BIOMIMETIC AND BIOINSPIRED MEMBRANES FOR NEW FRONTIERS IN SUSTAINABLE WATER TREATMENT TECHNOLOGY

Authored by **Amira Abdelrasoul, Huu Doan** and **Ali Lohi**

Biomimetic and Bioinspired Membranes for New Frontiers in Sustainable Water Treatment Technology
http://dx.doi.org/10.5772/65691
Authored by Amira Abdelrasoul, Huu Doan and Ali Lohi

Contributors

Amira Abdelrasoul, Huu Doan, Ali Lohi

Published by InTech
Janeza Trdine 9, 51000 Rijeka, Croatia

Publishing Process Manager Edi Lipovic
Technical Editor SPi Global
Cover InTech Design team

First published December, 2017
Printed in Croatia
Legal deposit, Croatia: National and University Library in Zagreb

Additional hard copies can be obtained from orders@intechopen.com

Biomimetic and Bioinspired Membranes for New Frontiers in Sustainable Water Treatment Technology,
Authored by Amira Abdelrasoul, Huu Doan and Ali Lohi
p. cm.
Print ISBN 978-953-51-3661-3
Online ISBN 978-953-51-3662-0

For EU product safety concerns:
IN TECH d.o.o., Prolaz Marije Krucifikse Kozulić 3, 51000 Rijeka, Croatia,
info@intechopen.com or visit our website at intechopen.com.

Meet the authors

Dr. Amira Abdelrasoul, P.Eng., is an assistant professor in the Chemical and Biological Engineering Department at the University of Saskatchewan. Prior to joining the University of Saskatchewan in 2017, she was a lecturer and postdoctoral research associate in the Department of Chemical Engineering at Ryerson University in Toronto, Ontario. Dr. Abdelrasoul received her PhD degree in Chemical Engineering in 2015 from Ryerson University with the Governor General's Academic Gold Medal and the C. Ravi Ravindran Outstanding Doctoral Thesis Award. Her academic excellence was recognized by numerous scholarships and academic awards from North American institutions. Her research interests focus on membrane science and technology for energy and water sustainability; process modeling, simulation, and optimization of complex systems; and nanomaterials for advanced technologies and clean energy. In addition, Dr. Abdelrasoul is a licensed Instructional Skills and Educational Development Workshops Facilitator and a fellow of the Staff and Educational Development Association.

Dr. Huu Dung Doan is a professor in the Department of Chemical Engineering at Ryerson University, Toronto, Ontario, Canada. He received his Bachelor of Engineering degree from Ryerson University, Toronto. After his graduation from Ryerson University, he worked in the engineering consulting industry for a few years and then went on for his Master of Science degree from the University of Guelph, Guelph, Ontario, and his PhD degree in Chemical Engineering from the University of Toronto, Toronto, Ontario. Dr. Doan joined Ryerson University in 1996. His research interests include membrane filtration, especially fouling and modeling, mass transfer in a packed bed, and industrial wastewater treatment by electrochemical methods.

Dr. Ali Lohi is a professor of Chemical Engineering at Ryerson University, Toronto, Canada. His research work covers topics in areas of process simulation/modeling, optimization of chemical processes, development of processes and models for reduction in energy consumption—remediation of membrane fouling, development of innovative and proficient green energy production processes, utilization of waste materials into new and value-added products including biofuel/bioenergy and synthesis gas (syngas) from agriculture waste, sequestration of CO2, oil/gas processing and production, transport phenomena in environmental systems, and application of CFD and ANN in modeling/simulation and optimization.

Contents

Preface

The world's water supply is one of the most critical issues today. According to a recent report from the National Intelligence Council, global water demand will exceed sustainable supply by 40% as early as 2030. The United Nations report highlights that almost 4,000 children die each day as a result of the diseases caused by ingestion of unsanitary water. Clean water is not available to 1.4–1.8 billion people globally. On the other hand, 75% of the earth is covered with water, and 97.5% of this water is in the oceans. Desalination of sea water is a common technique for the alleviation of the increasing shortage of fresh water in many areas of the world. In desalination, the salt water is forced through a reverse osmosis (RO) membrane at high pressures of 600 to 1000 psi. As a result, the energy associated with the desalination process is very large. Within the context of the continually rising energy cost and the impending exhaustion of conventional energy resources, the development of an efficient desalination technology is an attractive and promising direction.

Biomimetic and bioinspired membranes are the most promising type of membrane for multiple usage scenarios, includingcommercial separation applications as well as water and wastewater treatment technologies. In recent years, aquaporin biomimetic membranes (ABMs) for water purification have raised considerable interest. These membranes display uniquely favorable properties and outstanding performances, such as diverse interactions, varied selective transport mechanisms, superior stability, high resistance to membrane fouling, and distinct adaptability. Biomimetic membranes could make a significant contribution to alleviate water stress, environmental threats, and energy consumption.They have the potential to produce clean water more efficiently than RO membranes, while saving up to 88% of the energy used in desalination process.More than half of the 15,000 desalination plants around the world utilize RO technologies, and the implementation of biomimetic membranes on a large scale could save hundreds of millions of dollars in energy cost annually (potential savings of $1.45 million/year for a 100 ML/day desalination plant).

While there are still numerous challenges that need to be confronted, the biomimetic and bioinspired membranes may turn into the next-generation membranes that may open novel and efficient avenues leading toward advanced membrane technology. In this book, an overview of the current research and development of biomimetic and bioinspired membranes will be presented. The book attempts to provide a broad review of how aquaporins (AQPs), block copolymers, and polymer support structures interact and how these interactions can be characterized. The properties of aquaporins, their preparation, and characterization are also reviewed. In addition, the book provides an overview of various attempts to exploit the remarkable properties of aquaporin in membranes for desalination, including recent research developments in aquaporin-based membranes. Finally, the book will discuss challenges, opportunities, and future prospects of applying biomimetic membranes to desalination and water reuse.

Finally, we would like to acknowledge the support of the Department of Chemical Engineering at Ryerson University. Also, we would like to express our appreciation for the reviewers of InTech Publisher, making our contribution possible.

Last but certainly not least, the continual encouragement and support of our families and friends are deeply and sincerely appreciated.

Amira Abdelrasoul
Department of Chemical and Biological Engineering,
University of Saskatchewan,
Saskatoon, Saskatchewan, Canada

Huu Doan
Department of Chemical Engineering,
Ryerson University,
Toronto, Ontario, Canada

Ali Lohi
Department of Chemical Engineering,
Ryerson University,
Toronto, Ontario, Canada

Chapter 1

Sustainable Water Technology and Water-energy Nexus

Amira Abdelrasoul, Huu Doan and Ali Lohi

Additional information is available at the end of the chapter

http://dx.doi.org/10.5772/intechopen.71716

Abstract

As water scarcity continues expanding in regions around the globe, there is an ever-increasing need to augment municipal, industrial, and agricultural water supplies through the purification of unconventional water sources, such as seawater, industrial and municipal wastewater. Due to the inextricable linkage between water and energy consumption, often called the water-energy nexus, the augmentation of water supplies must not come with a high cost energy consumption. As such, the high energy efficiency and often superior efficacy of membrane-based technologies have gained widespread implementation in various water treatment processes. Membranes allow passage of water, but largely reject salt and most other solutes, play a critical role in the majority of these processes. These types of membranes lie at the heart of traditional reverse osmosis (RO) processes.

Keywords: water, desalination, energy, reverse osmosis, membrane modules

1. Introduction

Biomimetic and bioinspired membranes are those membranes that are fabricated with natural or natural-like (inorganic, organic or hybrid) materials via biomimetic and bioinspired approaches (biomineralization, bioadhesion, self-assembly, etc.) to tailor specific properties (sophisticated structures, hierarchical organizations, controlled selectivity, antifouling or self-cleaning properties).

The accountability for today's water supplies is one of the essential concerns since clean and reliable water resources are not readily or permanently available to the global population and the industrial sectors. The world's water supply is of the uttermost importance to future development and prosperity. Even in contemporary global research, a substantial effort is being directed toward searching for water solutions. As water scarcity emerges as a concern in regions around the globe, there are exponentially growing needs to increase the potential for municipal, agricultural, and industrial water supplies with the aid of purification processes and unconventional water resources. Such unconventional resources include seawater, industrial residue water, and municipal wastewater. Because of the implicit connection between water and energy consumption, or the water-energy nexus concerns, the expansion of water supplies must not be accompanied by a high cost of energy consumption. The higher energy efficiency potential and superior efficacy of membrane-based technologies have received widespread implementation in a variety of water treatment processes and water solution projects. Desalination of seawater is a common technique used to counter the increasing shortage of freshwater in many remote global areas. In particular, desalination membranes allow for the passage of water, however, tend to reject salt and other solutes that play a critical role in the majority of these processes. These types of filtration membranes are at the core of traditional reverse osmosis (RO) processes. During desalination, the saltwater is forced through a reverse osmosis (RO) membrane at high pressures of 600–1000 psi. Due to these high pressure requirements, the energy use related to this desalination process is extensive. The continually rising energy costs and the expected limitations in conventional energy resources undermine the use of such filtration technology and instead call for the development of efficient desalination technologies as the alternative viable direction.

2. Recent and projected future needs for clean water resources

The global demand for fresh, clean water has increased by a factor of six during the period from 1900 to 1995. This increase is more than double when comparing it to the concurrent increase in population [1, 2]. This trend has further accelerated after 1995 due to increase in water use in emerging and growing global economies. On the other hand, the freshwater supply availability is being constantly reduced by growing pollution around the globe and the accumulative effects of climate change. The facts dictate that even though the amount of planet's freshwater remained fairly constant over time due to the continuous recycling process in the atmosphere, the population and its demands have dramatically exploded. This implies that with every year, the competition for a clean, copious supply of water for drinking, cooking, bathing, and sustaining life intensifies. The growing lack of access to potable water and sanitation is a major cause of disease and an obstacle to sustainable growth for a large portion of the global population [3, 4]. Many developing countries are undergoing rapid industrialization without incorporating appropriate long-term wastewater management systems, and are now facing increasing water pollution issues while still struggling with poor water supply and sanitation concerns [5]. The World Health Organization (WHO) reports that there are more than 2.5 billion people, or around 40% of the world's population currently lack access to proper sewer sanitation systems [6]. A report from the National Intelligence Council further highlights that the global water demands will drastically exceed sustainable supplies by 40% by as early as 2030 [7]. The United Nations reports on the correlation between water access and living conditions indicate that almost 4000 children die each day as a result of the various diseases instigated by unsanitary water ingestion and an absence of viable clean water resources. Clean, drinkable water is not available to anywhere between 1.4 and 1.8 billion people globally [8]. Sustainable and long-term access to clean water resources is

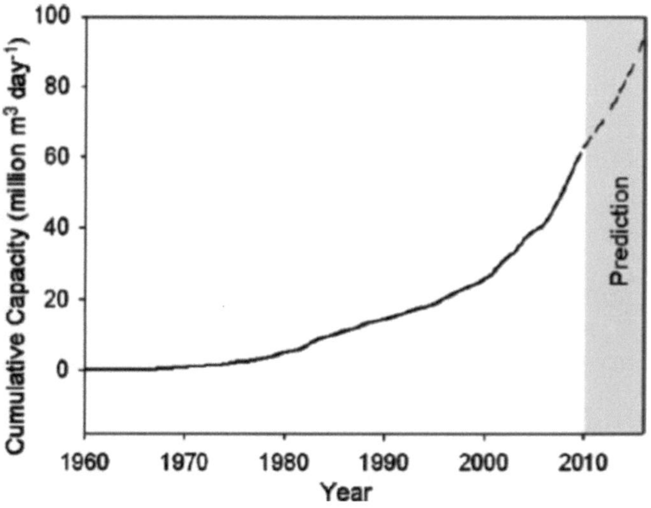

Figure 1. Cumulative desalination capacity from 1960 to 2016 [12, 13].

critical to all global economies and populations. The shift to biofuels may add further substantial water demands when it comes to crop irrigation, product manufacturing, and biorefining processes [9]. One of the primary reasons for the decrease in natural water supplies is the ongoing climate change and overexploitation of resources. The available remediation methods, including, water conservation, water transportation, and the construction of new dams, are insufficient for addressing or partially subduing with the exponentially growing demands. The most pressing global challenges include the production of potable water from salty or seawater as the world's largest water resource remaining, and the recycling or treatment of wastewater.

Desalination process is a technology that can convert saline water into usable water and can offer one of the most accessible solutions to the global water concerns [10]. A significant 75% of the Earth is covered with water and 97.5% of this water is located in the oceans. As indicated in **Figure 1**, the cumulative production capacity in desalination plants increased by 50% in 2010 to 60 million cubic meters per day, up from 40 million m^3 per day in 2006. In the first half of 2008, the growth in contract capacity was 39%, and the total global desalination production capacity was at around 50 million m^3 in 2009 [11–13].

3. Current state-of-the-art RO technology

The desalination membranes and their filtration potential are essential for the successful treatment of unconventional water sources such as seawater and wastewater. Arguably, desalination membranes can be the key to solving the concerns associated with water scarcity. There are multiple ways in which the process of seawater or brackish water desalination can be attained, and some of these ways, primarily rely on either thermal energy or the mechanical/electrical types of energy [14]. Because of its reliance on energy consumption, the process of seawater desalination is usually more costly than the process of filtering freshwater from rivers, groundwater, water recycling, and water conservation. **Table 1** outlines a summary of the technical and cost-related parameters for the major types of commercial and industrial desalination processes. The thermal processes for water filtration may include multistage flash (MSF), multiple effect distillation (MED), and thermal vapor compression (TVC) processes. The thermal-based filtration approaches have a record of high reliability accompanied by very high-energy consumption expectations. Processes that depend on mechanical/electrical energy include the mechanical vapor compression (MVC) and the reverse osmosis (RO) processes. The RO-based filtration process has the lowest energy consumption demands in comparison to other processes; however, its positives are undermined by a relatively lower reliability if compared to the other process types. In order for the seawater desalination to successfully occur, it is necessary to lower the total amount of the dissolved solids (TDS) content from 35,000–47,000 mg/L down to less than 500 mg/L [15].

Contemporary RO membrane technology is used as the primary methodology for desalination processes. RO membranes are introduced in a range of research and industrial applications that require desalination of saltwater and wastewater. Emerging shifts in energy recovery potential and innovations in pretreatment technology have dramatically enhanced the reliability and energy consumption demands in RO membrane technology. The expansive research and production efforts in RO desalination processes over the past four decades have

Energy used	**Thermal**		**Mechanical**	
Process	MSF	MED/TVC	MVC	RO
State of the art	Commercial	Commercial	Commercial	Commercial
World Wide Capacity 2004 (Mm^3/d)	13	2	0.6	6
Heat consumption (kJ/kg)	250–330	145–390	–	–
Electricity consumption (kWh/m^3)	3–5	1.5–2.5	8–15	2.5–7
Plant cost ($\$/m^3/d$)	1500–2000	900–1700	1500–2000	900–1500
Time of commissioning (months)	24	18–24	12	18
Production unit capacity (m^3/d)	<76,000	<36,000	<3000	<20,000
Conversion freshwater/seawater	10–25%	23–33%	23–41%	20–50%
Maximum top brine temperature (°C)	90–120	55–70	70	45 (max)
Reliability	Very high	Very high	High	Moderate (for seawater)
Maintenance (cleaning per year)	0.5–1	1–2	1–2	Several times
Pretreatment of water	Simple	Simple	Very simple	Demanding
Operation requirements	Simple	Simple	Simple	Demanding
Product water quality (ppm)	<10	<10	<10	200–500

Source: AQUA-CSP, DLR 2007.

Table 1. Major commercial desalination processes.

accounted for the 44% of the global desalination production capacity and for the 80% of the total number of desalination plants installed worldwide [16].

RO desalination has displaced the more conventional thermal types of technology, including the multistage flash (MSF) [17]. Furthermore, RO desalination technology is expected to maintain its dominance despite the arrivals and the proposals of new alternative technologies like membrane distillation [18], electrodialysis [19], capacitive deionization [20], and forward osmosis [21]. The commercial focus on RO technology is growing globally because of the constant improvement of the process and the respective cost reduction opportunities that arise from it. Some of the RO membrane advances feature innovation in membrane materials, membrane module design, feed pretreatment, process design, energy recovery, and reduction in energy consumption. Some of these improvements are quantitatively outlined in **Figure 2(a), (b)** and **(c)**. The significant sevenfold decrease in salt passage over the course of 30 years of development has greatly extended the potential range of saline feeds that can be treated to adhere the strict potable water standards. By improving mechanical, chemical, and biological strength of RO membranes, and by increasing their permeability, the developments have lowered the membrane cost per unit volume of water produced by more than 10 times since 1978. Combined efforts to lower fouling and concentration polarization, and to maximizing permeate flux and energy recovery potential, have decreased the energy consumption from 12 $kWhm^{-3}$ in the 1970s to less than 2 $kWhm^{-3}$ in 2006 [22].

The most extensive advantages to the water filtration process have come directly from the improvement of the membranes themselves. The key structural parameters, materials used, and morphology of RO membranes have been actively modified in order to advance membrane functionality, such as permeability and selectivity, and applicability, such as mechanical, chemical, and biological stability. The current RO membrane market is controlled by the thin film composite (TFC) polyamide membranes. These membranes consist of three layers that include a polyester web acting as a structural support (120–150 μm thick), a microporous interlayer (about 40 μm), and an ultrathin barrier layer located on its upper surface (0.2 μm) [23]. One of the possible concerns is that the polyester support web cannot itself provide the necessary support for the barrier layer since it is too porous and irregular. As a result, a microporous interlayer of polysulfonic polymer is added between the barrier layer and the support layer to enable the ultrathin barrier layer to endure the high-pressure compression. The barrier layer's thickness is lowered to minimize the opposition to the permeate transport process. The membrane pore size generally required to attain salt rejection higher than 99% on a consistent basis is less than 0.6 nm. This selective barrier layer is frequently created out of aromatic polyamide, for instance using interfacial polymerization of 1,3-phenylenediamine (also known as 1,3-benzenediamine) and the tri-acid chloride of benzene (trimesoyl chloride) [24]. Membrane featuring an improved chemical resistance and structural sturdiness can provide enhanced tolerance to impurities, better durability, and improved cleaning parameters [25, 26].

The most extensively used designs in RO desalination are the spiral wound membrane module configuration. This type of configuration allows for greater specificity of membrane's surface area, better scale-up operations, interchangeability, and low replacement costs. It is likewise the least expensive module configuration that can be used for production based on flat sheet TFC membranes [27, 28]. The spiral wound configuration was originally developed

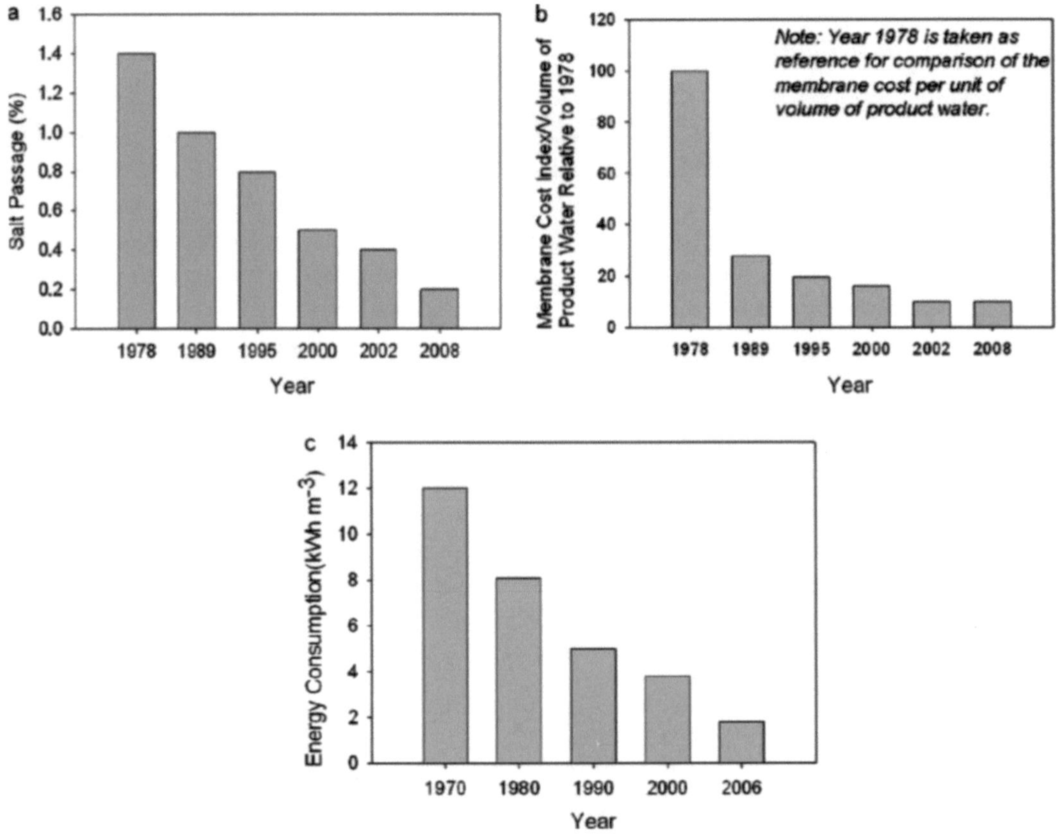

Figure 2. (a) Salt rejection enhancement potential, (b) membrane cost reduction, and (c) energy consumption minimization of RO [22].

over several decades ago and there has been extensive progress in the dimensions of spacers, feed channels, vessels, and the construction materials. These enhancements have helped to optimize the interconnection between module design and fluidic transport characteristics, which in turn decreased fouling and pressure losses. Polyamide spiral wound membranes control RO/Nanofiltration (NF) market sales with a 91% share majority, while the asymmetric cellulose acetate (CA) hollow fiber membranes in the far removed second spot [29]. CA membrane has a high chlorine resistance; hence, chlorine can be injected in the CA membrane unit to deter the growth of microorganisms and algae. On the other hand, RO/Nanofiltration (NF) has significantly higher salt rejection capacity and net pressure driving force potential [30]. Within the industrial sector, there are four primary membrane module suppliers that provide RO membranes to large-scale desalination plants, and these include DOW, Toray, Hydranautics, and Toyobo. State-of-the-art seawater desalination RO membrane modules from each respective supplier are shown in **Table 2** to provide a benchmark of current SWRO performance parameters. A more extensive comparison of these products has not been included since the research data corresponds to varying test or operating conditions [31–38]. Ongoing research on modular element design is currently concentrating on the optimization

Membrane module brand name	Material & module	Permeate flux (m^3 day^{-1})	Salt rejection (%)	Specific energy consumption[d] (kWh m^{-3})
DOW FILMTEC™ 8-in. SW30HRLE	TFC cross-linked fully aromatic polyamide spiral wound	28.0[a]	99.60–99.75[a]	3.40 (2.32)[e] at Perth SWRO Plant, Australia [32]
Hydranautics 8-in. SWC4+	TFC cross-linked fully aromatic polyamide spiral wound	24.6[b]	99.70–99.80[b]	4.17 (2.88)[e] at Llobregat SWRO Plant, Spain [32, 35]
Toray 8-in. TM820C	TFC cross-linked fully aromatic polyamide spiral wound	19.7–24.6[a]	99.50–99.75[a]	4.35 at the Tuas SWRO Plant, Singapore [33]
Toyobo 16-in. HB10255	Asymmetric cellulose tri-acetate hollow fiber	60.0–67.0[c]	99.40–99.60[c]	5.00 at Fukuoka SWRO Plant, Japan [34]

[a]Operating condition: 32 g L^{-1} NaCl solution, 55 bar, 25°C, pH 8 and 8% recovery [29, 35].

[b]Operating condition: 32 g L^{-1} NaCl solution, 55 bar, 25°C, pH 7 and 10% recovery [34].

[c]Operating condition: 35 g L^{-1} NaCl solution, 54 bar, 25°C and 30% recovery [36].

[d]These numbers are parameter specific and should not be compared to different desalination plants because of the different operating parameters (e.g., feed water quality, recovery, pretreatment processes, process design, etc.).

[e]The number in brackets is the energy consumption value of the RO membrane unit.

Table 2. Some of the state-of-the-art SWRO membrane modules in current application.

of hydrodynamics and the minimization of the concentration polarization effects. Further research directions have likewise focused on larger modular elements that are necessary for an improved desalination capacity.

One of the recent developments at Koch Membrane Systems includes a release of the 18-in. spiral wound modules featuring the MegaMagnum® trade name. Hydranautics and DOW (FILMTECTM) have 16-in. Modules piloted in cooperation with the Singapore National Water Agency. Research studies on the effects of the ongoing improvements indicate that such module designs can offer effective solutions when it comes to lowering the costs of desalination by up to 20% [39, 40].

4. RO desalination technology development and challenges

Research conducted by Sheikholeslami suggests that the upcoming challenges that the desalination industry will have to address are related to feed water characterization, process development, renewable energy source, materials development, stringent water standards, and brine management [41]. For instance, the largest SWRO plant in the world is located in Ashkelon, Israel, and features a production capacity of about 110 million m^3/year. If researchers consider the global average water consumption per capita of 1243 m^3/year (5% for domestic use, 85% for agricultural irrigation, and 10% for industrial use) [42], then even this plant can supply fresh drinkable water to less than 100,000 people. What this discrepancy in demands and expectations suggests is that mega-sized desalination plants are necessary and they need to be built within the foreseeable future if global multibillion populations are to be supplied with sufficient clean water.

From the perspective of global industrial complex, the greatest concern is how making RO desalination affordable in low-income countries with limited access to RO technology. Arguably, the capital investment and operating costs of RO plants must be further reduced in order to make this feasible. Electric energy, chemicals, and labor make up about 87% of the total RO cost [14]. Further advancement in membrane material and module optimization can dramatically contribute to the improvement of these aspects. The rejection of low-molecular weight compounds, especially boron species, is likewise necessary and would strongly contribute to lowering operational and production costs.

The membrane with the highest boron rejection potential currently available on the market can only achieve 93% boron rejection at optimum conditions. However, it has been reported that 99% of boron rejection is required in the Middle Eastern region for the one-pass RO process to comply with the WHO water drinking standards [43]. Higher salt rejection can potentially reduce the number of RO passes needed to achieve the desirable water quality. As a result, the decline in fouling, especially through the development of chlorine-tolerant membranes, is crucial since it reduces the costs of membrane replacement, backwashing chemicals, and energy needed to overcome the additional osmotic pressure. The operating pressure in many of the current systems is already nearing the thermodynamic limit and an additional reduction would have a relatively modest effect on the overall performance [44]. Nonetheless, a decrease in energy consumption would be substantial, as the energy costs signify half of the total water production costs. Higher permeability would further reduce the membrane area, which in turn will lead to a decrease in membrane replacement costs, smaller plant footprint, and a reduced use of harsh cleaning chemicals. From the perspective of industrial application, research potential, and commercial interest, all emerging membrane innovations need to outperform the materials and modules available on the market and are listed in **Table 2**.

Author details

Amira Abdelrasoul[1]*, Huu Doan[2] and Ali Lohi[2]

*Address all correspondence to: amira.abdelrasoul@usask.ca

1 Department of Chemical and Biological Engineering, University of Saskatchewan, Saskatoon, Saskatchewan, Canada

2 Department of Chemical Engineering, Ryerson University, Toronto, Ontario, Canada

References

[1] Werber JR, Deshmukh A, Elimelech M. The critical need for increased selectivity, not increased water permeability, for desalination membranes. Environmental Science & Technology Letters. 2016;**3**:112–120

[2] Comprehensive Assessment of the Freshwater Resources of the World. World Meteorological Organization-Geneva, CH, WMO; 1997. p. 9

[3] Fry LM, Mihelcic JR, Watkins DW. Water and nonwater-related challenges of achieving global sanitation coverage. Environmental Science & Technology. 2008;**42**:4298-4304

[4] Montgomery MA, Elimelech M. Water and sanitation in developing countries: Including health in the equation. Environmental Science & Technology. 2007;**41**:17-24

[5] Health and Environment in Sustainable Development: Five Years after the Earth Summit 1997;54-55

[6] Progress on Drinking Water and Sanitation: Special Focus on Sanitation; World Health Organization Report (WHO). 2008

[7] Global Water Security, Intelligence Community Assessment. 2012 ICA 2012-08, 2 February 2012

[8] Hanneke H. Water in Canada, A Resource in Crisis. Canada: Lone Pine Publishing. ISBN 10: 1926736044 2011

[9] Shannon MA, Bohn PW, Elimelech M, Georgiadis JG, Marinas BJ, Mayes AM. Science and technology for water purification in the coming decades. Nature. 2008;**452**:301-310

[10] Gleick PH. The World's Water 2008-2009 the Biennal Report on Freshwater Resources. Chicago: Island Press; 2008

[11] New Desalination Capacity, 1980-2009-Chart. United States: Global Water Intelligence; 2009. p. 10

[12] Desalination: A National Perspective. Washington, DC: National Academies Press; 2009

[13] The Big Dipper: Contracted Desalination Capacity Forecast-Chart. United States: Global Water Intelligence; 2009. p. 10

[14] Fritzmann C, Löwenberg J, Melin T, Wintgens T. State-of-the-art of reverse osmosis desalination. Desalination. 2007;**216**:1-76

[15] El-Nashar AM. Water and wastewater treatment technology—History and current status of membrane desalination process. Paris, France: Encyclopedia of Life Support Systems (EOLSS); 2010

[16] Greenlee LF, Lawler DF, Marrot BD, Mouli BP. Reverse osmosis desalination: water resources, and today's challenges. Water Research. 2009;**43**:2317-2348

[17] Energy Makes All the Difference: Desalination Operating Costs Compared Chart. Global Water Intelligence; 2007. p. 8

[18] Hsu ST, Cheng KT, Chiou JS. Seawater desalination by direct contact membrane distillation. Desalination. 2002;**143**:279-287

[19] Sadrzadeh M, Mohammadi T. Sea water desalination using electrodialysis. Desalination. 2008;**221**:440-447

[20] Oren Y. Capacitive deionization (Cdi) for desalination and water treatment past, present and future (a review). Desalination. 2008;**228**:10-29

[21] McGinnis RL, Elimelech M. Energy requirements of ammonia-carbon dioxide forward osmosis desalination. Desalination. 2007;**207**:370-382

[22] The International Desalination and Water Reuse Quarterly. Boca Raton, Fl: Lineal Pub. 2006;**16**:10-22

[23] Petersen RJ, Cadott JE. Thin film composite reverse osmosis membrane. In: Porter MC, editor. Handbook of Industrial Membrane Technology. New Jersey: Noyes Publication; 1990

[24] Cadotte JE. Reverse Osmosis Membrane. Patent Application No. 4039440. p. 1977

[25] Tarboush BJA, Rana D, Matsuura T, Arafat HA, Narbaitz RM. Preparation of thin-film-composite polyamide membranes for desalination using novel hydrophilic surface modifying macromolecules. Journal of Membrane Science. 2008;**325**:166-175

[26] Li L, Zhang S, Zhang X, Zheng G. Polyamide thin film composite membranes prepared from 3,4′,5-biphenyl triacyl chloride, 3,3′,5,5′-biphenyl tetraacyl chloride and M-phenylenediamine. Journal of Membrane Science. 2007;**289**:258-267

[27] Pearce G. Water and wastewater filtration: membrane module format. Filtration & Separation. 2007;**44**:31-33

[28] Polasek V, Talo S, Sharif T. Conversion from hollow fiber to spiral technology in large seawater RO systems—Process design and economics. Desalination. 2003;**156**:239-247

[29] Market Outlook for RO/NF and UF/MF Membranes Used for Large-Volume Applications. In: Water Executive. 2004. p. 9-11

[30] Kumano A, Fujiwara N. Cellulose triacetate membranes for reverse osmosis. In: Normam AGF, Li N, Ho WSW, Matsuura T, editors. Advanced Membrane Technology and Application. New Jersey: John Wiley & Sons; 2008. p. 21-43

[31] FilmtecTM Membranes—Filmtec Sw30hr Le-400 Seawater Reverse Osmosis Element. Dow Chemical Company.

[32] Laine JM. Design & Operation Considerations: Two Large-Scale Case Studies. Suez Environment; 2009

[33] Stover RL. Low Energy Consumption SWRO. in: Clean Technology 2008, Boston, Massachusetts, 2008

[34] Shimokawa A. Desalination plant with unique methods in Fukuoka, in: Japan-U.S. Governmental Conference on Drinking Water Quality Management and Wastewater Control, Las Vegas 2009

[35] Europe's Largest SWRO Plant Opens. Water Desalination Report 2009. p. 45

[36] Hydranautics introduces desalination membranes, Membrane Technology. 2003; **2003**(11):2-3

[37] Lee KP, Arnot TC, Mattia D. A review of reverse osmosis membrane materials for desalination-development to date and future potential. Journal of Membrane Science. 2011;**370**(1-2):1-22

[38] Kumano A, Fujiwara N. Cellulose triacetate membranes for reverse osmosis. In: Li NN, Fane AG, Ho WSW, Matsuura T, editors. Advanced Membrane Technology and Application. New Jersey: John Wiley & Sons; 2008. p. 39

[39] Yun TI, Gabelich CJ, Cox MR, Mofidi AA, Lesan R. Reducing costs for largescale desalting plants using large-diameter, reverse osmosis membranes. Desalination. 2006;**189**:141-154

[40] Ng HY, Tay KG, Chua SC, Seah H. Novel 16-inch spiral-wound RO systems for water reclamation—A quantum leap in water reclamation technology. Desalination. 2008;**225**:274-287

[41] Sheikholeslami R. Strategies for future research and development in desalination—Challenges ahead. Desalination. 2009;**248**:218-224

[42] Hoekstra AY, Chapagain AK. Globalization of Water: Sharing the Planet's Freshwater Resources. Oxford: Blackwell Publishing; 2008

[43] Li NN, Fane AG, Ho WSW, Matsuura T, editors. Advanced Membrane Technology Applications. New Jersey: John Wiley and Sons, Inc.; 2008. p. 3

[44] Zhu A, Christofides PD, Cohen Y. On RO membrane and energy costs and associated incentives for future enhancements of membrane permeability. Journal of Membrane Science. 2009;**344**:1-5

Chapter 2

Development of Conventional RO Membranes

Amira Abdelrasoul, Huu Doan and Ali Lohi

Additional information is available at the end of the chapter

http://dx.doi.org/10.5772/intechopen.71717

Abstract

Polymeric RO membranes have dominated commercial applications since the very first RO desalination plant became industrially active. Due to their technological development they offer low-cost fabrication, ease of handling, and improved performance in selectivity and permeability. One of the earliest reviews on polymeric RO membrane materials was reported by Cadotte. It focused on composite RO membranes and covered the period from the inception of composite RO membranes up to approximately 1985. In 1993 Petersen offered a comprehensive review of the same subject, specifically examining the chemistry of the membrane materials. This section will briefly highlight the early development of membrane chemistry and graphical illustrations are used to visualize the performance improvement potential in RO membranes. This chronological description provides the readers with a quick overview of RO membranes formed by different mechanisms and their respective impact on the desalination industry over the years. For a more complete study of the early RO membrane development, readers are advised to further refer to Petersen's work. In general, the development of membrane materials can be divided into two periods according to research activity: (i) the search for a suitable materials (chemical composition) and membrane formation mechanisms (1960s to late 1980s), and (ii) the evolution of more controlled conditions for membrane formulation as a way of enhancing membrane functionality and durability (late 1980s to date).

Keywords: asymmetric, thin-film composite, reverse osmosis, surface modification

1. Introduction

Polymeric RO membranes have dominated commercial applications since the very first RO desalination plant became industrially active. Because of the way they can be technologically developed, they allow for easier handling, lower cost of production, and superior performance capacities when it comes to permeability and selectivity parameters. Cadotte [1] produced one of the earliest reviews of the polymeric RO membrane materials and specifically focused on the composite RO membranes by examining the period from the composite RO membranes' inception until around 1985. At a later period, in 1993, Petersen [2] put together a more in-depth review of the same composite RO membranes, but this time prioritizing the membrane materials' chemistry. The following section will track earlier development and overall analysis of membrane chemistry and its effects, as well as utilize graphical illustrations to reflect and visualize the performance improvement potential that exists in RO membranes. Such a chronology-driven overview offers readers an effective summary of the RO membranes created with various mechanisms and their respective effects on the desalination industry throughout the last several decades. A substantially more rigorous assessment of the early RO membrane development can be found in Petersen's work [2]. Overall, the development of membrane materials may be strategically divided into two key periods that reflect the duality of research directions and activities: (1) the first is shaped by a search for more suitable materials, based on chemical compositions, and membrane formation mechanisms during the period between 1960s and late 1980s [1–4], and (2) the second is shaped by the development of controlled conditions for membrane creation as a method of improving membrane durability and functionality starting in the late 1980s and continuing today [3, 4].

2. Early membranes approaches and the development of asymmetric membranes

The Sea as a Source of Fresh Water report that came out in 1949 was the beginning of a progressive expansion into the salt-rejecting membranes research [5]. While the experimental project that came out of this report was not successful due to its limited membrane focus, it did offer the necessary grounds for future membrane study. Researchers Reid and Breton reported in the late 1950s that a hand-cast thin symmetrical cellulose acetate (CA) membrane was able to retain salt effectively by attaining a 98% rejection rate. The same experiment, however, showed that the critical permeate flux values were highly discouraging, at the order of <10 mL m^{-2} h^{-1} [6]. The development of the Loeb-Sourirajan CA membrane was the next vital step, since it was the first successful effort to create RO in practice [7]. A Loeb-Sourirajan CA asymmetric membrane was then produced, featuring a dense 200-nm thin layer over a thick microporous membrane body. As a result of these experiments, a new morphological approach resulted in water flux values of at least an order of magnitude greater than the initial symmetric membrane [8]. The specific molecular composition of these CA membranes is outlined in **Table 1**.

Figure 1 shows the primary developments in asymmetric RO membranes, starting with the early research initiatives and projects and up to 1980s. During the decade following Loeb-Sourirajan membrane creation, new research focused on CA materials was conducted with the particular intention of refining membrane transport properties, as well as improving its manufacturing process and then introducing this technology for a widespread industrial applications [1]. In the following research projects, the cellulose triacetate (CTA) membrane was created as an effective alternative that allows for increased stability in a much wider range of temperatures and pH values and offers greater resistance potential

Chemical type and description	Chemical structure
1. Cellulose acetate—Loeb-Sourirajan CA [8] Flux: 0.35 m^3 m^{-2} day^{-1} Salt rejection: 99% Test: > 100 bar, 4% NaCl solution	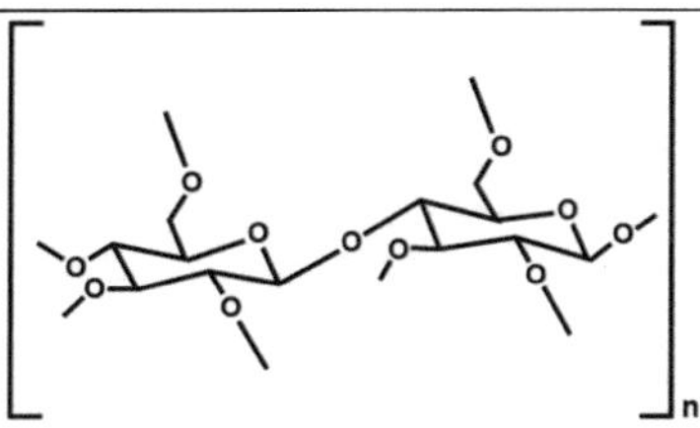
2. Aromatic polyamide: Polyamide-hydrazide [9] Flux: 0.67 m^3 m^{-2} day^{-1} Salt rejection: 99.5% Test: 30°C, > 100 bar, 3.5% NaCl solution	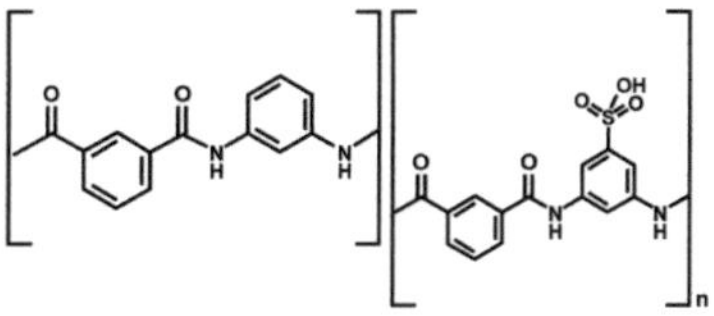
3. Polypiperzine-amide [10] Flux: 0.67 m^3 m^{-2} day^{-1} Salt rejection: 97.2% Test: > 80 bar, 0.36% NaCl solution	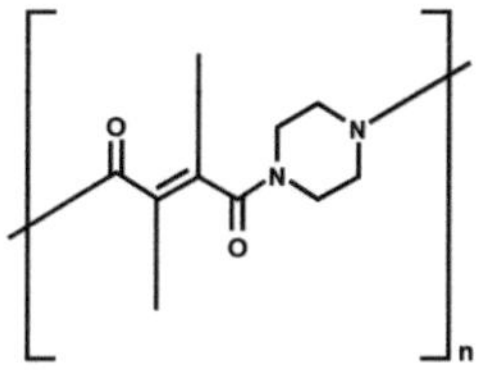
4. Polybenzimidazoline [11] Flux: 0.13 m^3 m^{-2} day^{-1} Salt rejection: 95% Test: > 6 bar, 0.105% NaCl solution	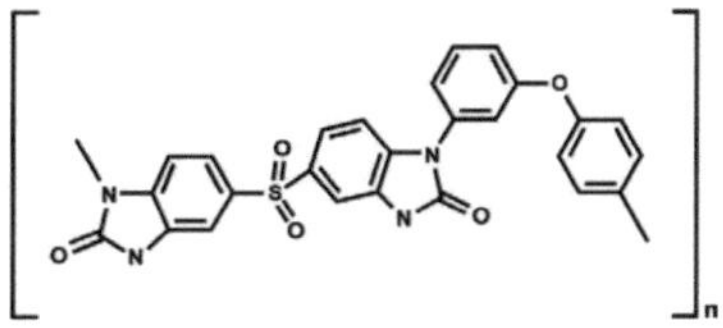
5. Polyoxadiazole [12] Flux: 0.07 m^3 m^{-2} day^{-1} Salt rejection: 92% Test: > 45 bar, 0.5% NaCl solution	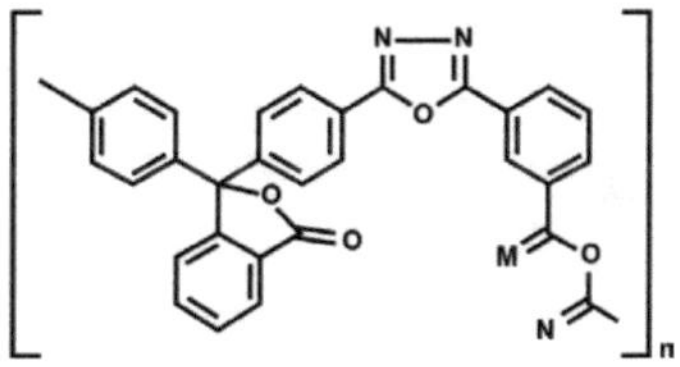

Note: The chemical structures listed represent segments of these membranes. It does not show all possible forms of the structure, for example, the CA structure shown is CDA, rather than CTA or mixed-CA.

Table 1. Asymmetric RO membranes.

of biological and chemical attacks when compared to the membranes composed of earlier cellulose diacetate (CDA) material. On the other hand, CTA membranes are predisposed to being affected by compaction and as a result may experience a substantial decrease in flux even at low operating pressures such as 30 bar or less [13]. The combination of CTA and CDA allowed for greater permeability and selectivity values than the CA membranes. Such a combination also offered greater resistance to the effects of compaction [14]. Additional research studies have been conducted with the aim to control degree of mixed ester substitution for the hydroxyl groups located in cellulose and to assess how well the CA membranes are working [15].

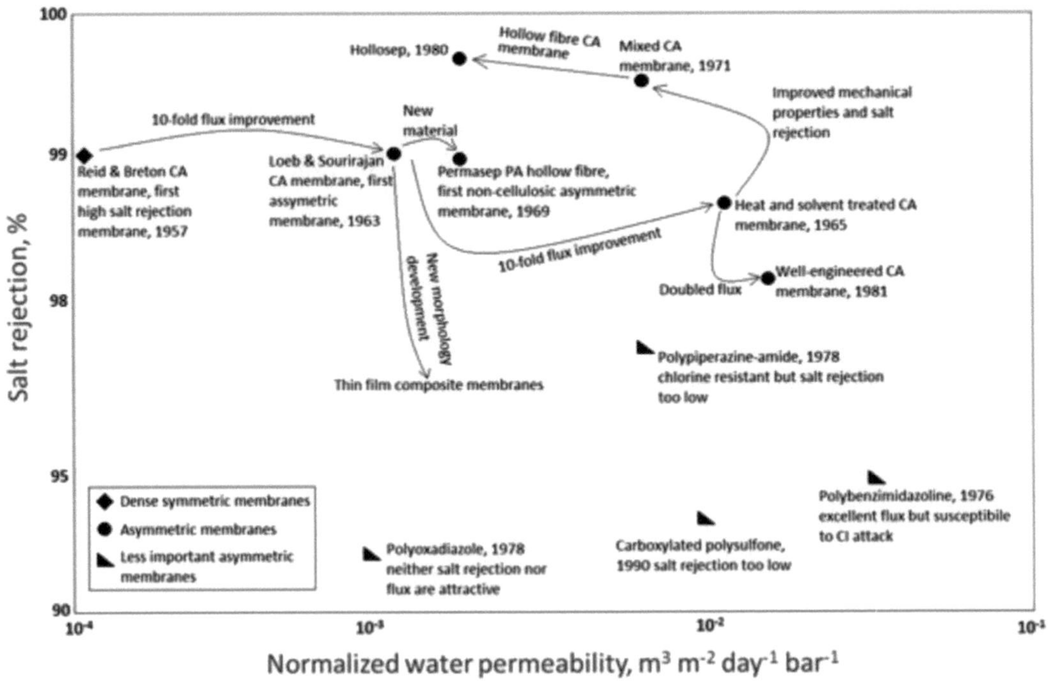

Figure 1. The development of asymmetric RO membranes [1].

The CA was the best available membrane material for the RO up to 1969, despite the extensive search for potential alternatives in membrane polymers. During this period, acetate group's susceptibility to hydrolysis, in both alkaline and acidic environments, and its relative sensitivity to microbial contamination had significantly limited its durability as well as its potential applications [16]. As a consequence, despite multiple alternative polymers examined in 1960s, a substantially sturdier material featuring greater chemical stability was severely needed.

Richter and Hoehn developed the first noncellulosic asymmetric membrane that received attention because it was comprised of an aromatic polyamide (PA) asymmetric hollow-fiber membrane [17]. This type of membrane was then commercialized by Du Pont with the trade name of B-9 Permasep® specifically for application during brackish water desalination processes. Although it features small flux and salt rejection values, it does offer stability, durability, and versatility which surpass those of aromatic polyhydrazides or CA [18]. Even though the low flux was problematic, it experienced relative commercial success due to its effectual packing of hollow fibers that helped it outperform the CA spiral wound elements, with respect to flux per unit module volume values. For a more in-depth discussion of the variation in reactants for the PA, asymmetric membranes refer to sources that specialize in this area [19].

The polyamides' predisposition to being attacked by disinfectants, such as ozone and chlorine (halogens), was noted after prolonged application of the B-9 Permasep® membranes. Subsequently, chlorine-resistant asymmetric membranes based on polypiperazine-amides were developed as a response [10, 20, 21]. These membranes have the permselectivity similar to the permselectivity of the asymmetric CA membrane. Lower quantities of amidic hydrogen in the membrane likewise improved its overall resistance to chlorine attacks [22]. Despite its advantages, this membrane was not commercialized because of its low salt rejection properties (≤95%) [23]. In this case, the phenyl and sulfonic groups in sulfonated polysulfone were projected to improve permeability, as well as chemical, biological, and mechanical stability; however, the salt rejection potential value was far lower than the acceptable level needed for industrial and commercial applications [24]. Carboxylated polysulfone likewise suffers from an uncompetitive salt rejection potential, even though it has promising flux values [25, 26]. On the other hand, polybenzimidazoline (PBIL) membranes created by Teijin reflect great permselectivity in harsh operating conditions, while at the same time remain susceptible to chlorine attacks and pressure compaction [11, 27, 28]. In another instance, polyoxadiazole has been exhibiting superior thermal and mechanical stability, while its permeability and salt rejection values fail to offer a commercially feasible or attractive option when it comes to industrial RO applications [12, 19].

3. Thin-film composite (TFC) membranes

Research indicates that only a small number of soluble polymers can successfully create asymmetric structures during a one-step casting process. Out of these soluble polymers, only a limited number are commercially attractive when it comes to the right combination of salt rejection capability and permeability potential. Furthermore, the densification in the middle transitional layer of the CA asymmetric membranes tends to occur under pressure [29]. These conditions have led to the creation of a two-step casting method that enables individual optimization of the materials being used for the barrier layer and for the microporous support film, the former for the optimal salt rejection potential and permeate flux and the latter for improved mechanical support. In addition, an extensive number of polymers may be tested for the support layer and barrier layer separately. Such anisotropic types of membranes are currently identified as the composite membranes.

3.1. Early development of TFC membrane

Figure 2 shows that the first TFC membrane was created by float-casting a CA ultrathin film onto the water surface and then followed by the process of laminating and annealing the film onto a preformed CA microporous support [30]. Membranes that were produced with the aid of this particular technique did not receive much commercial interest due to their asymmetric counterparts. An extensive empirical study into commercial membrane application suggested that polysulfone was the ideal material for the support layer because of its reasonable flux value, overall resistance to compaction, and its critical stability in acidic environmental settings. The latter key quality allows for the development of the TFC membrane using interfacial polymerization and acid polycondensation [31].

In order to overcome the scale-up issues in float-casting technology, a dip-coating methodology based on acid polycondensation of low molecular weight hydroxyl-containing compounds was

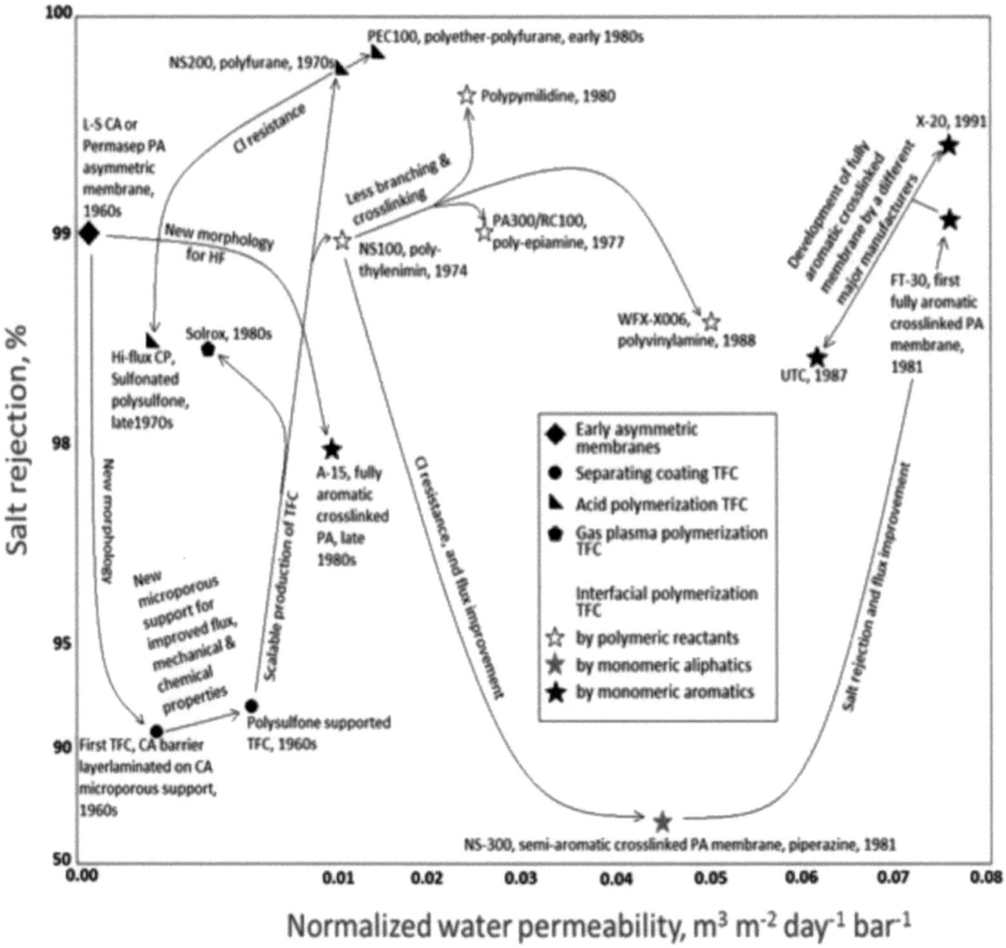

Figure 2. Development of thin-film composite RO membrane [30].

created [19, 32]. The original patented product based on this method was labeled NS-200 and entailed a product caused by the reaction of polyoxyethylene, sulfuric acid, and furfuryl alcohol (**Table 2**) [33]. Although NS-200 offered superior salt rejection capabilities, it was problematized by the irreversible swelling as well as hydrolysis of the sulfate linkage. Another example of the membrane created with the aid of acid polycondensation was the PEC-1000 TFC RO membrane produced by Toray Industries Inc., [34]. This membrane used 1,3,5-tris(hydroxyethyl)isocyanuric acid rather than polyoxyethylene.

While it did show a relatively high salt and organic compound rejection potential with sufficient flux levels, this membrane was vulnerable to chlorine attacks. In the next iteration of research, sulfonated polysulfone membranes were created so as to ensure a more uniform stability values in oxidizing environments [24]. However, substantial Donnan effects were still observed, thus suggesting that the shielding effect of divalent cations may drastically lower the monovalent ion rejection potential. A collective summary of key TFC RO membranes is listed in **Table 2**.

http://dx.doi.org/10.5772/intechopen.71717

Chemical type and description	Chemical structure
1. Polyfurane Name: NS-200 [33] Flux: 0.8 m^3 m^{-2} day^{-1} Salt rejection: 99.8% Test: > 100 bar, 3.5% NaCl solution	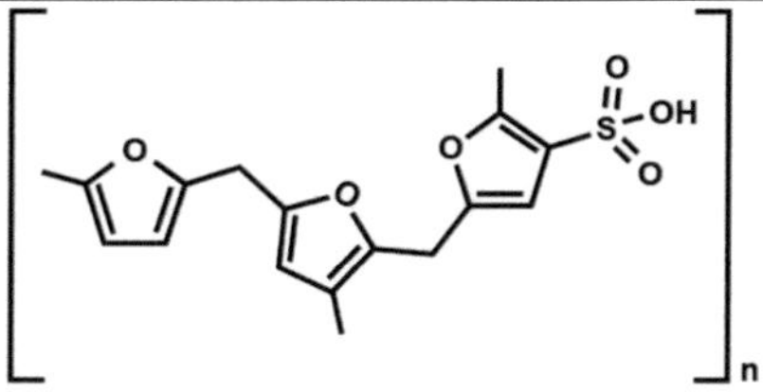
2. Polyether–Polyfurane Name: PEC-1000 [34] Flux: 0.5 m^3 m^{-2} day^{-1} Salt rejection: 99.9% Test: > 69 bar, 3.5% NaCl solution Excellent organic rejection	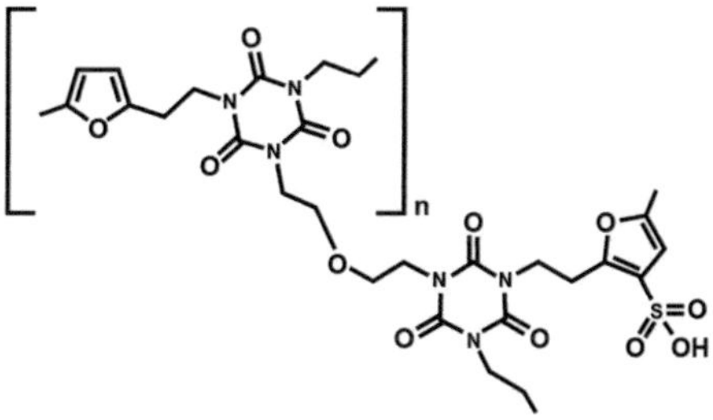
3. Sulfonated polysulfone Name: Hi-flux CP [67] Flux: 0.06 m^3 m^{-2} day^{-1} Salt rejection: 98% Test: > 69 bar, 3.5% NaCl solution Excellent chlorine resistance	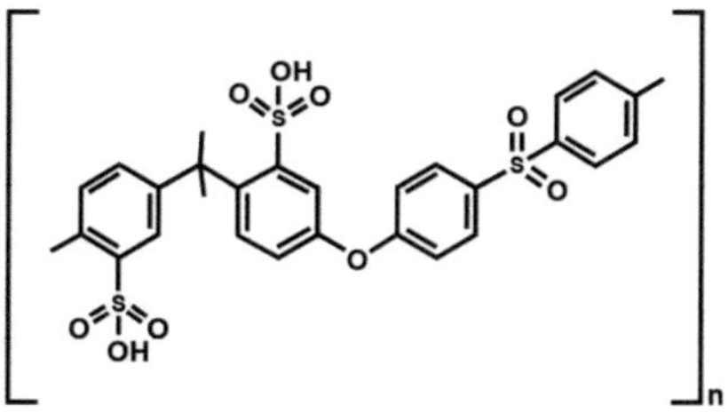
4. Polyamide via polyethylenimine Name: NS-100 [67] Flux: 0.7 m^3 m^{-2} day^{-1} Salt rejection: 99% Test: > 100 bar, 3.5% NaCl solution	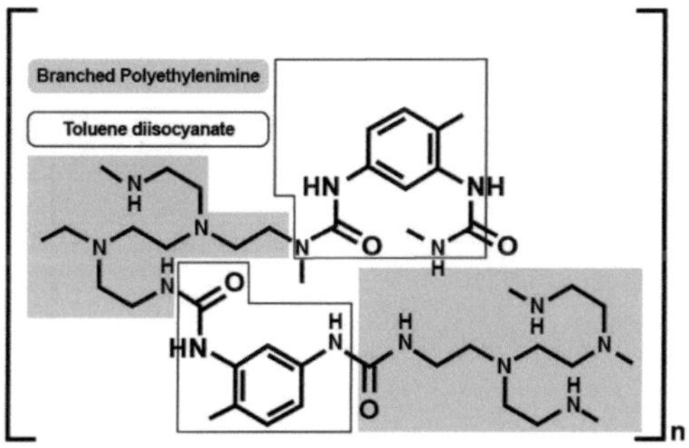
5. Polyamide via polyepiamine Name: PA-300 or RC-100 [45] Flux: 1.0 m^3 m^{-2} day^{-1} Salt rejection: 99.4% Test: >69% bar, 3.5% NaCl solution	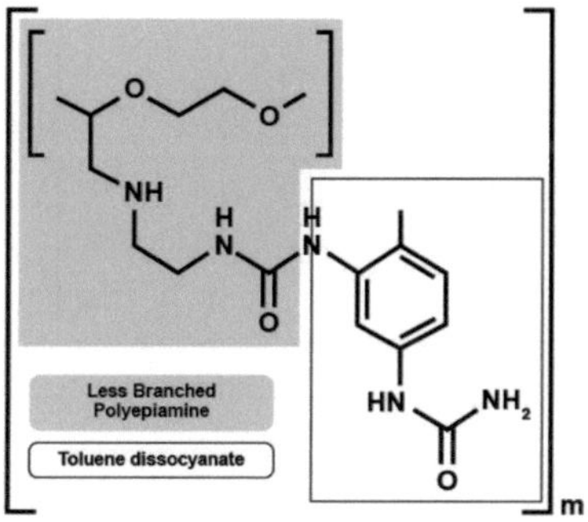

Chemical type and description	Chemical structure
6. Polyvinylamine – Name: WFX-X006 [68] Flux: 2.0 m^3 m^{-2} day^{-1} Salt rejection: 98.7% Test: > 40 bar, Conductivity: 5000 μScm^{-1}	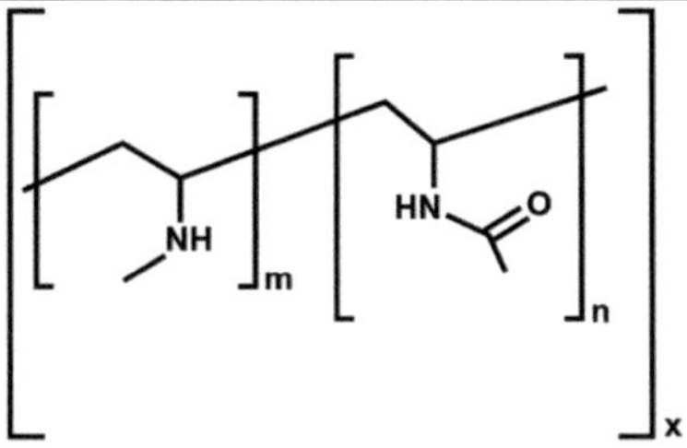
7. Polypyrrolidine [69] Flux: 0.8 m^3 m^{-2} day^{-1} Salt rejection: 99.7% Test: > 40 bar, 0.5% NaCl solution	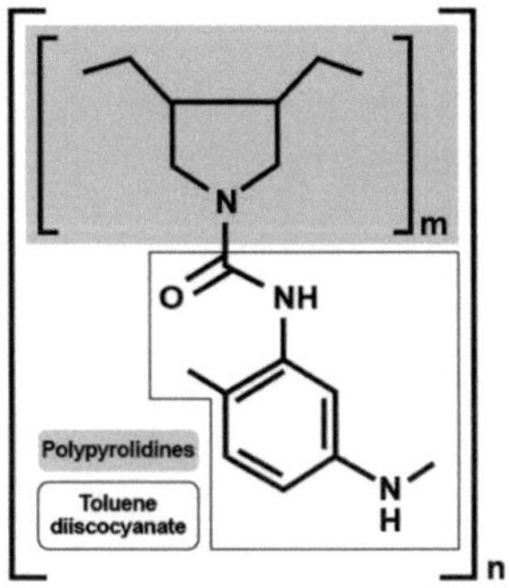
8. Polypiperazine-amine Name: NS-300 [52] Flux: 3.3 m^3 m^{-2} day^{-1} Salt rejection: 68% Test: > 100 bar, 3.5% NaCl solution	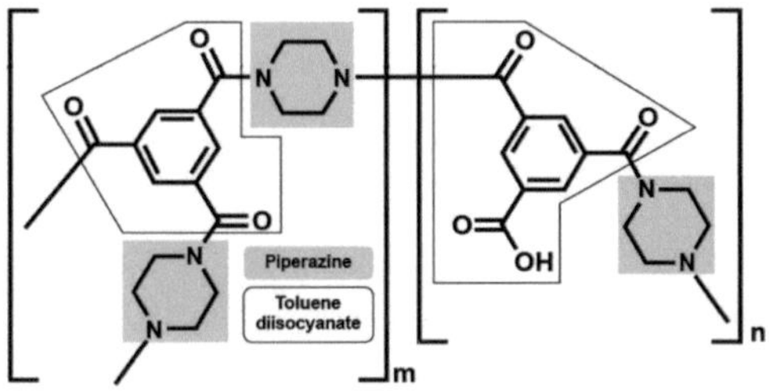
9. Cross-linked fully aromatic Polyamide: 1 Name: FT-30 [54] Flux: 1.0 m^3 m^{-2} day^{-1} Salt rejection: 99% Test: >15 bar, 0.2% NaCl solution	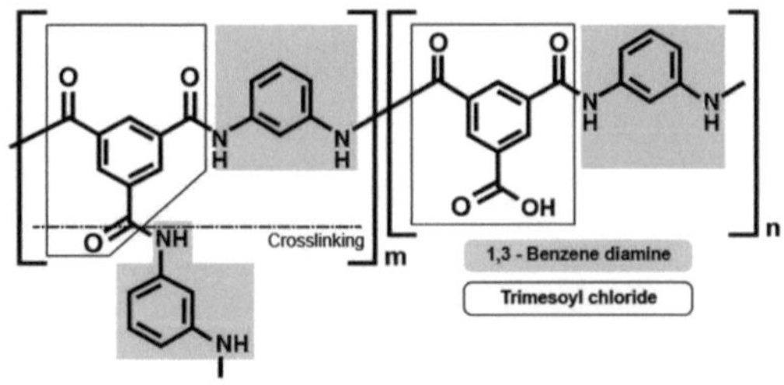
10. Cross-linked fully aromatic Polyamide: 2 Name: UTC series [70] Flux: 0.8 m^3 m^{-2} day^{-1} Salt rejection: 98.5% Test: > 15 bar, 0.5% NaCl solution	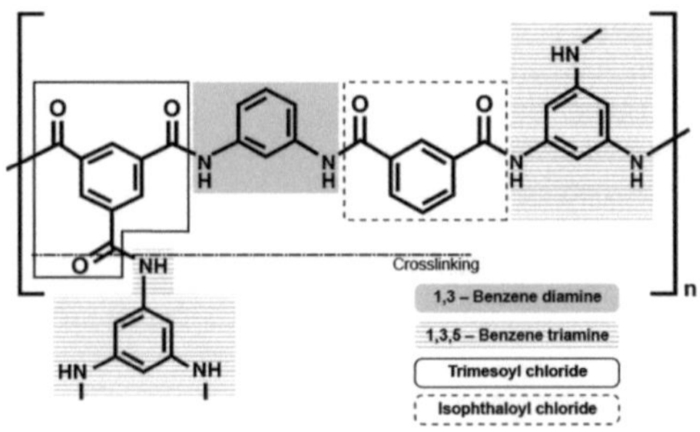

Chemical type and description	Chemical structure
11. Cross-linked aralkyl polymide Name: A-15 [64] Flux: 0.26% $m^3\ m^{-2}\ day^{-1}$ Salt rejection: > 98% Test: > 55 bar, 3.2% NaCl solution	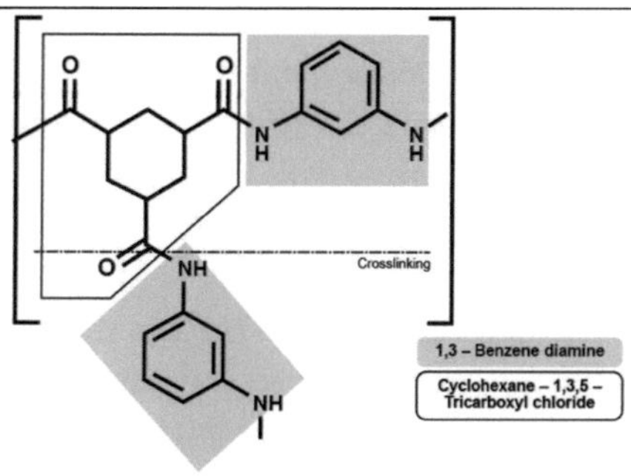
12. Cross-linked fully aromatic Polyamide: 3 Name: X-20 [71] Flux: 1 $m^3\ m^{-2}\ day^{-1}$ Salt rejection: 99.3% Test: > 15 bar, 0.2% NaCl solution	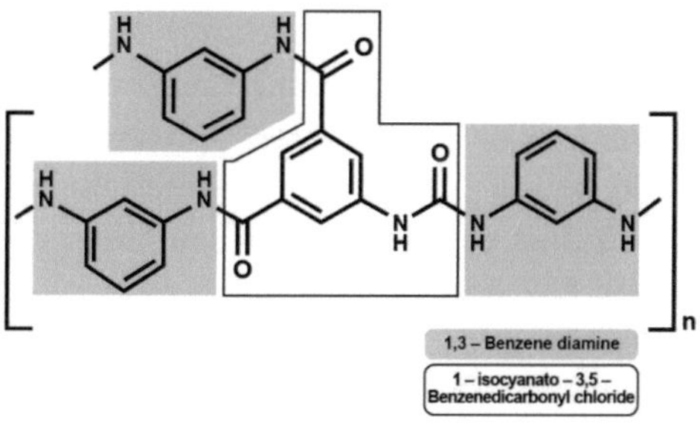

Table 2. TFC RO membranes.

The barrier layer may likewise be created using plasma polymerization during which the monomer vapor is incited with the help of the gas plasma method and the monomer's atomic polymerization is brought to the cool surface, frequently in the form of polysulfone support surface.

A variety of polymers have been tested as a part of membrane research, and adequate permselectivity levels have been achieved with vinylene carbonate/acrylonitrile, vinyl acetate/acrylonitrile, acetylene/water/ nitrogen, allylamine, as well as acetylene/water/carbon monoxide arrangements [35–41]. Yasuda's research collective was especially successful in plasma polymerization studies by producing a membrane created using water, acetylene, and nitrogen. This membrane performed exceptionally well during seawater desalination tests and showed 99% salt rejection with a flux of 1.5 $m^3\ m^{-2}\ day^{-1}$ at 100 bar operation parameters. Multiple research groups have conducted experiments with applications in gas separation and plasma-polymerized films [42]; however, only the Solrox membrane type of RO membrane has been successfully commercialized using this method. Plasma-polymerized RO membranes primarily feature low chlorine resistance potentials because of their nitrogen-enriched chemical configuration.

3.2. Interfacial polymerization of TFC membranes

Application of polysulfone as a potential support layer paved the way for production of RO membranes and interfacial polymerization because it could withstand the alkaline condition created by caustic soda applied as an acid acceptor throughout the interfacial polymerization processes. In the history of RO process, the production of NS-100, that is, polyethylenimine reacting with toluene diisocyanate (**Table 2**), is a critical developmental step in membrane research. In fact, this was the first successful instance of noncellulosic membrane featuring comparable monovalent and flux salt rejection potentials. This membrane likewise showcased improved stability at high temperatures, and advanced organic compound rejection capacity in acidic and alkaline environments [43, 44]. On the other hand, NS-100-type membranes provide very limited resistance to chlorine and show a distinct surface brittleness due to their overly cross-linked configuration. An alternative commercialized membrane product created using interfacial polymerization of polymeric amines is polyepiamine and specifically its two types PA-300 and RC-100 (**Table 2**) [45–47]. The PA-300-type membrane had enhanced permeate flux by 42.8% at about 1 $m^3 m^{-2} day^{-1}$ and with the salt rejection of 99.4% at 70 bar, if compared to the NS-100-type membranes. This noticeable improvement has contributed to the effective installation of PA-300 spiral wound modules at the TFC SWRO plant in Jeddah [48]. RC-100, on the other hand, exhibited high resistance to biofouling and was thus installed at Umm Lujj II as well as other desalination plants [49]. The two other notable interfacially polymerized TFC membranes are polyvinylamine and polypyrrolidone. While polyvinylamine offers higher flux capacity, polypyrrolidone is the type of membrane where the amino/carboxy groups may be controlled in a way that permits variable selectivity and amphotericity properties. Early efforts to use interfacial polymerization of monomeric amines, including aromatic and aliphatic diamines, with terephthaloyl chloride did not help facilitate the creation of membranes with the required salt rejection values [2]. Furthermore, once the polymerization conditions were improved, then the method was able to produce an improved type of membrane in the form of NS-300 [20, 21, 50]. This polypiperazine-amide membrane showed substantial Donnan exclusion effects because of the presence of anionic carboxylic groups at the membrane's surface. Due to this presence, the NS-300 membrane could produce outstanding rejection potential of divalent anions, including sulfate at higher flux values. Such a result makes NS-300 substantially more useful for practical industrial applications based on nanofiltration (NF) (**Table 2**). As a result, a variety of NF membranes based on this chemistry have been effectively commercialized, for example, NTR-7250 by NittoDenko [51], NF-40 series by DOW FILMTECTM [52], and UTC-20 by Toray Industries [53]. Research has shown that membranes with improved permselectivity can be created with the help of monomeric aromatic amines and aromatic acyl halides, while comprised of at least three carbonyl halide groups, with trimesoyl chloride allowing for the best results [54–56]. This method is further unlike other interfacial polymerization approaches since it helps to avoid the use of heat curing. Similarly, surfactants and acid acceptors were not compulsory as both cross-linking and polymerization were quick even when acyl halide was provided at low concentration values. FT-30-type membrane (**Table 2**) was created using the interfacial reaction between trimesoyl chloride and 1,3-benzenediamine and has resulted in distinct surface characteristics. This particular membrane can be defined through a "ridge and valley" configuration, instead of the slightly grainy and smooth surface achieved from the aliphatic amines [20]. Research studies have indicated that the rougher "ridge and valley" type of surface contributed to a larger surface area used specifically for water transport needs and as a consequence for the water flux [57]. During seawater desalination testing, FT-30 membrane produced fluxes of almost

1 m^3 m^{-2} day^{-1} and 99.2% salt rejection rate at 55 bar operation parameter. FT-30's aromatic polyamide structure offers a higher degree of thermal and chemical resistance, resistance to compression, and a much wider pH operating range. While the membrane was not entirely resistant to the effects of a chlorine attack, the FT-30 showed a substantial degree of tolerance to chlorine that proved to be enough to successfully endure accidental exposure to this kind of chemical substance [58]. DOW FILMTECTM has commercialized a number of products based on this membrane and its properties [59]. The commercialization of membrane such as this can have a substantial effect on the cost projections and design involved in RO desalination [60]. In fact, this was the first spiral wound-type membrane element that was able to rival the DuPont asymmetric hollow-fiber polyamide B-9 Permasep® membranes originally produced in 1972. The overall success of FT-30 contributed to the creation of a wide range of comparable membrane products [61], including the UTC-70 by Toray Industries [62] and CPA2 membrane produced by Hydranautics [63]. Furthermore, the Permasep A-15 TFC membrane (**Table 2**) was manufactured using the reaction of 1,3-benzenediamine with a saturated cross-linking agent, cyclohexane-1,3,5-tricarbonyl chloride, and thus facilitated an aralkyl polyamide membrane which offers improved flux capacity [64, 65]. The application of isocyanato aromatic acyl halides (e.g., 1-isocyanato-3,5-benzenedicarbonyl chloride) as cross-linking agents for 1,3-benzenediamine was likewise patented so as to design the type of membrane that includes urea and amide linkages, both of which improve salt rejection and flux properties (**Table 2**) [66]. This particular membrane was labeled as X-20 and has showcased noticeably better resistance to chlorine and fouling because of its stronger polyamide–urea bond linkage and comparatively neutral surface charge [66].

4. Postsynthesis modification and optimization of interfacial Polycondensation reaction

The research and development of innovative polymeric materials for RO membranes has declined drastically after the groundbreaking success that came with the market production of cross-linked fully aromatic polyamide TFC RO membranes. Existing membrane products from RO desalination membranes' major manufacturers are still being designed around the original chemistry developed in 1980s. As a result, current membranes rely on the interfacial polymerization of monomeric aromatic amines [72]. DOW FILMTECTM, as the largest manufacturers of desalination membranes, is currently marketing and selling membrane products based on FT-30. Other companies, like Toray Industries, are using UTC-70, while Hydranautics membranes are based on NCM1, a membrane that is similar to CPA2. Trisep membranes are still using X-20. Alternatively, asymmetric membrane products have remained unchanged and are still based on conventional CA materials. For instance, the Toyobo HollosepTM products use CTA while remaining the primary asymmetric RO membrane presently in usage.

Although there have been no original polymeric membranes commercialized recently, the overall performance of RO membranes was noticeably enhanced over time (**Figure 3**). For example, the water permeability capacity has been doubled and the freshwater recovery potential can be

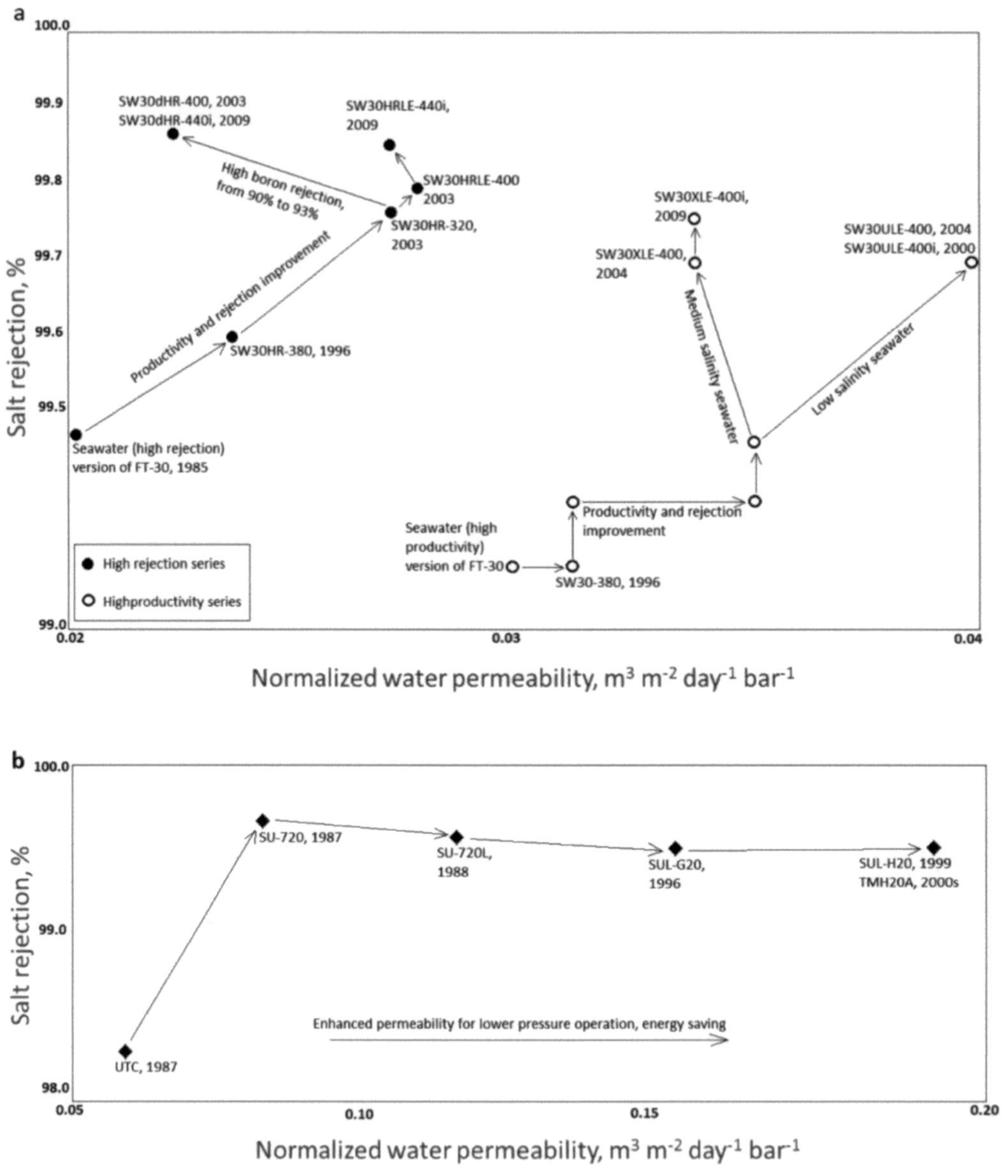

Figure 3. Development of RO membranes by reaction optimization and postsynthesis surface modifications: (a) Dow Filmtec seawater series and (b) Toray brackish water series.

as high as over 60%. Enhancements such as these are caused directly by surface modification as well as closer monitoring of interfacial polymerization reaction parameters. They are also linked to a substantially more effective module structure design [72, 73]. Furthermore, an in-depth research into the membrane structure has led to improvements in membrane characterization techniques [74]. Atomic force microscopy (AFM), for example, has remained a critical tool for confirming that membrane surface roughness properties may greatly improve permeability while at the same time ensuring higher salt rejection potential due to an increase in the effective membrane surface [75].

Tracking the development of commercially relevant RO membranes after 1990 has been difficult because of the low patenting activity among membrane manufacturers. To share posttreatment and chemical composition data that have been used with commercial RO membranes, researchers have chosen to merge the use of multiple methodologies and analytical techniques. One of these tools is the Rutherford backscattering spectrometry, and it can be used for elemental composition analysis of various layers and physicochemical characterization [76–79]. In order to gain a better understanding of chemical and physical membrane structure characteristics and how these characteristics are connected to the membrane's overall performance, a combination of analytical techniques, ATR-FTIR, XPS, TEM, and streaming potential measurement have likewise been applied [80–82]. The relevance of several analytical tools was also reviewed with respect to membrane characterization [83]. While there is ongoing proactive academic research conducted in the field, this review specialized in the engineering developments that offered the most innovation and were integrated into commercial products with direct industrial applications.

4.1. Surface modification

A key research area focused on membrane postsynthesis encompasses hydrophilization since it may help obtain an increase in chlorine resistance and permeability properties. Currently, the applied monomer reactants are not easily available and their preparation method is overly complex, despite the fact that there was some success in synthesizing membranes using monomer reactants that included hydrophilic groups, like eliminated amide hydrogen and carboxylate [84–87]. As a consequence, the preference is given to posttreatment that chemically modifies the membrane's surface properties, as well as numerous physical and chemical techniques established for this. Water-soluble solvents, including alcohols and acids, have been used for treating the membrane's surface. Mixtures of acid (hydrofluoric and hydrochloric acid) and alcohol (ethanol and isopropanol) in water have likewise been tested as a way to enhance rejection and flux performances because of the skin modification and partial hydrolysis initiated using acid and alcohol [88]. The existence of hydrogen bonds has been argued to instigate interactions between water and acid and thus to incite higher surface charges and enhance water flux and hydrophilicity values. For example, Mickols patented membrane surface posttreatment that includes alkyl or ammonia compounds, such as ethanolamine and ethylenediamine, that improved membrane salt rejection and flux capacity [89]. Specifically, a 70% flux improvement may be obtained when composite membranes are soaked in solutions containing a range of organic species, like sodium lauryl sulfate, triethylamine salt of camphorsulfonic acid, and glycerol [90, 91]. The membrane posttreatment base on an aqueous solution of poly(vinyl alcohol) (PVA) and a buffer solution can help improve the overall flux stability of the membrane as well as abrasion resistance potential [92, 93].

Hydrophilization has been produced by effectively coating the membrane's surface with additional hydrophilic compounds. The process of coating has been a favored method when it comes to addressing various fouling-related issues. A key example of this can be seen in the launch of the Hydranautics LFC series in 1996 [94], as well as the creation of the LFC3-LD membranes in 2005 [95] with the aim of targeting wastewater treatment/reclamation applications. Membranes such as these are neutrally charged and as such are designed to minimize the organic foulant adsorption. The relatively stable performance reported in research over time has been [96] connected to the poly(vinyl alcohol) (PVA) coating located on the surface of conventional fully aromatic polyamide membranes [81]. Coatings with poly(N,N-dimethylaminoethyl methacrylate) and PVA have likewise signaled greater resistance capabilities against chlorine attacks [97]. Recent research suggests that hydrophilic dendritic polymers were able to effectively modify the membrane's surface and enable fouling reduction [98, 99].

Substantial flow improvement was obtained when chemical treatment was used on a FT-30-type membrane. The FT-30 membrane was soaked in a 15% solution of hydrofluoric acid for seven consecutive days, after which period it showcased a slightly higher salt rejection capacity and a fourfold improvement in terms of flux. An analysis of the membrane's surface indicated that the fluorine ratio was higher due to the effects of the treatment. Furthermore, the etching of the surface helped to facilitate a thinner barrier layer [100]. While this method improved flux capacity without changing chemical structure, this approach is undermined over time by the leaching of hydrophilizing components that can cause the loss of any gained flux advantages [101].

Additional surface modification methods currently employed to covalently attach useful monomers onto the membrane's surface may include the applications of free radical-, redox-, radiation-, photochemical-, and plasma-induced grafting. Gas plasma treatment was similarly employed so as to encourage surface modifications. In this case, water permeability was enhanced using oxygen plasma treatment that relies on the addition of hydrophilic carboxylate groups, while the argon plasma treatment improved chlorine resistance potential by increasing the extent of crosslinking at the nitrogen sites [102, 103]. Recent research by Lin et al. suggests that the application of graft polymerization and atmospheric gas plasma surface activation on the conventional polyamide TFC membranes' surface is capable of significantly improving antifouling properties [104]. Once the gas plasma surface activation occurs, a polymeric brush layer will form with the help of the free radical graft polymerization based on acrylamide monomers or methacrylic acid. A brush layer of this type can successfully decrease the foulants' capacity to cling to the membrane's surface. This has been verified during multiple fouling tests, where such a membrane managed to outperform commercial low-fouling membrane LFC1, in particular during the mineral fouling tests. Both, graft polymerization and atmospheric gas plasma treatments can be easily adapted to large-scale membrane industrial applications and manufacturing.

4.2. Optimization of polymerization reactions

Optimization of interfacial polymerization reactions is another critical area of research, and it requires a controlled assessment of parameters such as kinetics, reaction time, reactant diffusion coefficients, solution composition, solvent solubility, polymer molecular weight range, curing time, nucleation rate, and other features of the microporous support [105–109]. Tomaschke and

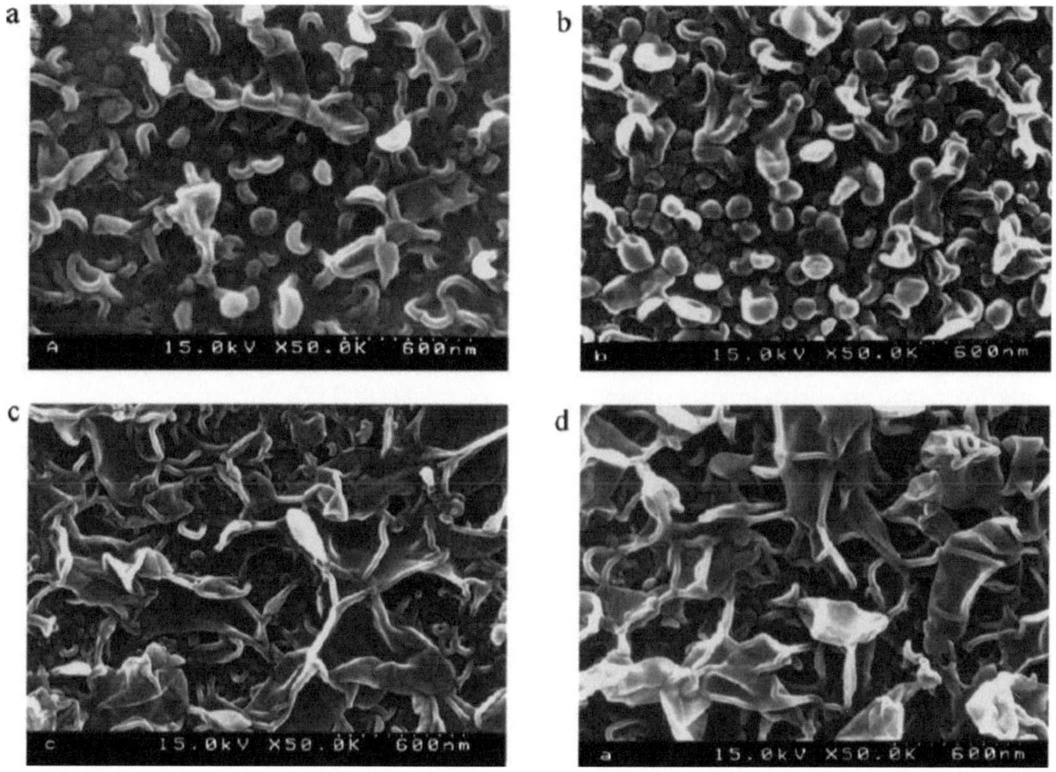

Figure 4. FE-SEM micrographs of RO membrane surface with various permselectivity. (reprinted with permission from Kwak et al., 1999) [117]. (a) Flux: 1.15 m^3 m^{-2} day^{-1}, salt rejection: > 96%. (b) Flux: 1.16 m^3 m^{-2} day^{-1}, salt rejection: > 99.1%. (c) Flux: 1.52 m^3 m^{-2} day^{-1}, salt rejection: > 98.7%. (d) Flux: 1.85 m^3 m^{-2} day^{-1}, salt rejection: > 98.4%. Note: Tested at 20°C, >15 bar for 0.2% NaCl solution. The scale bar is 600 nm for all figures.

Chau obtained early success when they experimented by introducing additives into the casting solution (amine reactants), which then incited extensive research into the application and addition of a range of additives [63, 110]. For instance, the application of amine salts, including the camphorsulfonic acid triethylamine salt, as an additive in an aqueous casting solution allowed for postreaction drying at temperatures greater than 100°C. As a consequence, a substantially more cross-linked membrane was created with improved characteristics in terms of the salt rejection properties and without the loss of flux capacities. Chau's research tried adding polar aprotic solvents, and particularly N,N-dimethylformamide, to the casting solutions. This experimental addition allowed for a greater carboxylate content in the membrane and subsequently improved water permeability.

The method of introducing additives into the casting solution can play a critical role when it comes to adjusting diffusivity, monomer solubility, protonation, and hydrolysis and to aiding the scavenging of reaction-inhibitory by-products [108]. A high number of patents reveal that the introduction of ethers, alcohols, polyhydric alcohol, water soluble polymers, or sulfur-containing compounds to the amine solution may enhance the membrane's overall permeability without substantively affecting its salt rejection capacity [111–115]. For instance, the miscibility of hexane and water was enhanced by the inclusion of dimethyl sulfoxide into the casting solution. In this case, the diffusion of amine monomers was improved and allowed to create a thinner barrier layer and better water flux potential [116]. **Figure 4** outlines the micrographs of RO membranes fabricated with the aid of various additives and that in turn facilitates diverse permselectivities [117]. Rather than mixing additives directly into the amine reactant solutions, the inclusion of a "complexing agent" into the acyl chloride solution, usually trimesoyl chloride, was patented. Research suggests that the phosphate-containing compounds are the ones used most frequently. These compounds include triphenyl phosphate that has the capacity to modify and reduce the repulsive interaction of acyl chloride with the other compounds by eliminating the halides that were formed during the formation of amide bonds. This minimizes concurrent hydrolysis and ensures that there is a sufficient reaction between amines and acyl halide, which improves membrane formation and helps increase permeate flux potential [118, 119].

Research conducted as part of recent membrane projects indicates that the successful addition of surface-modifying macromolecules into the reactants has been performed. During this method, additives can transfer toward the active surface area during polymerization and as a result change the surface chemistry. For example, the inclusion of hydrophilic surface-modifying macromolecules, like poly(ethylene glycol) end-capped oligomers, in the interfacial polycondensation reaction had overtime enhanced stability of salt rejection and membrane flux potential.

Author details

Amira Abdelrasoul[1]*, Huu Doan[2] and Ali Lohi[2]

*Address all correspondence to: amira.abdelrasoul@usask.ca

1 Department of Chemical and Biological Engineering, University of Saskatchewan, Saskatoon, Saskatchewan, Canada

2 Department of Chemical Engineering, Ryerson University, Toronto, Ontario, Canada

References

[1] Cadotte JE. Evolution of composite reverse osmosis membranes, in: Materials science of synthetic membranes. Symposium of the 187th Meeting of the American Chemical Society (ACS), St Louis, MO, USA 1985;273-294

[2] Petersen RJ. Composite reverse osmosis and nanofiltration membranes. Journal of Membrane Science. 1993;**83**:81-150

[3] Kumano A, Fujiwara N. Cellulose triacetate membranes for reverse osmosis. In: Li NN, Fane AG, Ho WSW, Matsuura T, editors. Advanced Membrane Technology and Applications. New Jersey: John Wiley and Shons, Inc; 2008

[4] Lee KP, Arnot TC, Mattia D. A review of reverse osmosis membrane materials for desalination - development to date and future potential. Journal of Membrane Science. 2011;**370**:1-22

[5] Hassler GL. The Sea as a Source of Fresh Water Department of Engineering Research Summary. Los Angeles: University of California; 1949

[6] Reid CE, Breton EJ. Water and ion flow across cellulosic membranes. Journal of Applied Polymer Science. 1959;**1**:133-143

[7] Loeb S, Sourirajan. *sea* Water demineralization by means of an osmotic membrane. In: Saline Water Conversion – II. Washington, D. C: American Chemical Society; 1963. p. 117-132

[8] Loeb S. The loeb-sourirajan membrane: How it came about. In: Synthetic Membranes. Washington, D.C: American Chemical Society (ACS); 1981. p. 1-9

[9] Hoehn HH. Aromatic polyamide membranes, in: Materials science of synthetic membranes. Based on a Symposium at the 187th Meeting of the American Chemical Society (ACS), St. Louis, MO, USA 1985;81-98

[10] Credali L, Baruzzi G, Guidotti V. Reverse Osmosis Anisotropic Membrane Based on Polypiperazine Amides. Patent Application No. 4129559 1978

[11] Senoo M, Hara S, Ozawa S. Permselective Polymeric Membrane Prepared from Polybenzimidazoles. Patent Application No. 3951920 1976

[12] Sekiguchi H, Sato F, Sadamitsu K, Yoshida K. Solute-Separating Membrane. Patent Application No. 4067804 1978

[13] Sudak RG. Reverse osmosis. In: Porter MC, editor. Handbook of Industrial Membrane Technology. New Jersey: Noyes Publication; 1990

[14] King WM, Cantor PA, Schoellenback LW, Cannon CR. High-retention reverse-osmosis desalination membranes from cellulose acetate. In: Turbak AF, editor. Membranes from Cellulose Derivatives. New York: Interscience Publisher; 1970

[15] Cannon CR, Cantor PA. Mixed Esters Pf Cellulose. Patent Application No. 3585126 1971

[16] Edgar KJ, Buchanan CM, Debenham JS, Rundquist PA, Seiler BD, Shelton MC, Tindall D. Advances in cellulose ester performance and application. Progress in Polymer Science. 2001;**26**:1605-1688

[17] Hoehn HH, Richter JW. Aromatic Polyimide, Polyester and Polyamide Separation Membranes. Patent Application No. RE30351 1980

[18] Beasley JK. The evaluation and selection of polymeric materials for reverse osmosis membranes. Desalination. 1977;**22**:181-189

[19] Kirsh YE, Popkov YM. New trends in the development of polymeric materials for reverse osmosis membranes. Russian Chemical Reviews. 1988;**57**

[20] Credali L, Parrini P. Properties of piperazine homopolyamide films. Poly. 1971;**12**:717-729

[21] Credali L, Chiolle A, Parrini P. New polymer materials for reverse osmosis membranes. Desalination. 1974;**14**:137-150

[22] Glater J, Hong SK, Elimelech M. The search for a chlorine-resistant reverse osmosis membrane. Desalination. 1994;**95**:325-345

[23] Parrini P. Polypiperazinamides: New polymers useful for membrane processes. Desalination. 1983;**48**:67-78

[24] Brousse C, Chapurlat R, Quentin JP. New membranes for reverse osmosis I. Characteristics of the base polymer: Sulphonated polysulphones. Desalination. 1976;**18**:137-153

[25] Guiver MD, Tremblay AY, Tam CM. Method of Manufacturing a Reverse Osmosis Membrane and the Membrane So Produced. Patent Application No. 4894159 1990

[26] Himeshima MKAY. The major developments of the evolving reverse osmosis membranes and ultrafiltration membranes. Polymer Journal. 1991;**23**:513-520

[27] Hara S, Mori K, Taketani Y, Noma T, Seno M. Reverse osmosis membranes from aromatic polymers. Desalination. 1977;**21**:183-194

[28] Murakami H, Igarashi N. Pbil tubular reverse osmosis. Application as low-energy concentrators. Industrial & Engineering Chemistry Product Research and Development. 1981;**20**: 501-508

[29] Congjie G. Development and extension of seawater desalination by reverse osmosis. Chinese Journal of Oceanology and Limnology. 2003;**21**:40-45

[30] Francis FS. Fabrication and evaluation of new ultra-thin reverse osmosis membranes. National Technical Information Services. 1966

[31] Rozelle LT, Cadotte JE, Corneliussen RD, Erickson EE. Final Report on Development of New Reverse Osmosis Membrane 1968

[32] Riley RL, Lonsdale HK, Lyons CR. Composite membranes for seawater desalination by reverse osmosis. Journal of Applied Polymer Science. 1971;**15**:1267-1276

[33] Reverse Osmosis Membrane. Patent Application No. 3926798 1975

[34] Kurihara M, Kanamaru N, Harumiya N, Yoshimura K, Hagiwara S. Spiral-wound new thin film composite membrane for a single-stage seawater desalination by reverse osmosis. Desalination. 1980;**32**:13-23

[35] Yasuda H. Reverse osmosis membranes formed by plasma polymerization of organic compounds. Applied Polymer Symposia. 1973:241-253

[36] Buck KR, Davar VK. Application of glow discharge polymerisation to the preparation of reverse osmosis. Membranes. 1970;**2**:238-239

[37] Peric D, Bell AT, Shen M. Effects of Deposition Conditions on Reverse Osmosis Characteristics of Composite Membranes Prepared by Plasma Polymerization of Allylamine, American Chemical Society, Polymer Preprints, Division of Polymer Chemistry. Vol. 171976. p. 534-538

[38] Yasuda H, Bumgarner MO, Marsh HC, Morosoff N. Plasma polymerization of some organic compounds and properties of the polymers. Journal of Polymer Science, Polymer Chemistry Edition. 1976;**14**:195-224

[39] Yasuda H, Marsh HC, Brandt ES, Reilley CN. Reactions of nitrogen and argon plasma with organic polymers, vol. 16. American Chemical Society, Polymer Preprints, Division of Polymer Chemistry. 1975:142-145

[40] Lai JY, Chao YC. Plasma-modified nylon 4 membranes for reverse osmosis desalination. Journal of Applied Polymer Science. 1990;**39**:2293-2303

[41] Yasuda H, Marsh HC. Preparation of composite reverse osmosis membranes by plasma polymerization of organic compounds. Iii. Plasma polymers of acetylene/Co/H2O. Journal of Applied Polymer Science. 1975;**19**:2981-2990

[42] Tran DT, Mori S, Suzuki M. Characteristics of polyimide-based composite membranes fabricated by low-temperature plasma polymerization. Thin Solid Films. 2008;**516**:4384-4390

[43] Bartels CR. A surface science investigation of composite membranes. Journal of Membrane Science. 1989;**45**:225-245

[44] Mysels KJ, Wrasidlo W. Strength of interfacial polymerization films. Langmuir. 1991;**7**: 3052-3053

[45] Riley RL, Fox RL, Lyons CR, Milstead CE, Seroy MW, Tagami M. Spiral-wound poly (ether/amide) thin-film composite membrane systems. Desalination. 1976;**19**:113-126

[46] Wrasidlo WJ. Semipermeable Membranes and the Method for the Preparation Thereof. Patent Application No. 4005012 1977

[47] Riley RL, Milstead CE, Lloyd AL, Seroy MW, Tagami M. Spiral-wound thinfilm composite membrane systems for brackish and seawater desalination by reverse osmosis. Desalination. 1966;**23**:331-355

[48] Hickman CE, Jamjoom I, Riedinger AB, Seaton RE. Jeddah seawater reverse osmosis installation. Desalination. 1979;**30**:259-281

[49] Light WG, Perlman JL, Riedinger AB, Needham DF. Desalination of nonchlorinated surface seawater using Tfcr membrane elements. Desalination. 1988;**70**:47-64

[50] Cadotte JE. Reverse Osmosis Membrane. Patent Application No. 4259183 1981

[51] Kamiyama Y, Yoshioka N, Matsui K, Nakagome K. New thin-film composite reverse osmosis membranes and spiral wound modules. Desalination. 1984;**51**:79-92

[52] Eriksson P. Water and salt transport through two types of polyamide composite membranes. Journal of Membrane Science. 1988;**36**:297-313

[53] Kurihara M, Himeshima Y. Major developments of the evolving reverse osmosis membranes and ultrafiltration membranes. Polymer Journal. 1991;**23**:513-520

[54] Cadotte JE. Interfacially Synthesized Reverse Osmosis Membrane. Patent Application No. 4277344 1981

[55] Koo JY, Petersen RJ, Cadotte JE. Esca characterization of chlorine-damaged polyamide reverse osmosis membrane, in: Polymer Preprints. Papers Presented at the Anaheim, California Meeting (ACS) Anaheim, CA, USA, 1986;391-392

[56] Cadotte JE, Petersen RJ, Larson RE, Erickson EE. A new thin-film composite seawater reverse osmosis membrane. Desalination. 1980;**32**:25-31

[57] Kwak SY, Jung SG, Yoon YS, Ihm DW. Details of surface features in aromatic polyamide reverse osmosis membranes characterized by scanning electron and atomic force microscopy. Journal of Polymer Science Part B: Polymer Physics. 1999;**37**:1429-1440

[58] Glater J, Zachariah MR, McCray SB, McCutchan JW. Reverse osmosis membrane sensitivity to ozone and halogen disinfectants. Desalination. 1983;**48**:1-16

[59] Larson RE, Cadotte JE, Petersen RJ. The Ft-30 seawater reverse osmosis membrane–element test results. Desalination. 1981;**38**:473-483

[60] Crowdus F. System economic advantages of a low pressure, spiral RO system using thin composite membranes. Ultrapure Water. 1984;**1**:29

[61] Light WG, Chu HC, Tran CN. Reverse osmosis TFC magnum elements for chlorinated/dechlorinated feedwater processing. Desalination. 1987;**64**:411-421

[62] Uemura TK, Yoshio H, Masaru K, Interfacially Synthesized Reverse Osmosis Membrane. Patent Application No. 4761234 1988

[63] Tomaschke JE. Interfacially Synthesized Reverse Osmosis Membrane Containing an Amine Salt and Processes for Preparing the Same. Patent Application No. 4948507 1990

[64] Sundet SA. Production of Composite Membranes. Patent Application No. 4529646 1985

[65] Sundet SA, Arthur SD, Campos D, Eckman TJ, Brown RG. Aromatic cycloaliphatic polyamide membrane. Desalination. 1987;**64**:259-269

[66] Jenkins M, Tanner MB. Operational experience with a new fouling resistant reverse osmosis membrane. Desalination. 1998;**119**:243-249

[67] Schiffer DK, Davis RB, Coplan MJ. Development of Composite Hollow Fiber Reverse Osmosis Systems, 1979

[68] Naaktgeboren AJ, Snijders GJ, Gons J. Characterization of a new reverse osmosis composite membrane for industrial application. Desalination. 1988;**68**:223-242

[69] Kawaguchi T, Minematsu H, Hayashi Y, Hara S, Ueda F. Amphoteric Ion-Permeable Composite Membrane. Patent Application No. 4360434 1982

[70] Uemura T, Himeshima Y, Kurihara M. Interfacially Synthesized Reverse Osmosis Membrane. Patent Application No. 4761234 1988

[71] Arthur SD. Multilayer Reverse Osmosis Membrane of Polyamide-Urea. Patent Application No. 5019264 1991

[72] Antrim B, Lesan R, Liu B, von Gottberg A. Worlds largest spiral element - history and development. Desalination 2005;**178**:313-324

[73] Bartels C, Hirose M, Fujioka H. Performance advancement in the spiral wound RO/NF element design. Desalination. 2008;**221**:207-214

[74] Matsuura T. Progress in membrane science and technology for seawater desalination - a review. Desalination. 2001;**134**:47-54

[75] Hirose M, Ito H, Kamiyama Y. Effect of skin layer surface structures on the flux behaviour of RO membranes. Journal of Membrane Science. 1996;**121**:209-215

[76] Mi B, Coronell O, BJ M˜n, Watanabe F, Cahill DG, Petrov I. Physicochemical characterization of NF/RO membrane active layers by Rutherford backscattering spectrometry. Journal of Membrane Science. 2006;**282**:71-81

[77] Zhang X, Cahill DG, Coronell O, Marinas BJ. Partitioning of salt ions in Ft30 reverse osmosis membranes. Applied Physics Letters. 2007;**91**:181903-181904

[78] Coronell O, BJ M˜n, Zhang X, Cahill DG. Quantification of functional groups and modeling of their ionization behavior in the active layer of Ft30 reverse osmosis membrane. Environmental Science & Technology. 2008;**42**:5260-5266

[79] Coronell O, Gonzaˇılez MI, BJ M˜a, Cahill DG. Ionization behavior, stoichiometry of association, and accessibility of functional groups in the active layers of reverse osmosis and nanofiltration membranes. Environmental Science & Technology. 2010;**44**:6808-6814

[80] Tang CY, Kwon YN, Leckie JO. Effect of membrane chemistry and coating layer on physiochemical properties of thin film composite polyamide RO and NF membranes: I. Ftir and Xps characterization of polyamide and coating layer chemistry. Desalination. 2009;**242**:149-167

[81] Tang CY, Kwon YN, Leckie JO. Effect of membrane chemistry and coating layer on physiochemical properties of thin film composite polyamide RO and NF membranes: Ii. Membrane physiochemical properties and their dependence on polyamide and coating layers. Desalination. 2009;**242**:168-182

[82] Tang CY, Kwon YN, Leckie JO. Probing the nano- and micro-scales of reverse osmosis membranes–a comprehensive characterization of physiochemical properties of uncoated and coated membranes by Xps, tem, Atr-Ftir, and streaming potential measurements. Journal of Membrane Science. 2007;**287**:146-156

[83] Cahill DG, Freger V, Kwak SY. Microscopy microanalysis of reverse osmosis and nanofiltration membranes. Material Research Society. 2008;**33**:27-32

[84] Kim CK, Kim JH, Roh IJ, Kim JJ. The changes of membrane performance with polyamide molecular structure in the reverse osmosis process. Journal of Membrane Science. 2000;**165**:189-199

[85] Li L, Zhang S, Zhang X, Zheng G. Polyamide thin film composite membranes prepared from isomeric biphenyl tetraacyl chloride and Mphenylenediamine. Journal of Membrane Science. 2008;**315**:20-27

[86] Roh IJ, Park SY, Kim JJ, Kim CK. Effects of the polyamide molecular structure on the performance of reverse osmosis membranes. Journal of Polymer Science Part B: Polymer Physics. 1998;**36**:1821-1830

[87] Moon EJ, Seo YS, Kim CK. Novel composite membranes prepared from 2,2 bis [4-(2-hydroxy-3-methacryloyloxy propoxy) phenyl] propane, triethylene glycol dimethacrylate, and their mixtures for the reverse osmosis process. Journal of Membrane Science. 2004;**243**:311-316

[88] Mukherjee D, Kulkarni A, Gill WN. Chemical treatment for improved performance of reverse osmosis membranes. Desalination. 1996;**104**:239-249

[89] Mickols WE. Method of Treating Polyamide Membranes to Increase Flux. Patent Application No. 5755964 1998

[90] Kuehne MA, Song RQ, Li NN, Petersen RJ. Flux enhancement in TFC RO membranes. Environmental Progress. 2001;**20**:23-26

[91] Pinnau I, Freeman BD. Formation modification of polymeric membranes: Overview. (ACS) Symp. Ser. 1999;**744**:1-22

[92] Tran CN, Maldonado AC, Somanathan R. Thin-Film Composite Membrane. Patent Application No. 5234598 1993

[93] Hachisuka H, Ikeda K. Composite Reverse Osmosis Membrane Having a Separation Layer with Polyvinyl Alcohol Coating and Method of Reverse Osmotic Treatment of Water Using the Same. Patent Application No. 6177011. p. 2001

[94] Gerard R, Hachisuka H, Hirose M. New membrane developments expanding the horizon for the application of reverse osmosis technology. Desalination. 1998;**119**:47-55

[95] Hydranautics' Lfc3-Ld Makes its Debut, Membrane Technology. Vol. 20052005. p. 3-3

[96] Wilf M, Alt S. Application of low fouling RO membrane elements for reclamation of municipal wastewater. Desalination. 2000;**132**:11-19

[97] Kang GD, Gao CJ, Chen WD, Jie XM, Cao YM, Yuan Q. Study on hypochlorite degradation of aromatic polyamide reverse osmosis membrane. Journal of Membrane Science. 2007;**300**:165-171

[98] Sarkar A, Carver PI, Zhang T, Merrington A, Bruza KJ, Rousseau JL, Keinath SE, Dvornic PR. Dendrimer-based coatings for surface modification of polyamide reverse osmosis membranes. Journal of Membrane Science. 2010;**349**:421-428

[99] Louie JS, Pinnau I, Ciobanu I, Ishida KP, Ng A, Reinhard M. Effects of polyether-polyamide block copolymer coating on performance and fouling of reverse osmosis membranes. Journal of Membrane Science. 2006;**280**:762-770

[100] Mukherjee D, Kulkarni A, Gill WN. Flux enhancement of reverse osmosis membranes by chemical surface modification. Journal of Membrane Science. 1994;**97**:231-249

[101] Kulkarni A, Mukherjee D, Gill WN. Flux enhancement by hydrophilization of thin film composite reverse osmosis membranes. Journal of Membrane Science. 1996;**114**:39-50

[102] Wu S, Xing J, Zheng C, Xu G, Zheng G, Xu J. Plasma modification of aromatic polyamide reverse osmosis composite membrane surface. Journal of Applied Polymer Science. 1997;**64**:1923-1926

[103] Gil'man AB. Low-temperature plasma treatment as an effective method for surface modification of polymeric materials. High Energy Chemistry. 2003;**37**:17-23

[104] Lin NH, NH KMM, Lewis GT, Cohen Y. Polymer surface nano-structuring of reverse osmosis membranes for fouling resistance and improved flux performance. JMCh. 2010;**20**:4642-4652

[105] Song Y, Sun P, Henry LL, Sun B. Mechanisms of structure and performance controlled thin film composite membrane formation via interfacial polymerization process. Journal of Membrane Science. 2005;**251**:67-79

[106] Karode SK, Kulkarni SS, Suresh AK, Mashelkar RA. New insights into kinetics and thermodynamics of interfacial polymerization. Chemical Engineering Science. 1998;**53**:2649-2663

[107] Dhumal SS, Wagh SJ, Suresh AK. Interfacial polycondensation – Modeling of kinetics and film properties. Journal of Membrane Science. 2008;**325**:758-771

[108] Ghosh AK, Jeong BH, Huang X, Hoek EMV. Impacts of reaction and curing conditions on polyamide composite reverse osmosis membrane properties. Journal of Membrane Science. 2008;**311**:34-45

[109] Ghosh AK, Hoek EMV. Impacts of support membrane structure and chemistry on polyamide-polysulfone interfacial composite membranes. Journal of Membrane Science. 2009;**336**:140-148

[110] Chau MM, Light WG, Chu HC. Dry High Flux Semipermeable Membranes. Patent Application No. 4983291 1991

[111] Koo JY, Kim N. Composite Polyamide Reverse Osmosis Membrane and Method of Producing the Same. Patent Application No. 6015495 2000

[112] Hirose M, Ikeda K. Method of Producing High Permeable Composite Reverse Osmosis Membrane. Patent Application No. 5576057 1996

[113] Hirose MS, Hiroki I, Masatoshi M, Kazuo T. Highly Permeable Composite Reverse Osmosis Membrane, Method of Producing the Same, and Method of Using the Same, Patent Application No. 5614099 1997

[114] Koo JY, Yoon YS. Composite Polyamide Reverse Osmosis Membrane and Method of Producing the Same. Patent Application No. 6063278 2000

[115] Kwak SY, Jung SG, Kim SH. Structure-motion-performance relationship of flux-enhanced reverse osmosis (RO) membranes composed of aromatic polyamide thin films. Environmental Science & Technology. 2001;**35**:4334-4340

[116] Kim SH, Kwak SY, Suzuki T. Positron annihilation spectroscopic evidence to demonstrate the flux-enhancement mechanism in morphology-controlled thin-film-composite (TFC) membrane. Environmental Science & Technology. 2005;**39**:1764-1177

[117] Kwak SY, Woo Ihm D. Use of atomic force microscopy and solid-state Nmr spectroscopy to characterize structure-property-performance correlation in high-flux reverse osmosis (RO) membranes. Journal of Membrane Science. 1999;**158**:143-153

[118] Mickols WE. Composite Membrane and Method for Making the Same. Patent Application No. 6878278 2005

[119] Mickols WE. Composite Membrane and Method for Making the Same. Patent Application No. 6337018 2002

Chapter 3

Novel Desalination RO Membranes

Amira Abdelrasoul, Huu Doan and Ali Lohi

Additional information is available at the end of the chapter

http://dx.doi.org/10.5772/intechopen.71719

Abstract

Since the initial operation of the first reverse osmosis (RO) desalination plants, only polymeric membranes have been employed for industrial use. As described in the previous chapter, the various advancements in the conventional polymeric RO membranes have been rather limited since the late 1990s, especially in the membrane permeability issue. Although new membrane modules have been released, however most of them are improved through a method that relies on increasing the membrane area per module. Recently, advances in nanotechnology have led to the development of nanostructured materials which may form the basis for new RO membranes. Li and Wang have included inorganic membranes and thin film nanocomposite membranes in a recent review, whereas Mauter and Elimelech have discussed the potential of carbon nanotube membranes for use as high flux membrane filters. In this chapter, the development of membranes that have been discussed in the previous two reviews will be briefly highlighted with a particular focus on the possibility of them being engineered into commercial RO membranes. At the same time this chapter includes a discussion about structured polymeric membranes synthesized via a new course featuring carbon-derived nanoporous membranes and biomimetic membranes. The coverage of all proposed novel desalination RO membranes in this section is aimed to provide a general overview of these materials and to draw a fair comparison of them possibly being developed into commercial RO membranes.

Keywords: polymeric membrane, reverse osmosis, mixed matrix membranes, inorganic membrane, nanoporous membranes

1. Introduction

Since the initial operation of the first reverse osmosis (RO) desalination plants, only polymeric membranes have been employed for industrial use [1, 2]. As described in the previous chapter, the various advancements in the conventional polymeric RO membranes have been rather limited since the late 1990s, especially in the membrane permeability issue. Although

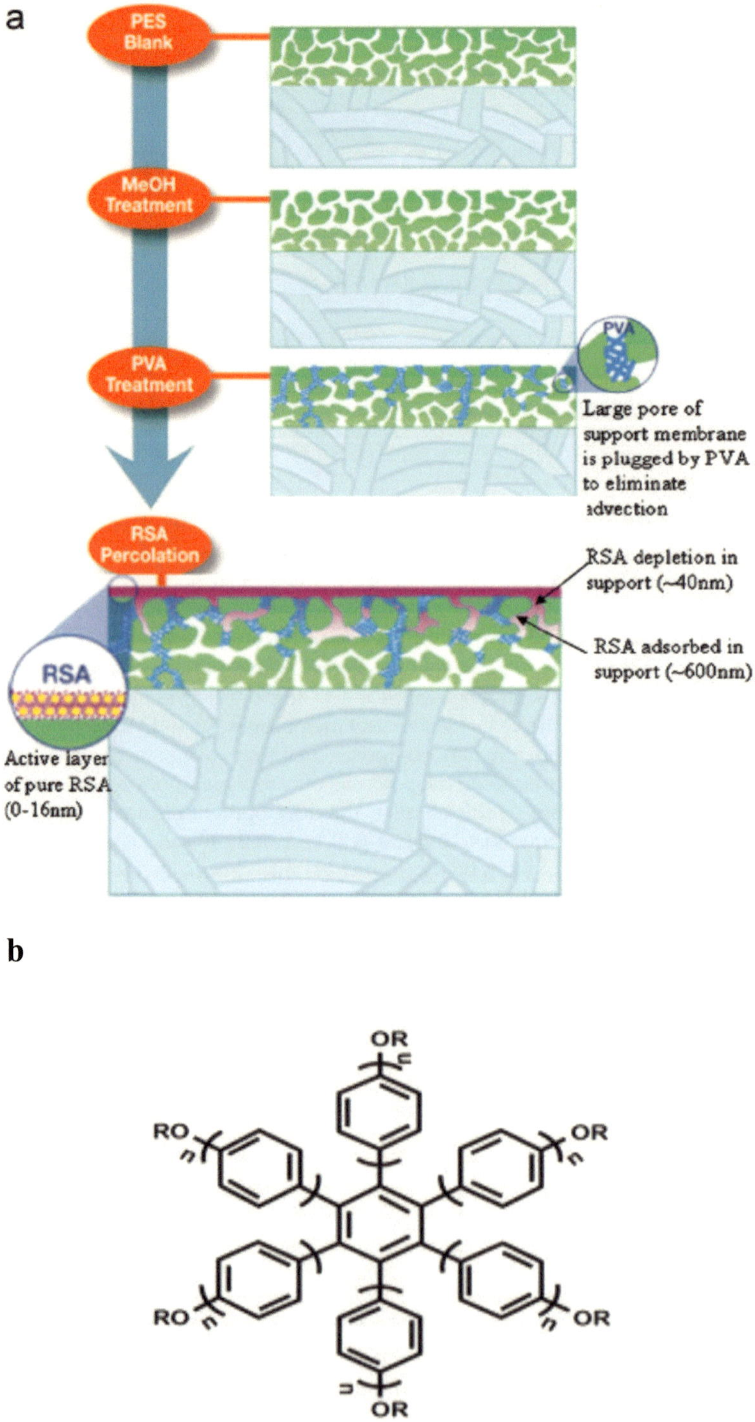

Figure 1. (a) RSA membrane synthesis process and (b) one of the RSA molecules tested [3].

a number of new membrane modules have been designed and tested, the majority of these are enhanced using a methodology that is reliant on the augmentation of the membrane's area per module. Developments in recent nanotechnology research have contributed to the creation of nanostructured materials that may prove to be a more effective basis for new types of RO membranes. For instance, Li and Wang research collaborative have focused on thin-film nanocomposite membranes and inorganic membranes in a recent overview, while Mauter and Elimelech examined at the possibilities of implementing carbon nanotube (CNT) membranes as high-flux membrane filters [1, 2]. This chapter outlines the evolution of membranes that have been discussed in the previous two reviews and briefly interrogates the possibility of these designs being engineered into successful commercial RO membranes. Furthermore, this chapter begins a critical discussion on the structured polymeric membranes synthesized using innovative course based on biomimetic membranes and carbon-derived nanoporous membranes. In this section, the covered methods of proposed desalination of RO membranes aim to offer a broader summary of membrane materials and outline the comparative potential of them being developed into viable commercial RO membranes.

2. Polymeric membrane created using rigid star amphiphiles

Recent research report has noted the development of a nanofiltration (NF) membrane based on rigid star amphiphiles (RSAs) [3, 4]. As shown in **Figure 1(a)**, nanofiltration membranes were created using percolation of methanol solutions of the RSAs and through an asymmetric polyethersulfone support that had been conditioned earlier with crosslinked poly-vinyl alcohol and methanol. **Figure 1(b)** illustrates one of the RSA molecules synthesized with the help of diverse cyclization methods as membrane-building block materials.

During atomic force microscopy (AFM) and scanning electron microscopy (SEM) analysis, the membrane showed an exceedingly smooth surface with an average roughness values within the range of 1–2 nm. Notably, this roughness potential is quite distinct from the roughness values found in commercially available NF membranes (20–70 nm). The RSA membrane barrier layer is extremely thin and features a thickness of around 20 nm. This composite multi-layer dendrimer arrangement helps regulate the narrower pore size distribution. When compared to the commercially available NF membranes, these membranes have exhibited similar contaminants rejection performances, but with doubled flux values. This newer approach to polymeric membrane synthesis can provide a superior alternative for tuning membrane structure, when considering polymeric NF and RO membranes' morphological similarities. Further research is still crucial when it comes to definitively verifying the method's suitability for the RO process. Specifically, the membrane's salt rejection potential has remained unspecified.

3. Ceramic/inorganic membranes

The ceramic types of membranes are primarily created using silica, alumina, zirconia, titania, or a combination of these materials. Because of their substantially higher manufacturing costs, the application of ceramic membranes is presently limited to cases where polymeric membranes cannot be properly used, such as in processes involving radioactives/heavily contaminated feeds, highly reactive environments, and elevated operating temperatures [5]. In most cases, ceramic membranes are based on a meso- or micro-porous active layer and a macro-porous support layer. The current top techniques for ceramic membrane production include paste extrusion for its supports, as well as powder suspensions slip-casting or the sol–gel processing of colloidal suspensions for active layer deposition. Ceramic membrane's foundational elements have been created using tubular modules that have been converted into monolithic honeycomb structures capable of providing much higher packing potential and efficiency. At this point, commercialized ceramic membranes are implemented extensively in micro- and ultra-filtration applications. On the other hand, ceramic membranes that can be used for successful nanofiltration still require further development and testing [6].

The industrial-scale application of ceramic-type membranes in domestic water production settings is uncommon; however, the membrane's overall process sturdiness has gained attention of researchers for purposes of pervaporation [9] and membrane distillation [7, 8]. A group of researchers from the New Mexico Institute of Mining and Technology have recently reported on the potential applications of ceramic membranes for RO desalination processes [10]. Due to the advantages of desalting oil field water, and inspired by molecular dynamic simulation results that showcased 100% of ion rejection potential in all-Si ZK-4 zeolite-type membranes [11], this research collaborative has conducted an experiment-based examination of the RO separation mechanism as well as assessed the viability of using ceramic membranes. The sub-nm inter-crystalline pores of the zeolite structure allowing for the rejection of the salt and passage of water molecules are outlined in **Figure 2** [12]. Theoretical sets of calculations suggest that ions can be completely rejected by zeolite membranes if they have pore sizes smaller than the hydrated ion's size. A-type zeolite membranes have 0.4-nm size pores and MFI-type membranes of 0.56-nm pore diameters. The first experimental attempt at developing an RO of an NaCl solution with the aid of the MFI silicalite-1 zeolite membrane represented a 77% salt rejection potential and a water flux as low as 0.003 $m^3 m^{-2} day^{-1}$ at 21 bar. It was similarly noted that the rejection potential of bivalent cations was greater than for monovalent ions in a test environment that implemented a feed including mixed ion species. As a consequence, the rejection of sodium ions in a mixed ion solution was smaller than the one occurring in a pure solution of NaCl. These data indicate that the filtration mechanism relies on, both, the size exclusion potential and the Donnan exclusion created by the charged double layer produced by adsorbed ions on the intercrystalline walls or the pore [13].

New research has been undertaken to improve the results by modifying the zeolite structure, despite the fact that the first RO tests with a zeolite membrane failed and, both, water flux and salt rejection values were far too low for practical industrial applications. The Si/Al ratio, which tends to dictate the membrane's surface charge and wettability parameters, has been enhanced to offer better salt rejection and flux. The Al content within the membrane can noticeably modify the surface hydrophilicity and, as a consequence, its affinity with water [14]. The flaws of the crystal structure can be minimized using a secondary growth of a zeolite layer on top of the zeolite placed onto a porous α-alumina substrate [15]. A combinatory approach such as this produced a significant improvement during testing of a 2-μm thick

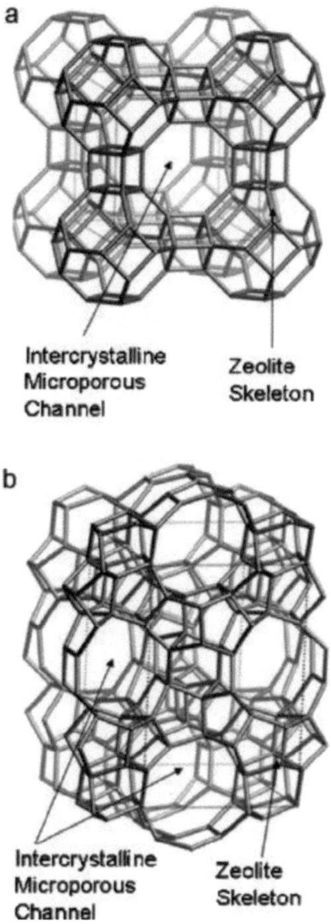

Figure 2. Micro-porous ceramic membrane structure: micro-porous channel in the crystalline structure (a) Type A zeolite and (b) MFI zeolite (reprinted with permission from Baerlocher et al. [12]).

zeolite membrane with 50:50 Si/Al ratio that rejected 92.9% of sodium ions and offered a water flux value of 1.129 $kgm^{-2} h^{-1}$ at 28 bar [16]. The same research tests were able to further reduce the thickness of the membrane to 0.7 μm, thus ensuring outstanding salt rejection (97.3%) and organic values (>99%), and obtaining a permeate flux improved by almost four times [17, 18].

Although innovation in the zeolite membranes research has shown substantial progress in the last 10 years, their economical viability and overall performance do not yet compare to that of the polymeric membranes. On the downside, the zeolite membrane thickness value is at least three times greater than the current quality polymeric RO membrane, and this creates higher overall resistance to permeate flux. As a consequence, ceramic-type membranes require a membrane area that is at least 50 times larger, than would be necessary in the polymeric counterpart, in order to reach similar production capacity. This membrane area parameter may even be greater if the lower-packing effectiveness and increased density are taken into consideration. Although zeolite membranes are intended to allow for higher organic rejection potential, organic fouling has produced almost 25% loss in flux capacity after only 2 h of operation, even though full recovery of flux was eventually achieved with the help of chemical washing [18]. The higher-salinity feeds are projected to cause shrinkage of the double layer because of the counter ions screening effect on the surface charge values. Thus, an unwanted growth in effective intercrystalline pore sizes can incite ion transport and then lower the rejection efficiency. These experimental tests were conducted with a low NaCl concentration (0.1%) and standard seawater desalination tests at 3.5% NaCl and must be further examined so as to assess their potential application for desalination of oil field seawater.

Another possible candidate for the formation of sub-nm porous materials is carbon. A carefully controlled carbide-derived carbon (CDC) materials' pore size distribution has been formerly reported in research studies [19]. There are several advantages to carbon since CDC allows for enhanced shape, control of pore size, and uniformity. For instance, these advantages can be seen in the manipulation of chlorination temperatures as indicated in **Figure 3**. Likewise noted was the synthesis of CDC membranes through the creation of a thin CDC film on top of the porous ceramic support [20]. This initial research study offers a method of developing asymmetric CDC membranes with an average pore size of about 0.7 nm, as well as of showcasing the potential for effective monovalent salt exclusion. Further research testing is crucial in order to assess the practicality and feasibility of CDC membranes for applications in long-term RO desalination settings.

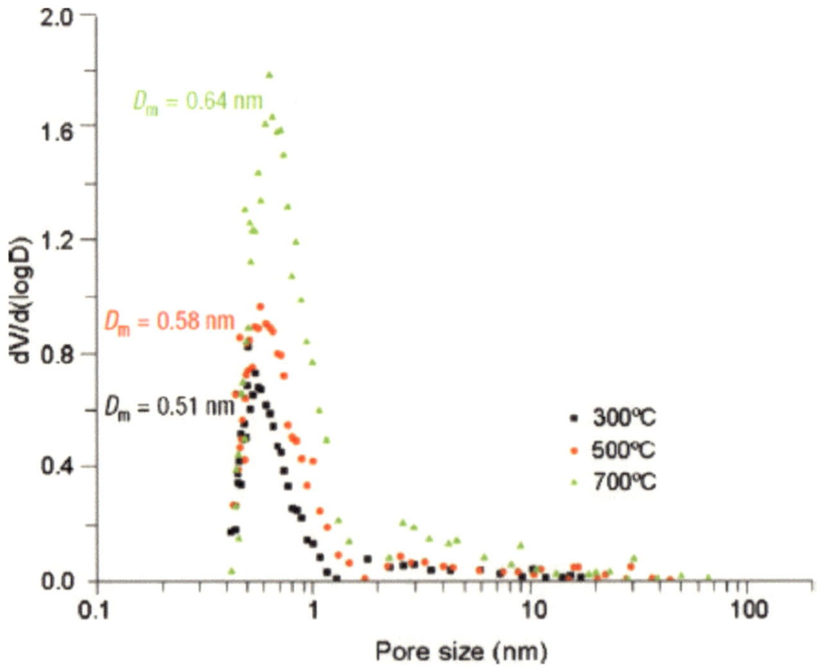

Figure 3. Differential pore size distribution for CDC-synthesized membrane at varying chlorination temperatures measured by methyl chloride adsorption (reprinted with permission from Gogotsi et al. [19]).

4. Mixed matrix membranes

The theoretical framework behind mixed matrix membranes (MMM), is the mixing of inorganic and organic materials and is not quite recent. In 1980, Universal Oil Product (UOP) created a silicalite-cellulose acetate MMM for the purposes of gas separation which showed advanced selectivity potential when compared to the more conventional polymeric membranes [21]. Even though MMMs were made for water/ethanol separation using pervaporation in the 1990s, the inclusion of inorganic materials as part of the organic RO thin-film composite (TFC) membranes and TFC membranes began in the early part of 2000s [22]. The primary aim of MMM is to efficiently merge the potential benefits offered by either material. In particular, there is a need to combine advantages such as good permselectivity, long operational experience, and higher packing density of the polymeric membranes, together with the improved biological, thermal, and chemical stability offered by the inorganic membranes [23].

4.1. Nanoparticle/polymeric membranes

In a wide range of industries, the synthetic membranes are becoming a key technology in separation processes and their applications. In most cases, the synthetic membranes are made out of organic materials like polymers, and inorganic materials like ceramics. Currently ongoing membrane research primarily looks at polymeric membranes because of their greater flexibility, smaller spaces required for installation, improved control over the pore-forming mechanisms, and comparatively smaller costs. In combination, these properties make the polymeric membranes a more suitable material in a variety of applications. However, frequent issues that researchers focus on when it comes to polymeric membranes are exposure to biofouling, lower fluxes, inferior mechanical strength, and higher hydrophobicity. The adding of nanoparticles into polymeric membranes became a key trend in the recent membrane research studies. Arguably, the insertion of nanosized materials may be able to incite synergistic effects when combined with diverse material types [24].

Iron is the most abundant transition metals as well as the fourth most available element in the earth's crust, qualities that make it the pillar of modern infrastructure. In fact, as a type of nanoparticle, iron has been overlooked in comparison to oxides and other metals like nickel, gold, platinum, and cobalt. Although this is an unfortunate limitation, there is an explicit reason for it. While iron's reactivity can be a key quality for macroscopic applications, especially rusting, it becomes a concern at the nanoscale levels. The fact that finely divided iron is pyrophoric is one of the main reasons responsible for iron nanoparticles not being more researched or applied. Such a severe reactivity potential has habitually made iron nanoparticles challenging to study and difficult to apply practically. On the other hand, iron has many properties to offer at the nanoscale, such as its catalytic and strong magnetic properties. In the recent research studies, iron's potential is being explored more rigorously, especially in relation to the membrane separation processes, which are discussed here.

As already noted, excessive reactivity of the iron metal makes it unfitting for application as pure metal nanoparticles [24]. As a result, instead of the pure iron nanoparticles, iron compounds are integrated into the polymeric membranes. The inclusion of iron compounds nanoparticles helps improve membrane performance with regard to some of their specific uses. Nafion is one of the polymeric membranes integrated with iron compounds. Specifically, Nafion is a sulfonated tetrafluoroethylene-based fluoropolymer-copolymer, and a first type in the synthetic polymers class featuring ionic properties, namely ionomers. Due to its superior thermal and mechanical stability, Nafion received much attention as a proton conductor used in proton exchange membrane (PEM) fuel cells. Films based on Nafion are effective as proton-conducting membranes in the case of direct methanol fuel cells (DMFCs). On the other hand, unmodified Nafion membranes usually have higher methanol permeability and this effectually makes them unsuitable for uses for the production of commercial-type fuel cells [25]. Therefore, the inclusion of nanoparticles featuring highly acidic and inorganic materials into Nafion membranes was shown to be the most effectual method for lowering methanol permeability potential [26–32]. Incorporating nanomaterials with greater surface acidity helps produce higher proton conductivity composite membranes, and the particle blocking of pores decreases methanol transport potential.

Inside the pores of Nafion membranes, the sol–gel synthesis of inorganic phases (SiO_2, TiO_2, ZrO_2) has been found as a successful modification direction that allows obtaining higher selectivity values [32–35]. Zeolites are also used as modifiers in Nafion membranes, primarily because of their high surface acidity, greater water intake values, and intrinsically narrow pore size distributions [36]. Recent research project reported results of methanol transport properties in a number of Nafion-composite membranes integrating micro- and nanosized zeolite particles (Fe-silicate-1), in situ crystallized Fe-silicalite-1, and amorphous silica, comparable to the unmodified commercial nafion-115-type membranes [25]. During the development of the composites, a supercritical carbon dioxide treatment was applied to some of the membranes before the integration of the inorganic phase. In this case, two approaches to zeolites deposition were used, specifically, direct in situ synthesis inside the pores of Nafion membrane and deposition from colloid or suspension solution. Relatively low methanol permeability potential was achieved in composite-type membranes created with the aid of the colloidal intercalation route (from colloidal Fe-sillicate-1 and silica solution), as well as using the in situ synthesis of Fe-silicate-1 inside the Nafion membrane's pores. In order to change its structure, supercritical CO_2 activation of a Nafion membrane before the zeolites deposition was implemented. The developed Nafion-zeolite composite membranes indicated a substantial reduction in methanol permeability, if colloidal rather than suspended Fe-silicalite-1 particles were used for deposition. Furthermore, there was a 19-fold greater selectivity value if compared to pure commercial Nafion-115 membrane or composite membranes made without previous supercritical treatment. The approach of in situ synthesis of zeolite inside the membrane's pores was determined to be quite successful for the production of composites and ensured a sixfold greater selectivity potential for the composite membrane, if compared to pure Nafion [25]. Iron compounds are likewise integrated into other polymeric membranes so as to create proton exchange membranes, in addition to iron's use with Nafion membranes and the reduction of methanol permeability. Further research has been recently done on the production of ion-conducting membranes by self-assembly of surface-charged nanoparticles [37]. Researchers suggest that the membranes created using closely packed nanoparticles offered noticeably greater proton conductivity if compared to solution-cast films of similar ion-exchange capacity (IEC) and composition. Nonetheless, there was a degree of limitation on maximum IEC, since high-IEC membranes showcased extreme swelling in water and thus substantially deterred the testing for proton conductivity. Membranes featuring proton-conducting particles constructed in a suitable matrix may be designed in a manner that avoids these issues. In particular, the particles could be carefully aligned to gain the percolation necessary for proton conduction, while the swelling in methanol or water can be supervised through the selection of a water-resistant matrix. Another research project outlines the synthesis of composite particles with sulfonated crosslinked polystyrene (SXLPS) intended for use in the proton exchange membranes and ribbed fuel cells [38]. In this instance, the approach implemented for polymerization was comparable to the mini-emulsion polymerization described by Ramirez et al. [39]. Admittedly, a number of procedure modifications were necessary for the production of functional and crosslinked polymer-iron oxide composites. Researchers likewise reported a membrane production process that requires the alignment of synthesized particles in a high-performance sulfonated poly(etherketoneketone) (SPEKK) matrix [40] and offers a number of properties of PEMs for fuel cell uses. The membrane's ultimate properties depended on a range of aspects, including the matrix, size of particles, and the particle's IEC. The primary aim of this research project was to show an applicable membrane-fabrication method that can be used to improve the PEMs' overall conductivity. As noted [38], the composite ion-conducting nanoparticles were synthesized with the help of emulsion polymerization. The particles' polymeric component contained sulfonated crosslinked polystyrene as well as the inorganic component of γ-Fe_2O_3. As a result of the experimental runs, a particle synthesis method was compiled so as to elucidate the abnormal morphologies of composite particles. The distribution breadth increased with sulfonated content, while the average synthesized particles' diameter value correlation to changing feed compositions were within the range of 230–340 nm (**Figure 4**).

Titanium oxide (TiO_2) is a relatively common photocatalytic material that has been frequently employed for decomposition and disinfection of various organic compounds [41]. This potentially dual property makes TiO_2 a compelling anti-fouling coating option. The anatase TiO_2 nanoparticles (<10 nm) are prepared using the process of controlled hydrolysis of titanium tetraisopropoxide. These particles are then dip-coated as an interfacially polymerized fully cross-linked polyamide TFC membrane with a surface layer further functionalized with carboxylate groups [42]. Notably, these carboxylate groups are essential for the TiO_2 self-assembly within the barrier layer and require the use of an adsorption mechanism. Experimental testing based on an *Escherichia coli* containing feed water has shown improved anti-biofouling properties. In fact, these improvements aided with UV excitation and did not compromise the salt rejection and flux performance of the original membrane and its properties. There was no substantial loss of TiO_2 nanoparticles from the membrane notes after a continuous 7-day RO research trial run [42, 43].

Moreover, zeolite nanoparticles have been used similarly during the preparation of MMMs (**Figure 5**). The first step in this process is for the zeolite nanoparticles to be synthesized with the help of a templated hydrothermal reaction. What follows are a series of difficult processes that involve carbonization, sodium exchange, template removal, and calcination [44]. The formed NaA-type zeolite particles appear to be within the size range of 50–150 nm and feature an Si/Al

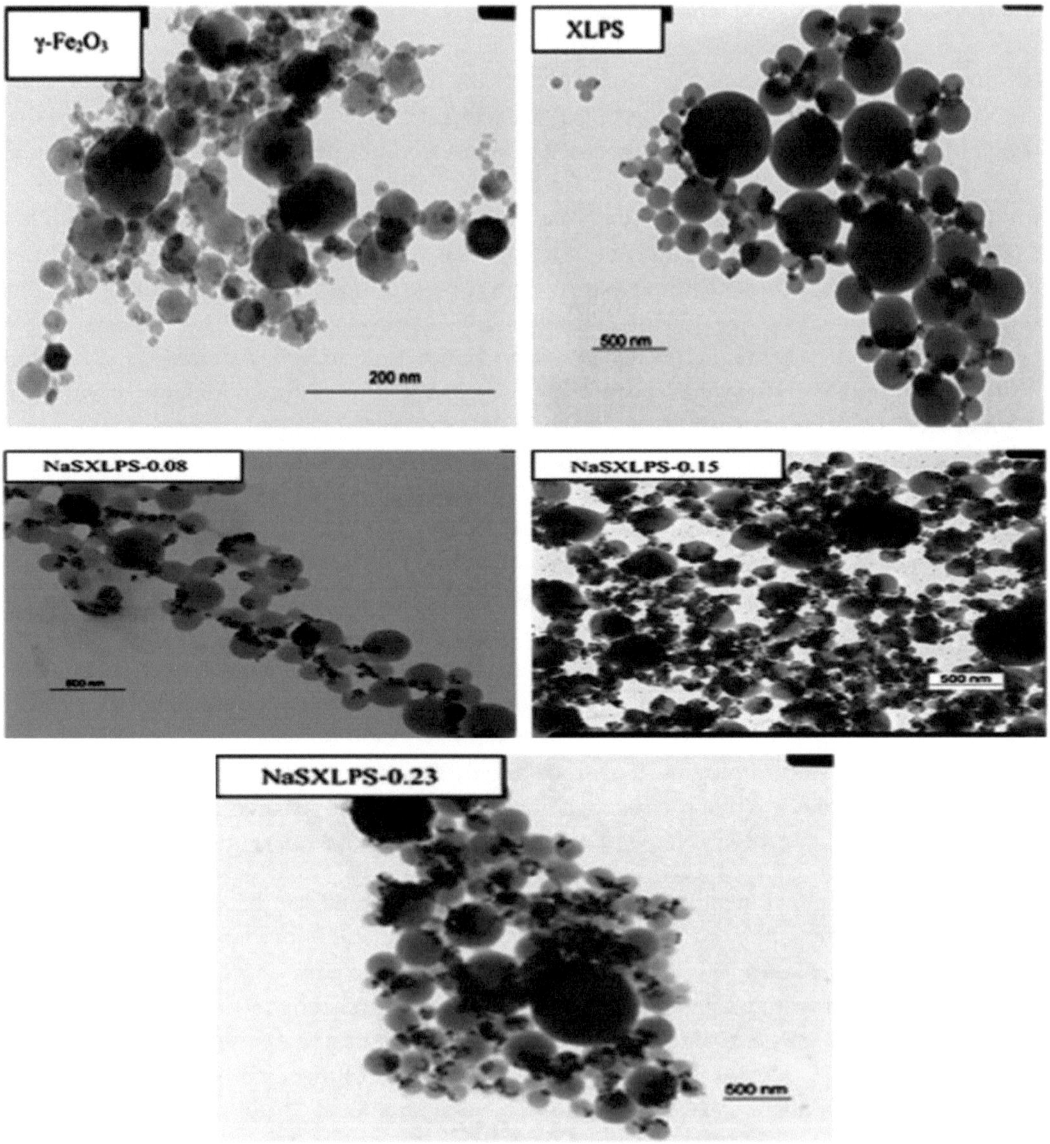

Figure 4. TEM images of the starting material γ-Fe2O3 and synthesized crosslinked polystyrene γ-Fe2O3 particles. Adapted from Ref. [24].

ratio of 1.5. Reports suggest that these particles are highly hydrophilic (contact angle of <5°) and have negatively charged 0.4-nm pores that repulse anions. Before the interfacial polycondensation reaction can occur, the zeolite nanoparticles are dissolved into a crosslinking agent solution, such as trimesoyl chloride dissolved in hexane. This method differs from the approach based on dipping the previously made membrane into a nanoparticle-containing solution, as occurs with the TiO_2 nanocomposite-type membranes. In this case, a homogeneous zeolite particle dispersion is created using ultrasonication, prior to the enactment of the standard interfacial polymerization

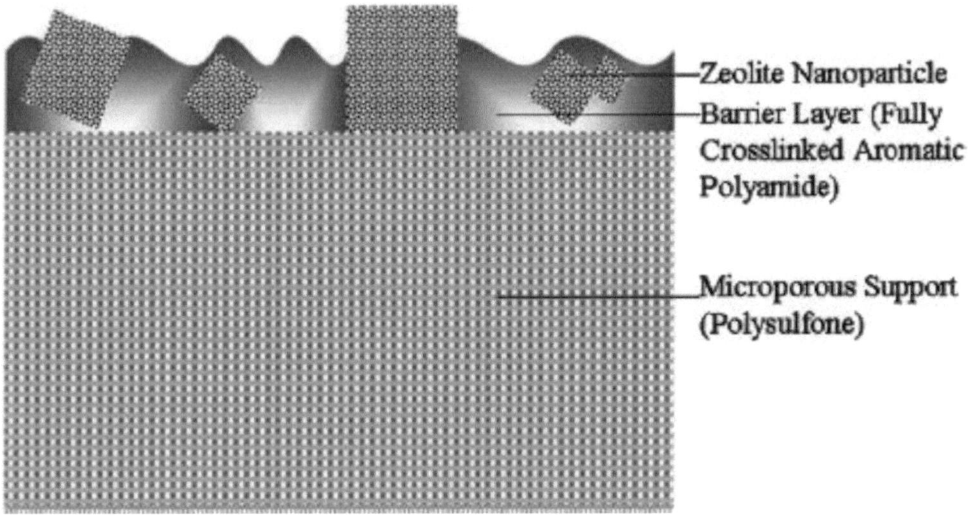

Figure 5. Schematic cross-section of zeolite nanocomposite membrane (reprinted with permission from Jeong et al. [44]).

process. During this experiment, the RO membranes with a variety of zeolite-loading values were prepared and the subsequent changes in the membrane's properties were examined. For instance, the membranes became smoother in quality, more negatively charged with increasing nanoparticle loading, and more hydrophilic. Relative to the hand-cast TFC membranes without zeolite nanoparticles, the MMM membranes showed 90% of flux and a minor increase in the salt rejection potential. Researchers indicate that this could be due to the changes in membrane morphology and improved Donnan exclusion with regard to zeolite particles [44, 45]. The altered surface properties and membrane's new separation performance facilitated by the variations in zeolite nanoparticle loading are shown in **Figure 6**.

4.2. Carbon nanotube/polymeric membranes

A number of researchers begun to pay closer attention to carbon nanotubes because of the resemblance between their fluid transport characteristics and the water transport channels in biological type of membranes [46]. Successful experimental instance of fluid flow in a CNT membrane was first observed in 2004 [47]. For this experiment, well-aligned multi-wall CNT membranes were created with the aid of catalytic chemical vapor deposition (CCVD) situated on the surface of quartz substrates. These membranes were then spun coated with polystyrene in order to effectively cover inter-tube gaps, while plasma etching was employed to open the CNTs' tips. An experimental assessment of water transport in a solid polystyrene film membrane that includes 7-nm diameter multi-wall CNTs indicates that the detected flow velocity was four to five orders of magnitude greater than the rate expected from the Hagen–Poiseuille equation governing macroscale hydrodynamics [48]. A research report notes another fluid flow experiment based on a CNT membrane synthesized with the help of nanofabrication methodologies (**Figure 7(a)**). This type of membrane had a double-wall CNTs with <2-nm diameter and exhibited flow velocity rates of three to four orders of magnitude greater than the values calculated theoretically [49].

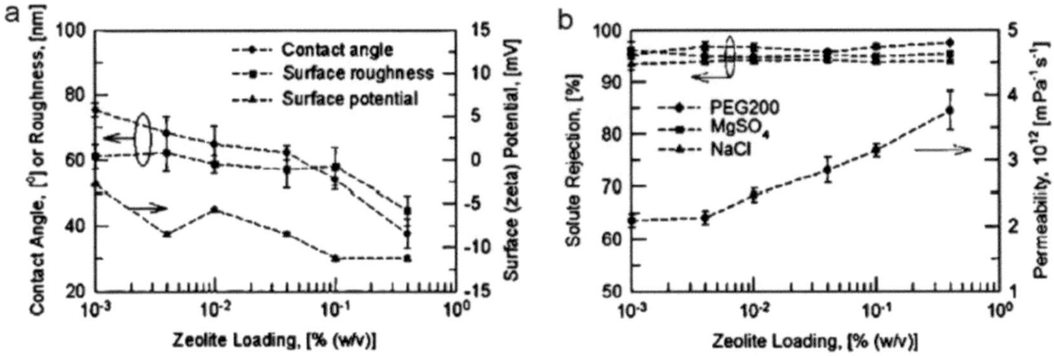

Figure 6. Effect of zeolite-loading dosage on (a) surface properties and (b) separation performance (reprinted with permission from Jeong et al. [44]).

A new investigation focused on the water transport potential passing through template-grown car bon nanopipes (CNPs) of around 44 nm in diameter. In this case, the CNPs were designed based on non-catalytic CVD that yield an amorphous, or turbostratic, graphitic structure (**Figure 7(b)**). The existence of the template helps eliminate structural imperfections, from tortuosity, and including, to branching and pore misalignment. However, the flow enhancement factor of over the theoretically predicted value is around 20, which is substantially less than was observed in earlier cases [50]. Researchers argue that this is caused by the varying structure and surface chemistry, if compared to the CNTs described in the previous articles and prepared by CVD.

A number of ongoing scientific discussions have emerged from the experimental results showcasing fast water transport [46, 51, 52]. The critical source of this exceptionally fast water transport in CNTs is not entirely defined and multiple, often contradictory, justifications have been observed in studies exploring this property [53–55]. One of the explanations suggests that the formation of a robust hydrogen-bonding network between atomically smooth hydrophobic inner nanotube wall and water molecules can facilitate spontaneous inhibition. This in turn leads to the creation of a vapor layer between the surface and the bulk flow, the latter responsible for water transport in a projected flow [56, 57]. Alternatively, an argument can be made that the frictionless water flow is caused by the creation of a layer of liquid water molecules on the CNT walls. This layer can in theory offer a type of "shielding" to the bulk water molecules in a way that forces them to flow at a faster rate [58].

The ion transport through CNT channels has been examined from, both, computational and experimental angles. Transport of ions with different valences has been assessed in double-walled-type CNTs of 1–2-nm diameter and functionalized with negatively charged groups. Despite the ion rejection potential not being suitably high for the purposes of desalination, this research study suggests that CNTs' ion exclusion mechanism is controlled by electrostatic interactions (Donnan exclusion), instead of the steric effects. This analysis is grounded in the conclusion that the electrostatic screening length and solution pH considerably alter the ion rejection potential [49, 59]. Majumder's group also showed two specific methods of changing the selectivity of various ion species, specifically, the pore size modification using CNT tip functionalization and the voltage-based gate control approach [60, 61]. While the monovalent salt rejection potential is not verified in both instances, these research studies illustrate the potential of effectively

http://dx.doi.org/10.5772/intechopen.71719

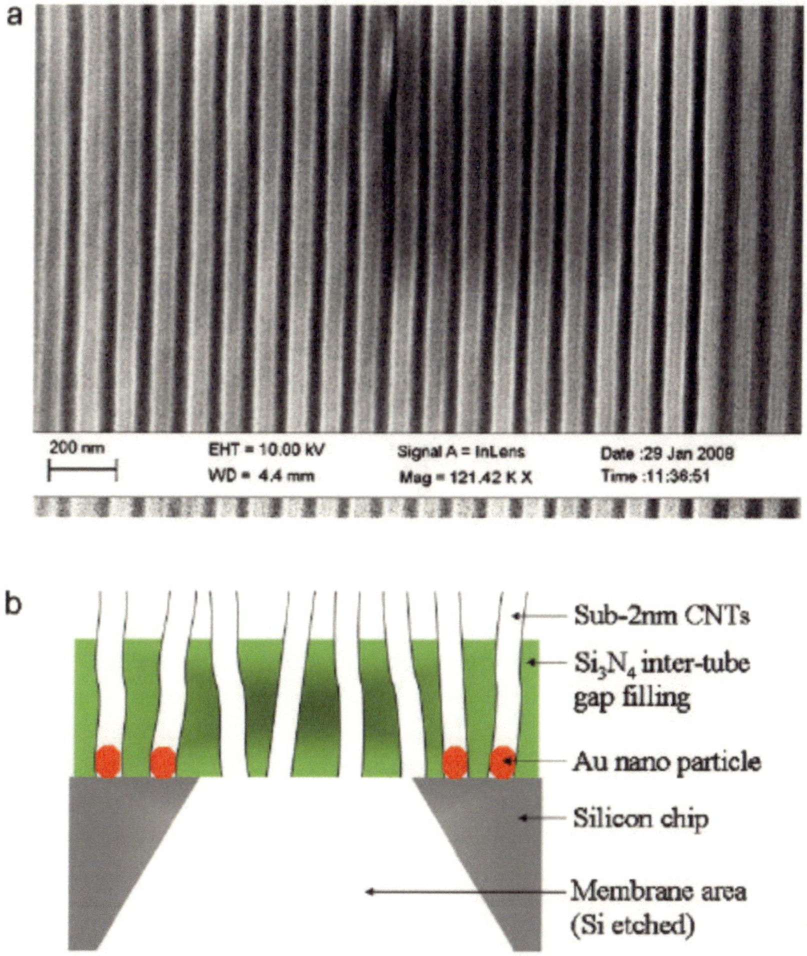

Figure 7. (a) Schematic of CNT membrane reported in [49]; and (b) SEM micrograph showing cross-section of CNP membrane [50].

altering pore characteristics and as a consequence improving selectivity. A molecular dynamic simulation run based on RO using CNTs has been conducted in connection to the physical size exclusion mechanisms. This experiment-based testing indicated that 0.8-nm CNTs can entirely reject the salt, while at the same time offer at least a fourfold flux enhancement over the current high-quality TFC RO membranes, and contingent on the expected CNT packing density values [62, 63]. This simulation did not consider the influence of charge functionalities that frequently occur at both ends of CNTs. Their occurrence may be capable of expanding the CNT size regime, since the creation of charge double layer could help raise the total salt rejection values [51].

Mauter and Elimelech combined the preceding desalination developments in CNT membranes in order to prepare the key data for the next generation in CNT membrane production [2]. Their study suggests that although CNT membranes are a highly encouraging area of research when it comes to flux improvement, a substantially more in-depth research work is necessary for the successful creation of synthesis methodologies. In particular, they argue that there is a need for synthesis methodologies that would be able to align arrays of single-walled CNTs with sub-nanometer diameters, as well as improve tip functionalization and offer a more effective salt rejection potential. The creation of the CNT/polymeric membranes from the experimental studies outlined above [47–49], and in various other gas separation studies, requires a multiplicity of complex steps, including polymer filling of the inter-tube spaces, substrate removal, catalytic growth of CNTs onto expensive substrates, and CNT tip opening using etching. Moreover, the CNT diameter size distribution is not small enough to align with the simulation studies being conducted. To transcend these concerns, a patent has revealed a dynamic where there is a mixing of CNTs into solutions, primarily crosslinking agent solutions like trimesoyl of isophthalic chlorides, for the construction of successful composite polymeric membranes (**Figure 8**). This alternative option may ensure that the CNTs can be efficiently fixed into the barrier layer created using conventional interfacial polymerization on a micro-porous polyethersulfone support [64]. In this instance, the CNTs must be functionalized in order to ensure improved organic solvent solubility, and in this particular patent the CNTs are functionalized with octadecylamine. The membrane design in this manner can be effortlessly customized for RO systems and current filtration needs, with the CNTs measuring 0.8 nm in diameter. The water then permeates through the membrane by, both, the embedded CNT pathways and the conventional polymeric barrier layer. A test revealed as part of the patent shows a comparison between membranes developed with and without embedded CNTs, so as to illustrate the improved flow capacity achieved using the CNT pathways. With CNTs present in the experiment, a marginally higher salt rejection potential was obtained (97.69% as compared with 96.19%), as well as an almost a doubled water flux value (44 L m^{-2} day^{-1} bar^{-1} as compared with 26 L m^{-2} day^{-1} bar^{-1}). Although the resulting data are highly encouraging, the synthesized membrane disc was only 47 mm in diameter and as such still leaves a research gap. Extended

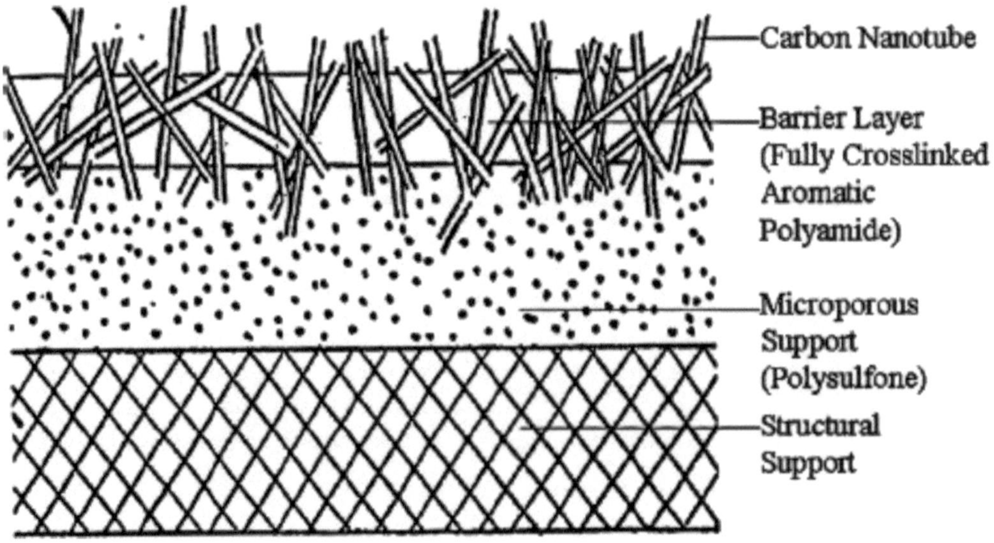

Figure 8. Schematic cross-section of CNTs-embedded TFC membrane [64].

http://dx.doi.org/10.5772/intechopen.71719

studies using membranes with much larger surface areas are needed before large-scale manufacturing methodologies can be successfully established, tested, and implemented.

5. Development of robust organosilica membranes for reverse osmosis

The hybrid organically bridged silica membranes have attracted considerable attention due to their successful performances in a variety of applications. The development of robust reverse osmosis membranes that can withstand aggressive operating conditions is still a major challenge. A new type of microporous organosilica membrane has been developed and applied in reverse osmosis. Sol–gel-derived organosilica RO membranes reject isopropanol with a rejection potential higher than 95%, demonstrating superior molecular-sieving ability for neutral solutes of low-molecular weight. Due to the introduction of an inherently stable hybrid network structure, this membrane withstands higher temperatures in comparison to the commercial polyamide RO membranes and is resistant to water to at least 90°C with no obvious changes in its filtration performance. Furthermore, both an accelerated chlorine-resistance test and Fourier transform infrared analysis confirm excellent chlorine stability for this material, a quality that demonstrates promise for a new generation of chlorine-resistant RO membrane materials [65]. Organosilica membranes were prepared using the sol–gel technique via a polymeric route with $(EtO)_3Si$- CH_2CH_2-$Si(OEt)_3$ (BTESE) as a single precursor. The synthesis of nanometer-sized BTESE polymer sol was performed as previously reported [66]. The organosilica membrane composition is presented in **Figures 9** and **10**.

Figure 9. Cross-sectional SEM image of the BTESE-derived organosilica membrane [66]; Copyright 2011, Journal of Membrane Science.

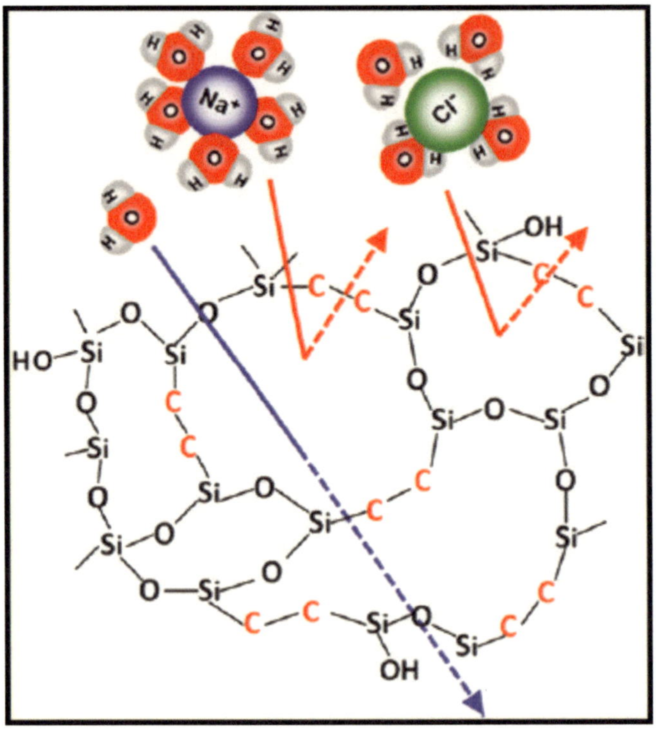

Figure 10. Hybrid organically bridged silica membranes [66]; Copyright 2011, Journal of Membrane Science.

The organosilica membranes derived from BTESE exhibit superior molecular-sieving abilities for neutral solutes of low-molecular weight. Exceptional hydrothermal stability has been obtained due to the introduction of an inherently stable, organically bridged silica network structure, significantly broadening the application fields of the organosilica membranes. Furthermore, these organosilica RO membranes already show excellent chlorine stability under a wide range of chlorine concentrations. For example, after a total chlorine exposure of up to 35,000 ppm for 3 h, there was no noticeable change in separation performance.

6. Biomimetic RO membranes

Superior water transport characteristics found in biological membranes have directed researchers toward the study of membranes that incorporate aquaporins or proteins acting as water-selective channels in biological cell membranes [67]. Membranes that include bacterial aquaporin Z proteins have been shown to offer enhanced water transport capabilities if compared to conventional RO membranes [68]. For example, aquaporins have been integrated into the walls of self-assembled polymer vesicles consisting of triblock copolymer, poly(2-methyl-2-oxazoline)-block-poly(dimethylsiloxane)-block-poly(2-methyl-2-oxazoline). Stopped-flow light-scattering experiments were carried out as an initial permeability test for the aquaporin-triblock polymer vesicles. The experimental data from this test indicate that there was at least one order of magnitude detected in permeability improvement, if compared to the standard commercially available TFC RO-type membranes [68]. While a salt separation test was not included in the experiment, an exceedingly high salt rejection potential is anticipated from the aquaporins due to their functional biological performance, which only permits the passage of water molecules. As a result, these conditions embody a superior prospect for the creation of ultra pure water [68–70]. Studies such as these have a narrow focus that limits them to examining water permeability properties through a barrier layer composed of triblock polymers and aquaporins. Numerous key issues, including the understanding of the membrane's fouling resistance, identifying appropriate support materials, and locating the suitable range of operating conditions, have to be dealt with before this membrane can be effectively developed for industrial and practical applications. The relevance of the NF membranes as a biomimetic membrane support has been likewise observed [71]. By applying the vesicle fusion approach, a continuous phospholipid bilayer was formed on an NTR-7450 and was able to fully cover the membrane. Additional research studies must be undertaken with the aim to include aquaporins into the phospholipid bilayer for practical applications in water purification processes. In Denmark in 2005, a company Aquaporin was formed specifically to advance these membranes for long-term industrial applications. The company has recently received a patent for the methodology of building membranes that successfully include aquaporins (**Figure 11**) [72]. As part of this patented approach, instead of using triblock polymers, the aquaporins are reorganized into lipid bilayers developed with the help of the Langmuir–Blodgett method, using spin coating, or vesicle fusion method. Aquaporin's patent likewise notes two distinct membrane orientations: (1) first orientation where a lipid bilayer including the aquaporins is inserted between two hydrophilic porous support layers, like polysulfone, cellulose, or mica; and (2) second orientation where a lipid bilayer including aquaporins is constructed over a hydrophobic porous support membrane, like the porous PTFE film. Although severe concentration polarization and fouling are reported, the patent does not offer significant numerical data with regard to the membrane's salt rejection and flux potential values.

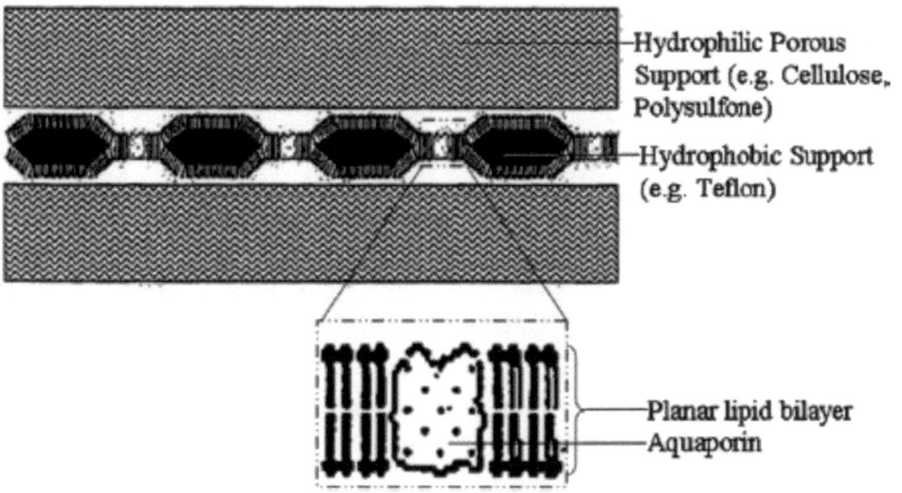

Figure 11: Schematic cross section of aquaporin-embedded membrane [72].

Author details

Amira Abdelrasoul[1]*, Huu Doan[2] and Ali Lohi[2]

*Address all correspondence to: amira.abdelrasoul@usask.ca

1 Department of Chemical and Biological Engineering, University of Saskatchewan, Saskatoon, Saskatchewan, Canada

2 Department of Chemical Engineering, Ryerson University, Toronto, Ontario, Canada

References

[1] Li D, Wang H. Recent developments in reverse osmosis desalination membranes. JMCh. 2010;**20**:4551-4566

[2] Mauter MS, Elimelech M. Environmental applications of carbon-based nanomaterials. Environmental Science & Technology. 2008;**42**:5843-5859

[3] Lu Y, Suzuki T, Zhang W, Moore JS, Mariñas BJ. Nanofiltration membranes based on rigid star amphiphiles. Chemistry of Materials. 2007;**19**:3194-3204

[4] Suzuki T, Lu Y, Zhang W, Moore JS, Mariñas BJ. Performance characterization of nanofiltration membranes based on rigid star amphiphiles. Environmental Science & Technology. 2007;**41**:6246-6252

[5] Siskens CAM. Applications of ceramic membranes in liquid filtration. In: Burggraaf AJ, Cot L, editors. Membrane Science and Technology. Elsevier; 1996. Vol. **13**. p. 619-639

[6] Pabby AK, Rizvi SSH, Sastre AM. Handbook of Membrane Separations: Chemical, Pharmaceutical Food and Biotechnological Applications. Boca Raton: CRC Press; 2009

[7] Gazagnes L, Cerneaux S, Persin M, Prouzet E, Larbot A. Desalination of sodium chloride solutions and seawater with hydrophobic ceramic membranes. Desalination. 2007;**217**:260-266

[8] Duke MC, Mee S, da Costa JCD. Performance of porous inorganic membranes in non-osmotic desalination. Water Research 2007;**41**:3998-4004.

[9] Kujawski W, Krajewska S, Kujawski M, Gazagnes L, Larbot A, Persin M. Pervaporation properties of fluoroalkylsilane (Fas) grafted ceramic membranes. Desalination. 2007; **205**:75-86

[10] Lia L, Dong J, Nenoff TM, Lee R. Reverse osmosis of ionic aqueous solutions on Amfi zeolite membrane. Desalination. 2004;**170**:309-316

[11] Lin J, Murad S. A computer simulation study of the separation of aqueous solutions using thin zeolite membranes. Molecular Physics: An International Journal at the Interface Between Chemistry and Physics. 2001;**99**:1175-1181

[12] Baerlocher C, McCusker LB, Olson DH. Atlas of Zeolite Framework Types. 6th ed. Amsterdam: Elsevier; 2007

[13] Li L, Dong J, Nenoff TM, Lee R. Desalination by reverse osmosis using Mfi zeolite membranes. Journal of Membrane Science. 2004;**243**:401-404

[14] Jareman F, Hedlund J, Sterte J. Effects of aluminum content on the separation properties of Mfi membranes. Separation and Purification Technology. 2003;**32**:159-163

[15] Duke MC, O'Brien-Abraham J, Milne N, Zhu B, Lin JYS, Diniz da Costa JC. Seawater desalination performance of Mfi type membranes made by secondary growth. Separation and Purification Technology. 2009;**68**:343-350

[16] Li L, Liu N, McPherson B, Lee R. Enhanced water permeation of reverse osmosis through Mfi-type zeolite membranes with high aluminum contents. Industrial & Engineering Chemistry Research. 2007;**46**:1584-1589

[17] Liu N, Li L, McPherson B, Lee R. Removal of organics from produced water by reverse osmosis using Mfi-type zeolite membranes. Journal of Membrane Science. 2008;**325**:357-361

[18] Lu J, Liu N, Li L, Lee R. Organic fouling and regeneration of zeolite membrane in wastewater treatment. Separation and Purification Technology. 2010;**72**(2):203-207

[19] Gogotsi Y, Nikitin A, Ye H, Zhou W, Fischer JE, Yi B, Foley HC, Barsoum MW. Nanoporous carbide-derived carbon with tunable pore size. Nature Materials. 2003;**2**:591-594

[20] Hoffman EN, Yushin G, Wendler BG, Barsoum MW, Gogotsi Y. Carbidederived carbon membrane. MCP. 2008;**112**:587-591

[21] Kulprathipanja S, Neuzil RW, Li NN. Separation of Fluids by Means of Mixed Matrix Membranes, Patent Application No. 4740219 1988

[22] Okumus E, Gurkan T, Yilmaz L. Development of a mixed-matrix membrane for pervaporation. Separation Science and Technology. 1994;**29**:2451-2473

[23] Ismail AF, Goh PS, Sanip SM, Aziz M. Transport and separation properties of carbon nanotube-mixed matrix membrane. Separation and Purification Technology. 2009;**70**(1):12-26

[24] Nga LY, Mohammada AW, Leob CP, Hilal N. Polymeric membranes incorporated with metal/metal oxide nanoparticles: A comprehensive review. Desalination. 2013;**308**:15-33

[25] Gribov EN, Parkhomchuk EV, Krivobokov IM, Darr JA, Okunev AG. Journal of Membrane Science. 2007;**297**:1

[26] Arico AS, Baglio V, Di Blasi A, Antonucci V. Electrochemistry Communications. 2003;**5**:862

[27] Arico AS, Baglio V, Di Blasi A, Creti P, Antonucci PL, Antonucci V. Solid State Ionics. 2003;**161**:251

[28] Lee K, Nam JH, Lee JH, Lee Y, Cho SM, Jung CH, Choi HG, Chang YY, Kwon YU, Nam JD. Electrochemistry Communications. 2005;**7**:113

[29] Baglio V, Arico AS, Di Blasi A, Antonucci V, Antonucci PL, Licoccia S, Traversa E, Fiory FS. Electrochimica Acta. 2005;**50**:1241

[30] Bauer F, Willert-Porada M. Journal of Membrane Science. 2004;**233**:141

[31] Lin CW, Fan KC, Thangamuthu R. Journal of Membrane Science. 2006;**278**:437

[32] Daiko Y, Klein LC, Kasuga T, Nogami M. Journal of Membrane Science. 2006;**281**:619

[33] Xu W, Lu T, Liu C, Xing W. Electrochimica Acta. 2005;**50**:3280

[34] Jalani NH, Dunn K, Datta R. Electrochimica Acta. 2005;**51**:553

[35] Ren S, Sun G, Li C, Liang Z, Wu Z, Jin W, Qin X, Yang X. Journal of Membrane Science. 2006;**282**:450

[36] Chen Z, Holmberg B, Li W, Wang X, Deng W, Munoz R, Yan Y. Chemistry of Materials. 2006;**18**:5669

[37] Gao J, Lee D, Yang Y, Holdcroft S, Frisken BJ. Macromolecules. 2005;**38**:5854

[38] Brijmohan SB, Shaw MT. Journal of Membrane Science. 2007;**303**:64

[39] Ramirez LP, Landfester K. Macromolecular Chemistry and Physics. 2003;**204**:22

[40] Gasa JV, Boob S, Weiss RA, Shaw MT. Journal of Membrane Science. 2006;**269**:177

[41] Sunada K, Kikuchi Y, Hashimoto K, Fujishima A. Bactericidal and detoxification effects of TiO2 thin film photocatalysts. Environmental Science & Technology. 1998;**32**:726-728

[42] Kim SH, Kwak SY, Sohn BH, Park TH. Design of TiO_2 nanoparticle selfassembled aromatic polyamide thin-film-composite (TFC) membrane as an approach to solve biofouling problem. Journal of Membrane Science. 2003;**211**:157-165

[43] Kwak SY, Kim SH, Kim SS. Hybrid organic/inorganic reverse osmosis (RO) membrane for bactericidal anti-fouling. 1. Preparation and characterization of TiO2 nanoparticle self-assembled aromatic polyamide thin-film-composite (TFC) membrane. Environmental Science & Technology. 2001;**35**:2388-2394

[44] Jeong BH, Hoek EMV, Yan Y, Subramani A, Huang X, Hurwitz G, Ghosh AK, Jawor A. Interfacial polymerization of thin film nanocomposites: A new concept for reverse osmosis membranes. Journal of Membrane Science. 2007;**294**:1-7

[45] Lind ML, Ghosh AK, Jawor A, Huang X, Hou W, Yang Y, Hoek EMV. Influence of zeolite crystal size on zeolite-polyamide thin film nanocomposite membranes. Langmuir. 2009;**25**:10139-10145

[46] Noy A, Park HG, Fornasiero F, Holt JK, Grigoropoulos CP, Bakajin O. Nanofluidics in carbon nanotubes. Nano Today. 2007;**2**:22-29

[47] Hinds BJ, Chopra N, Rantell T, Andrews R, Gavalas V, Bachas LG. Aligned multiwalled carbon nanotube membranes. Sci. 2004;**303**:62-65

[48] Majumder M, Chopra N, Andrews R, Hinds BJ. Nanoscale hydrodynamics: Enhanced flow in carbon nanotubes. Nature. 2005;**438**:44

[49] Holt JK, Park HG, Wang Y, Stadermann M, Artyukhin AB, Grigoropoulos CP, Noy A, Bakajin O. Fast mass transport through Sub-2-nanometer carbon nanotubes. Science. 2006;**312**:1034-1037

[50] Whitby M, Cagnon L, Thanou M, Quirke N. Enhanced fluid flow through nanoscale carbon pipes. Nano Letters. 2008;**8**:2632-2637

[51] Holt JK. Carbon nanotubes and nanofluidic transport. Advanced Materials. 2009;**21**: 3542-3550

[52] Mattia D, Gogotsi Y. Review: Static and dynamic behavior of liquids inside carbon nanotubes. Microfluidics and Nanofluidics. 2008;**5**:289-305

[53] Striolo A. The mechanism of water diffusion in narrow carbon nanotubes. Nano Letters. 2006;**6**:633-639

[54] Joseph S, Aluru NR. Why are carbon nanotubes fast transporters of water? Nano Letters. 2008;**8**:452-458

[55] Thomas JA, McGaughey AJH. Reassessing fast water transport through carbon nanotubes. Nano Letters. 2008;**8**:2788-2793

[56] Hummer G, Rasaiah JC, Noworyta JP. Water conduction through the hydrophobic channel of a carbon nanotube. Nature. 2001;**414**:188-190

[57] Kalra A, Garde S, Hummer G. Osmotic water transport through carbon nanotubes membranes. PNAS. 2003;**100**:10175-10180

[58] Kotsalis EM, Walther JH, Koumoutsakos P. Multiphase water flow inside carbon nanotubes. International Journal of Multiphase Flow. 2004;**30**:995-1010

[59] Francesco Fornasieroa HGP, Holta JK, Stadermanna M, Costas P, Grigoropoulosc NA, Bakajina O. Ion exclusion by sub- 2-nm carbon nanotube pores. Proceedings of the National Academy of Sciences of the United States of America. 2008;**105**:17250-17255

[60] Majumder M, Chopra N, Hinds BJ. Effect of tip functionalization on transport through vertically oriented carbon nanotube membranes. Journal of the American Chemical Society. 2005;**127**:9062-9070

[61] Majumder M, Zhan X, Andrews R, Hinds BJ. Voltage gated carbon nanotube membranes. Langmuir. 2007;**23**:8624-8631

[62] Corry B. Designing carbon nanotube membranes for efficient water desalination. The Journal of Physical Chemistry B. 2007;**112**:1427-1434

[63] Suk ME, Raghunathan AV, Aluru NR. Fast reverse osmosis using boron nitride and carbon nanotubes. Applied Physics Letters. 2008;**92**:133120

[64] Ratto TV, Holt JK, Szmodis AW. Membranes with Embedded Nanotubes for Selective Permeability patent application No. 20100025330 2010

[65] Subramani A, Voutchkovb N, Jacangelo JG. Desalination energy minimization using thin film nanocomposite membranes. Desalination. 2014;**350**:35-43

[66] Lee KP, Arnot TC, Mattia D. A review of reverse osmosis membrane materials for desalination—Development to date and future potential. Journal of Membrane Science. 2011;**370**:1-22

[67] Agre P. Membrane water transport and aquaporins: Looking back. Biology of the Cell. 2005;**97**:355-356

[68] Kumar M, Grzelakowski M, Zilles J, Clark M, Meier W. Highly permeable polymeric membranes based on the incorporation of the functional water channel protein aquaporin Z. PNAS. 2007;**104**:20719-20724

[69] Gonzaˇılez-Peˇırez A, Stibius KB, Vissing T, Nielsen CH, Mouritsen OG. Biomimetic triblock copolymer membrane arrays: A stable template for functional membrane proteins. Langmuir. 2009;**25**:10447-10450

[70] Taubert A. Controlling water transport through artificial polymer/protein hybrid membranes. PNAS. 2007;**104**:20643-20644

[71] Kaufman Y, Berman A, Freger V. Supported lipid bilayer membranes for water purification by reverse osmosis. Langmuir. 2010;**26**:7388-7395

[72] Jensen PH, Keller D, Nielsen CH. Membrane for Filtering of Water. Patent Application No. EP1885477 2010

Chapter 4

Fabrication of Biomimetic and Bioinspired Membranes

Amira Abdelrasoul, Huu Doan and Ali Lohi

Additional information is available at the end of the chapter

http://dx.doi.org/10.5772/intechopen.71718

Abstract

Biomimetic and bioinspired membranes are those membranes that are fabricated with natural or natural-like (inorganic, organic, or hybrid) materials via biomimetic and bioinspired approaches (bio-mineralization, bio-adhesion, self-assembly, etc.) to tailor-specific properties (sophisticated structures, hierarchical organizations, controlled selectivity, antifouling or self-cleaning properties, etc.). With the support of knowledge on mechanisms, models and functions from many scientific disciplines, research activity on biomimetic and bioinspired membrane during the last decade has increased rapidly.

Keywords: biomimetic, bioinspired, natural prototypes, membrane proteins, fabrication

1. Natural prototypes for bioinspired membranes

1.1. Cell membranes

Figure 1 illustrates the state-of-the-art examples that has shown great diversity of biomimetic and bioinspired membranes based on imitation of compositions (zwitterion and glycosyl), structures (biological channel), formations (biomineralization, bioadhesion, and self-assembly), and functions (self-cleaning) of the natural prototypes.

Among the natural prototypes, cell membranes are the most important due to their excellent abilities in mass transfer, energy transformation, and signal transduction. Cell membranes separate the cell interior from the outside environment and play a crucial role in almost all cellular phenomena. Each cell consists of ~63,000 μm^2 membrane area and a human body with 10^{14} cells that total to 10^7 m^2 of membrane area [1]. Cell membranes have a high degree of sophistication, miniaturization, and multi-functionalization. The present understanding of the cell membrane functions and complex membrane structures is primarily dependent on the fluid lipid bilayer and the proteins embedded within it (**Figure 2**). As such, cell

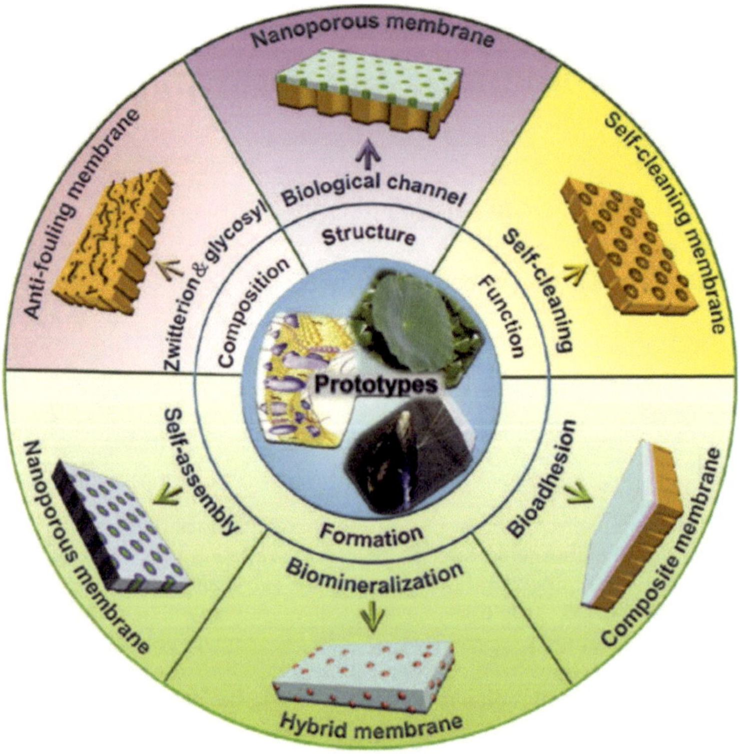

Figure 1. Overview of biomimetic and bioinspired membranes prepared by the imitation of natural prototypes.

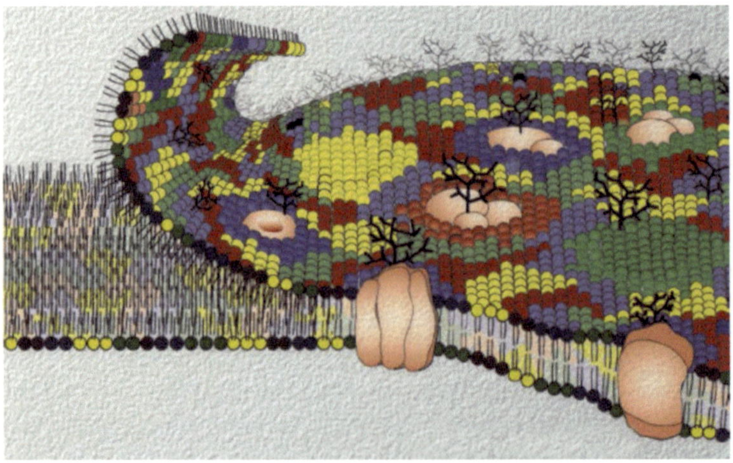

Figure 2. The fluid mosaic model with different lipid species shown in different colors. Source: Ref. [2], Copyright 2003; reproduced with permission from the Nature Publishing Group.

membranes are created using amphipathic lipids (phospholipids, cholesterols, glycolipids, and cholesterol esters), carbohydrates (oligosaccharides and polysaccharide), and membrane proteins (lipid anchored proteins, peripheral proteins, and integral proteins).

1.1.1. Lipid bilayer

The lipid bilayer is a universal component of all cell membranes. Its role is critical because its structural components provide the barrier that marks the boundaries of a cell. The structure is called a "lipid bilayer" because it is composed of *two* layers of *fat* cells organized in two sheets. The lipid bilayer is typically about five nanometers thick and surrounds all cells, providing the cell membrane structure.

Framed by the amphipathic nature of phospholipid molecules, the development of cell membranes occurs as a self-assembly type process. The phospholipids' nonpolar groups are included into planar bilayers with the aid of the hydrophobic effect. For instance, in a planar lipid bilayer, the nonpolar groups are mostly submerged into the bilayer's hydrophobic interior, while the polar head regions are positioned with respect to the external aqueous phase. Generally, the lipid bilayer is very fluidic and features assemblies of amphiphilic proteins (or lipoproteins) and lipids within the lipid's matrix. Furthermore, the interactions between the membrane lipids and exogenous proteins and peptides can incite a number of key biological processes at the level of the cell membrane [3]. As the primary phospholipid on the exterior surface, zwitterionic phosphatidylcholine shows superior nonfouling and nonthrombogenic qualities [3, 4]. This effective array of cell membranes offers excellent as well as rare instances of antifouling membranes' rational design.

1.1.2. Membrane proteins

Membrane proteins are proteins that interact with, or are part of, biological membranes. They include integral membrane proteins that are permanently anchored or part of the membrane and peripheral membrane proteins that are only temporarily attached to the lipid bilayer or to other integral proteins.

Cell membranes exhibit outstanding selectivity that allows certain substances permeating through them. Water as well as various smaller size molecules may move in and out of cells through active transport, facilitated diffusion, and direct diffusion. Small types of molecules, such as oxygen, water, ethanol, urea, and carbon dioxide, may easily move through cell membranes using simple diffusion mechanisms because of their higher solubility properties in the lipid bilayers' oily interior phase. These types of molecules move straight through the lipid bilayer or through the pores produced by essential membrane proteins. Alternatively, substances such as small organic molecules or ions move through cell membranes with the help of facilitated diffusion or active transport featuring protein-mediated carriers.

The facilitated or active diffusion is a diffusion using a carrier or channel proteins in the cell membrane that assists in the movement of molecules across a concentration gradient. All these processes play a crucial role in regulating the movement of solutes and water. The major intrinsic protein (MIP) is an important type of integral membrane proteins. MIPs are primarily divided as either aquaporins (AQPs) that can be only permeable to water or aquaglyceroporins (GLPs) that assist the diffusion of solutes like urea and glycerol [5]. As part of this, water channels have become the focus of a more rigorous research due to their effective transport mechanism. Specifically, there are several distinct water channel varieties. The AQP1 water channels permit water to travel bidirectionally and easily using osmosis across cell membranes; however, this is not the case with other small inorganic and organic molecules as well as ions [6]. The overall rate of water transport through AQP1 (3×10^9 water molecules per sub-unit per second) is substantially higher than that of the channels [7]. The dynamic and crystallographic structures of AQP1 allow for a rapid water transport process. Experimental runs show that the AQP1 selectivity filter is relatively hydrophobic and covered with hydrophilic nodes, a series of six completely spanning α-helices and a junction of two shorter helices from the channel [8]. **Figure 3a** indicates the way in which partial charges from the helix dipoles constrain the positioning of the water molecules moving through the restricted area of the pore. The interactions between Asn 192 and Asn 76 amino acids hold this junction together and create a hydrophilic water gate that allows for the AQP1 selectivity. In this case, the water molecule's oxygen atoms construct hydrogen bonds with the amide groups (**Figure 3b**), while the assembling of the water's molecular orbital produces well-tuned water dipole rotation (**Figure 3c**). Furthermore, the overall diameter of the narrowest

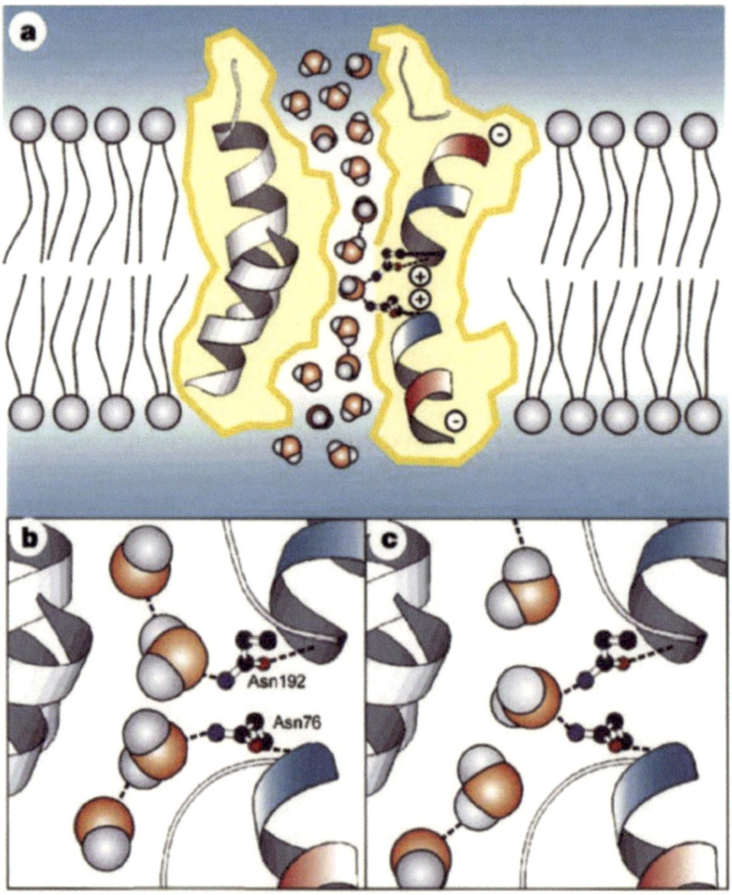

Figure 3. Schematic diagram of the water molecules transports in AQP1. (a) How partial charges from the helix dipoles restrict the orientation of the water molecules passing through the constriction of the pore. (b and c) The interactions and the hydrogen bonding of a water molecule with Asn 76 and/or Asn 192. Source: Ref. [9], Copyright 2000; reproduced with permission from the Nature Publishing Group.

point present is around 0.28 nm, and this likewise poses a steric obstacle for other molecules. The hydrophilic nodes, narrow size of the constriction region, and hydrophobic channel wall in combination contribute to the quick and accurate water molecule transport process [6].

The ion channels are a succession of pore-forming proteins that help control the voltage gradient throughout the membranes of living cells. Gating and selective ion conduction are two of the essential features attributed to ion channels. The selective ion conduction controls the performance of the channel, and how well it can choose specific ionic species among the available species in the cellular environment and then catalyze them using a prompt flow through [10]. Alternatively, the gating process controls the ion channel activity by being turned on and off. For instance, potassium (K^+) channels feature a selective filter close to the pore's extracellular side as well as a gate close to the intracellular side (**Figure 4**) [11, 12]. Whenever K^+ ion arrives into the selective filter, it is completely dehydrated. Such an unusual selectivity in K^+ channels is caused by the main chain atoms that have a stack of modified polar oxygen cages, which in turn allow for a series of closely spaced sites of appropriate dimensions to carefully arrange the process of K^+ ion dehydration. The hydrogen bonding and extensive van der Waals interactions ensure that the protein packing around the selective filter expands outward radially so as to keep the pore open at its appropriate diameter. Four helix dipoles and their electrostatic influence guarantee the cation selectivity by creating a negative electrostatic, or cation attractive, potentially close to the entry into the narrow selectivity filter [11]. The amino acid sequence preservation offers a frequent structural basis for the gating of K^+ channels, and the gating stimulus itself is caused by the membrane electric field and ligand binding [12].

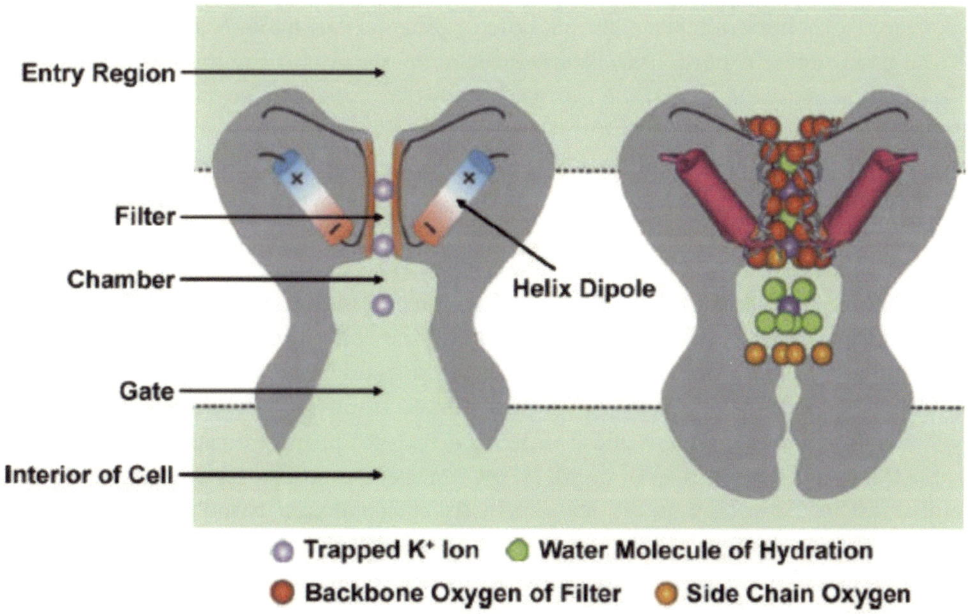

Figure 4. Cross-section of K^+ channels. (a) Wide open intracellular vestibule and pore helix dipoles; (b) high resolution structure for a closed channel. Source: Ref. [11], Copyright 2002; reproduced with permission from the Nature Publishing Group.

These delicate K^+ channel structures guarantee that K^+ ion is capable of diffusing from one site to the next within a relatively short distance. Moreover, it also restricts the accommodation of other ions or the rapid conduction within the high-selectivity dynamic. In nature, ion transport likewise happens with the aid of ion pumps. In terms of structure, the ion pumps are the large protein complexes that have their central channel portion spanning into the cell membrane [13]. Ion pumps are active transporters that are responsible for fulfilling a range of functions, unlike the ion channels that solely encourage the downhill movement of ions. The pumps effectively transport ions against their electrochemical gradient by coupling the "uphill" transport process with an energy source. An example of such a source can be found in the form of adenosine triphosphate (ATP) hydrolysis or the "downhill" movement of a substrate molecule or another ion [14]. A cell membrane and its functionality are one of the most effective designs available in nature. Cell design has inspired the creation of artificial as well as synthetic membranes with tailored structures, designed components, targeted performance, and specialized functions that offer a variety of applications in many fields. Applications like the complex lipid components and structures, multisubunit assemblies in cell membranes, and membrane proteins provide innovative solutions using new chemically and physically controlled mechanisms for artificial membrane designs requiring particular hierarchical structures and components. Uniquely advantageous cell membrane characteristics like self-healing, controllable permeability, and antifouling may offer promising directions in the use and exploration of artificial membranes [14].

1.2. Biomineralization

The concept of biomineralization shows how organisms can make hard materials in green and mild conditions. In particular, biomineralization stands for the mineral-formation process in organisms during which the inorganic elements collect on specific organics from the external surroundings and then form minerals under the modulating and inducing organics. The key feature of biomineralization is that biomolecules, such as polysaccharide protein and peptide, secreted by cells dictate the creation of minerals with a defined size, orientation, structure, and shape. This occurs because of the ordered collections of biomolecules and the interactions between inorganic and organic phases [15]. Living organisms are well known for utilizing minerals' material properties, when developing organic-inorganic hybrid materials for a range of applications [16]. In fact, in nature, biomineralization phenomena exist in each of the five major organism groups. So far, around 70 different types of biominerals have been classified, for example, calcium carbonate in the invertebrate skeletons, calcium phosphate in the bones and teeth of the vertebrate, iron oxide and iron sulfide in the magnetotactic bacteria, and the silica in diatoms [17]. Of these, the silicon-based and calcium-based minerals exist in the largest quantities, especially since calcium-based mineral accounts for about half of the biominerals [18]. If they are created under diverse circumstances, materials featuring identical chemical compositions may offer different morphologies. For example, calcium carbonate created in the leaves of plants is identified as amorphous, and the same calcium carbonate is calcite in the mollusk's shell [18]. **Figure 5** provides a rough overview of the roles that the organic and inorganic constituents have during the biomineral formation process. In most cases, insoluble organic matrix and inorganic mineral reactants are the key

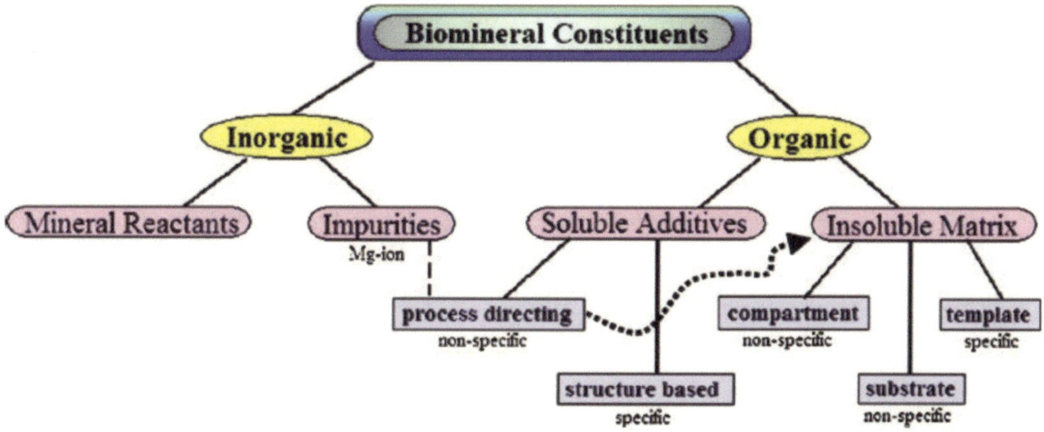

Figure 5. A simplistic view of the roles the inorganic and organic constituents played in biomineral formation process. Source: Ref. [19], Copyright 2008; reproduced with permission from the American Chemical Society.

factors, since the latter provides the necessary inorganic elements, and the former offers the substrate and functions as a template and an inducer of the mineral deposition. Furthermore, the crystallographic control can be regulated through the inclusion of organic additives and/or inorganic impurities. Although it is difficult to assess the intricate mechanisms leading to the formation of each biomineral, there are certain common strategies for manipulating mineralization. These strategies include spatial control, morphological control, structural control, constructional control, and chemical control [18]. Materials created using biomineralization in most cases have a substantially more complex structure and hierarchical organization than artificially synthesized materials. This ensures that they have improved physicochemical properties necessary for the molecular level control of organisms over the microstructure and nanobiominerals [20, 21]. For example, the ordered brick-and-mortar organization of $CaCO_3$ tablets and proteins in seashell nacre combines the strength of $CaCO_3$ together with the elasticity of proteins, thus ensuring that the seashell nacre exhibits strength, toughness, and hardness that exceed most manmade ceramics [22]. Furthermore, the physiological environment of the living organisms guarantees that the biominerals may be effectively synthesized in conditions that are environmentally friendly and mild, with almost neutral pH, aqueous environment, atmospheric pressure, and room temperature [23]. As a process, biomineralization joins superior properties, environmentally friendly conditions, and unique morphology, all of which are appealing features, when it comes to material synthesis. As a consequence, the idea of simulating biomineralization processes has remained an effective and promising methodology for synthesis and design of sophisticated organic-inorganic hybrid and inorganic materials using low energy and green approaches [24].

1.3. Bioadhesion

The area of bioadhesion stands for the ways in which natural materials adhere to a range of solid surfaces in a strong and quick manner. When it comes to natural phenomena, there are numerous examples of rare and exciting bioadhesion phenomena. This is particularly applicable to marine organisms such as sandcastle worms, limpets, starfish, tube worms, giant clams, sea cucumbers, barnacles, kelp, and mussels. For example, marine mussels are capable of secreting adhesive proteins all along the ample threads fanning out from the sides of the shells and then terminating each thread from the external coating of the thread and the adhesive plaques [25]. Adhesive proteins can stick to solid surfaces and then harden during short periods of time in order to create a solidified layer in water. This process of solidification ensures that mussels may be firmly attached to almost any type of substrates, like rocks, ship hulls, and even wave prompt habitats [26, 27]. **Figure 6b** shows the attachment of mussels to glass using an adhesive system based on plaques and threads and called "beard" or "byssus." Another relevant example is the sandcastle worm (*Phragmatopoma californica*) and its related species of marine polychaetes. These marine organisms can secrete cement from their "building organ" located on their thoraces, which allow them to glue particles such as shell fragments and sand grains together and then build a tube-like shelter [28, 29]. **Figure 6a** reflects the community of mussels fixed to rocks, and **Figure 6c** outlines the chemical formula embodying the byssus of mussel.

As **Figure 7** indicates that whenever a part of the worm's tube is removed and if the building blocks like glass beads are available in abundance, the worm will carefully go through the gluing process in order to repair its tube section. Adhesive systems listed earlier have several key similarities, when it comes to composition. Research studies show that the mussels' adhesive capabilities may be caused by the proteins located close to the plaque-substrate interface, like Mytilu edulisfoot protein 3 (Mefp-3) and Mefp-5, both of which contain sufficient 3,4-dihydroxy-l-phenylalanine (DOPA), with 21 and 27 mole%, respectively [26, 30, 31]. DOPA has a critical role in adhesive proteins, since it participates in the reactions that bring about the bulk adhesive proteins' hardening. DOPA also helps to form durable noncovalent and covalent connections with substrates because of the chemically multifunctional characteristics of catechol groups in relation to DOPA [26]. Furthermore, metal ions in nonmineral forms are necessary for a range of bioadhesive processes. The iron-DOPA complexes are created in the byssus of the mussel (**Figure 6c**) and feature at least two important functions [25]. The first function allows to simultaneously enhance extensibility and solidity of the threads using the reversible formation of iron-DOPA bonds. The second, key function permits inducing the

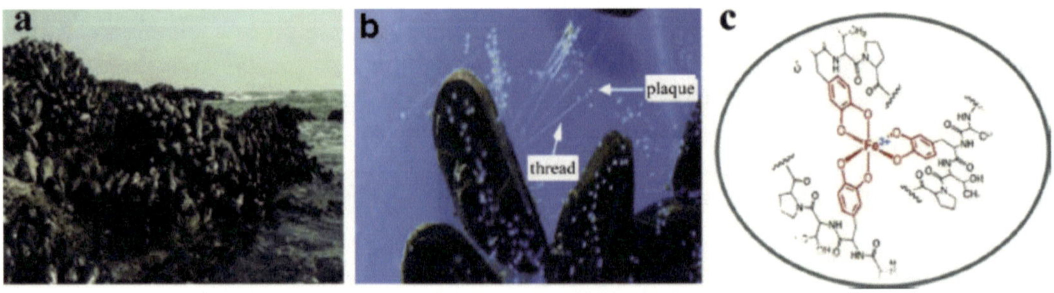

Figure 6. (a) A community of mussels affixed to rocks. (b) Mussels adhering to glass. The picture shows their byssus adhesive system consisting of threads and plaques. (c) An [Fe(DOPA)3] complex. Source: Ref. [25], Copyright 2010; reproduced with permission from Wiley-VCH Verlag GmbH & Co. KGaA.

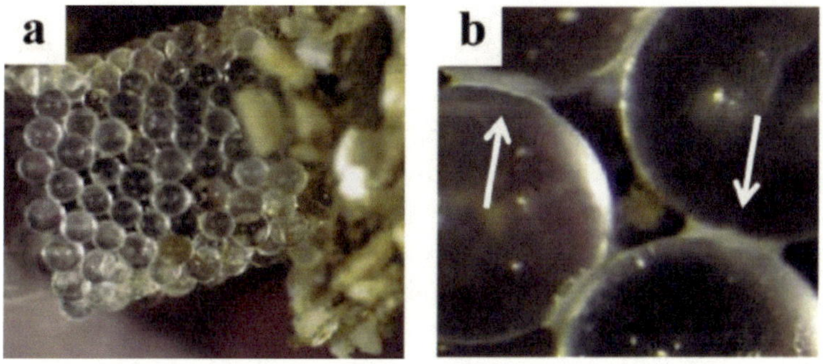

Figure 7. Sandcastle glue. (a) A tube rebuilt on top of the natural tube with 0.5 mm glass beads in the laboratory. (b) Close up of the rebuilt tube. Source: Ref. [29], Copyright 2011; reproduced with permission from Elsevier Ltd.

oxidation and the following DOPA reactions, which in turn helps to achieve the creation of the adhesive plaques and outer coating of threads. When comparing them to various synthetic adhesives, bioadhesives offer substantially more gains, such as durability, superior strength, quicker formation process, nontoxicity, milder formation conditions, and universality [25, 30]. In addition, all the bioadhesion processes occurring in living organisms happen in the presence of water, while underwater adhesion has been a constant roadblock for the majority of man-made types of adhesives. As a result, bioadhesion phenomena and the mechanisms they use have drawn a lot of attention in the last decade. Some researchers have tried to synthesize or screen models analogous to bioadhesives by simulating their properties and constitutions, since substantial difficulties arise in relation to the costs and processes of obtaining purified natural bioadhesives. For example, dopamine (DA) has been commonly used as an adhesive because of its similar properties and structure resembling DOPA [32, 33]. The growing research area of bioadhesion mechanisms has been exploring new innovative directions, including biomimetic adhesion strategies that can have extensive applications in the development and design of composite membranes with robust interfaces and uses [34].

1.4. Self-assembly

The phenomenon of self-assembly conveys the way in which organisms can create a wide range of complex structures featuring a high level of intricacy and precision. The definition of self-assembly is that it is a process of spontaneous organization of molecules in specific thermodynamic equilibrium conditions and into well-defined structural arrangements. In nature, there are a number of ingenious designs for structurally compatible and chemically complementary constituents capable of molecular self-assembly. Examples of these include deoxyribonucleic acid/ribonucleic acid DNA/RNA, polysaccharides, and peptide/proteins. The degree of ubiquity of the self-assembly phenomenon which occurs in nature, at either macroscopic or microscopic scales, reflects the capacity to spontaneously combine different individual entities into well-defined structures and cohesive organizations using nonspecific as well as specific intra/intermolecular relations [35, 36]. The cell membrane and its structure is one such characteristic example of molecular self-assembly occurring in nature. The lipid bilayer configuration has the capacity to show complex morphological changes using the phospholipids assembly. For instance, primitive cells can sustain the basic cellular functions such as division and growth with the help of lipid assembly [37]. The lipid bilayer likewise has a key role in the organization and assembly of amphiphilic transmembrane proteins, since they are guided by hydrophobic, or hydrophilic, interactions. Natural proteins and peptides may self-assemble into ordered molecules due to their evolutionarily fine-tuned functions and unique structures. A widely known instance of this occurs in the spider silk, which is famous for remarkable flexibility and strength [38]. Spiders are capable of manufacturing different types of spider silks using amphiphilic silk proteins, or spidroins, that have repetitive hydrophobic and hydrophilic amino acid stretches bordered by carboxy terminal and conserving nonrepetitive (NR) amino-terminal regions [39–41]. In this case, the assembling of charged N-terminal domain may be controlled with the aid of pH, since the pH gradient of spider silk glands can help to regulate the silk formation process. Next, the C-terminal domain, which is indifferent to pH changes, can regulate silk formation process by ordering the assembly of repetitive segments into actual fibers [39, 40]. The larger hydrophilic NR terminal regions make these silk protein molecules surfactant-like and make sure that they have the capacity to form micelles or hexagonal columns. This is followed by larger globular structures that are elongated due to the changes in their shear forces and extensional flow, thus creating the precursors to the subsequently produced spider silk fibers (**Figure 8**) [42]. As a common characteristic of extracellular organic matrix macromolecules, self-assembly depends on specific intermolecular interactions. In fact, the formation process of natural inorganic-organic composites begins with the careful assembly of extracellular matrix, then followed by selective transportation of inorganic ions to the organized compartments, subsequent mineral nucleation, and, finally, to the mineral growth defined by the confined cellular compartments [43–45]. As a consequence, the process of self-assembly in protein scaffolds has a vital role when it comes to the composite seashells' rich diversity [46]. Self-assembly

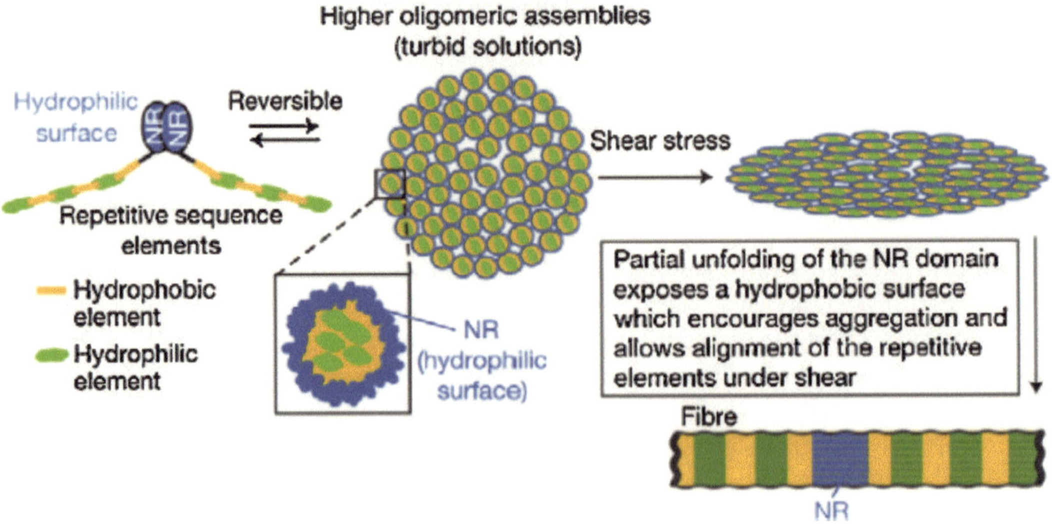

Figure 8. High oligomeric assemblies from silk proteins. Source: Ref. [39], Copyright 2010; reproduced with permission from the Nature Publishing Group.

has been proposed as an intelligent and bioinspired strategy for producing membranes with controlled architecture and composition and highlighted for incorporating a variety of building blocks into artificial/synthetic membranes.

1.5. Self-cleaning

The phenomenon of self-cleaning reflects how the surfaces of an organism can show a low-adhesion potential for a wide range of foulants occurring during the fluid flow. The qualities of biological surfaces, ranging from interplay between chemistries to surface morphologies, have a key role when it comes in defining specific wettability of biological materials. For example, superhydrophobic nonwetting quality is an essential property of standard self-cleaning biological surfaces. In the case of plants, this self-cleaning phenomenon is generally referred to as the "Lotus effect". Drops of water accumulated on the lotus leaves bead up when experiencing a high contact angle and then roll off, collecting dirt along the way in a mechanism of self-cleaning [47]. Plant surface nanostructures and microstructures play an intrinsic role in self-cleaning processes. Certain plant surfaces become hyper self-cleaning and hydrophobic because of the hydrophobic epicuticular waxes and hierarchical roughness. As one of the typical biological objects, the lotus leaf is well known for the combinatory use of hydrophobic epicuticular wax and the micro/nanoscale hierarchical architectures on its surface [48, 49]. In this case, the first structure is made out of microlevel mound-like protrusions featuring papillose epidermal cells, while the second structure is made out of nanoscale branch-like growths happening in the epidermal cells (**Figure 9a** and **b**) [50, 52]. This hierarchical roughness produced by randomly oriented hydrophobic wax tubules and convex cell papillae is essential for the preservation of the lotus leaf's self-cleaning characteristics (**Figure 9c**) [51, 53]. Particles contaminating the lotus leaves are picked up by the water droplets and then removed as the droplets slide off [54]. Plant surfaces tend to appear as rather diversified types of surface structures, as indicated in **Figure 10**. Distinct structures in two scales are helpful for lowering surface energy, forming the self-cleaning surfaces, and trapping air [57]. Furthermore, the physical adhesion forces that exist between the structured surfaces and contaminating particles can be significantly reduced. Within the realm of nature, self-cleaning processes and

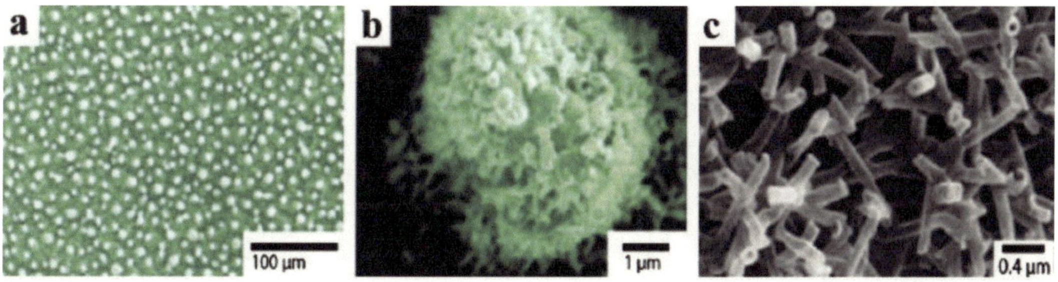

Figure 9. (a) Large-area SEM image of the lotus leaf's surface. Every epidermal cell creates a papilla and has a dense layer of epicuticular waxes superimposed on it. (b) Enlarged overview of a single papilla from panel [50]. (c) SEM image of 3D epicuticular wax tubules on lotus leaf surfaces, which create nanostructures [51]. Source: Refs. [50, 51], Copyright 2002 and 2009; reproduced with permission from Wiley-VCH Verlag GmbH & Co. KGaA and the American Chemical Society, respectively.

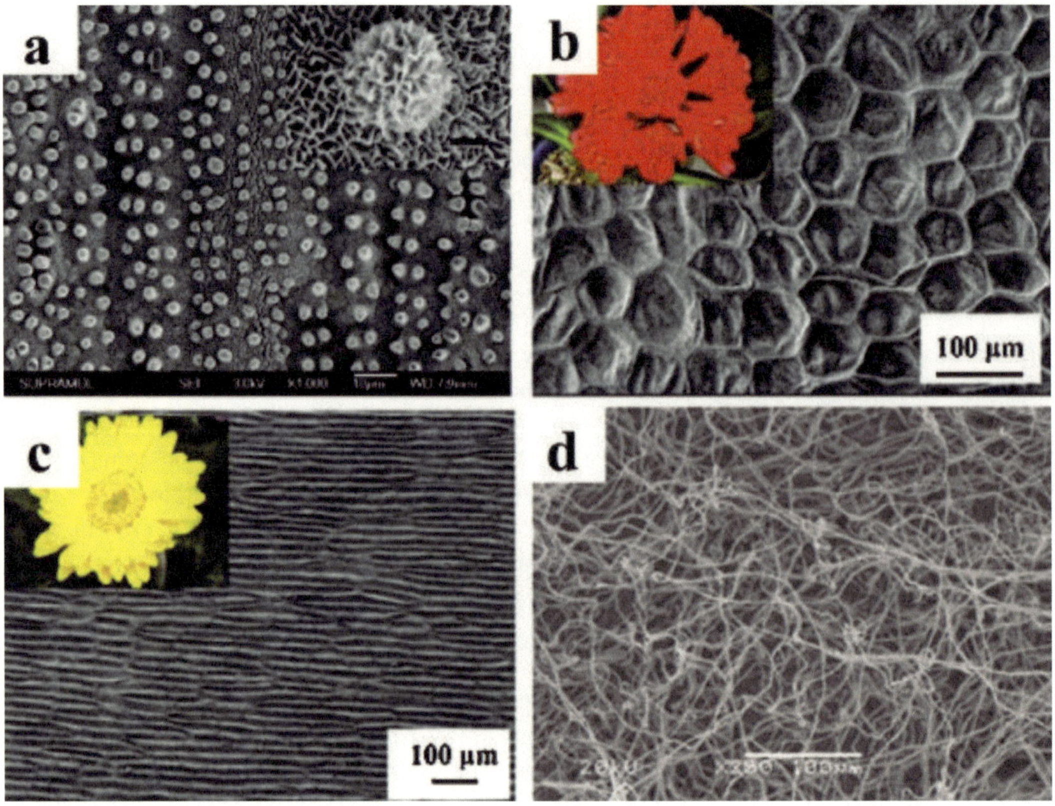

Figure 10. SEM images of the surface of (a) hierarchically structured papillae arranged in quasi-one-dimensional order parallel to the leaf edge [55], (b) periodic array of close-packed hexagons and strips on Chinese Kafir lily petal [56], (c) periodic array of parallel lines and helices on sunflower petal [56], and (d) unitary web of micro-fibers on ramee rear face [57]. Source: Refs. [55–57], Copyright 2010, 2008, and 2007; reproduced with permission from the American Chemical Society and Elsevier Ltd., respectively.

mechanisms are not limited only to plant surfaces. A wide range of self-cleaning surfaces have likewise been identified in water strider legs, insect eyes, insect wings, shark skin, gecko feet, spider silks, bird feathers, fish scales, and other types of surfaces [49, 58].

In the case of the Morpho butterfly wings, multiscale as well as ordered photonic structures improve self-cleaning and superhydrophobicity characteristics (**Figure 11**) [59, 60]. This directional easy-cleaning quality of the Morpho butterfly wings can be explained by its unique direction-dependent alignment of flexible nanotips on top of the lamella-stacked nanostripes and microscales overlapped on top of the wings [61]. Another example is found in gecko's feet, as they can engage in the process of self-cleaning, while the walking occurs with sticky toes. This exciting self-cleaning quality can be caused by the nanostructure, or single seta with a branched structure terminating in hundreds of spatula tips, and microstructure, that is setae on overlapping lamellar pads in uniform arrays. It seems that nonadhered lamellar surfaces can be quite nonwettable, and the particles contacting unloaded surface would be easily washed away when water becomes present. Furthermore, gecko feet that have been contaminated with microspheres may likewise retrieve their capacity to cling after a few steps on a dry

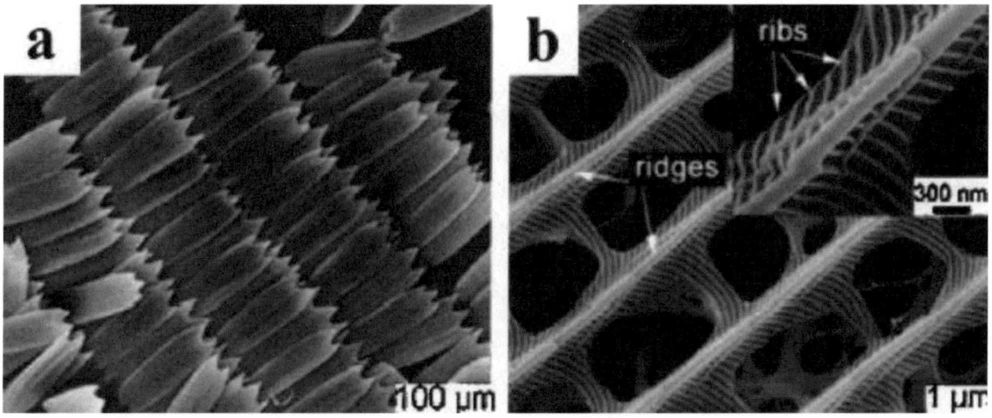

Figure 11. Hierarchical microstructures and nanostructures on the surface of the Morpho butterfly wings. (a) Secondary electron image of overlapping scales possesses an overall rectangular shape with pointed tips. (b) Secondary electron image of the porous architecture of the scale with parallel microscale ridges aligning along the scale length and nanoscale ribs lying on each ridge. Source: Ref. [59], Copyright 2008; reproduced with permission from Wiley-VCH Verlag GmbH & Co. KGaA.

surface, such as the one offered by clean glass. The process of self-cleaning is a consequence of the energetic disequilibrium that occurs between the adhesive forces that attract a dirt particle to the substrate, and those that attract the same particle to one or more spatula [62, 63].

In addition to the superior hydrophobic surfaces that occur in the air, nature likewise produces low adhesive surfaces in water environments. Examples of these adhesive surfaces that act as an inspiration for developing underwater self-cleaning surfaces are the surfaces of fish made up of tough scales and hydrophilic flexible mucus. The sector-like carp scales are created by orienting micropapillae with nanostructures of 30–40 μm in width and 100–300 μm in length and assembled in a radial direction (**Figure 12**) [63]. Whenever such fish scales come

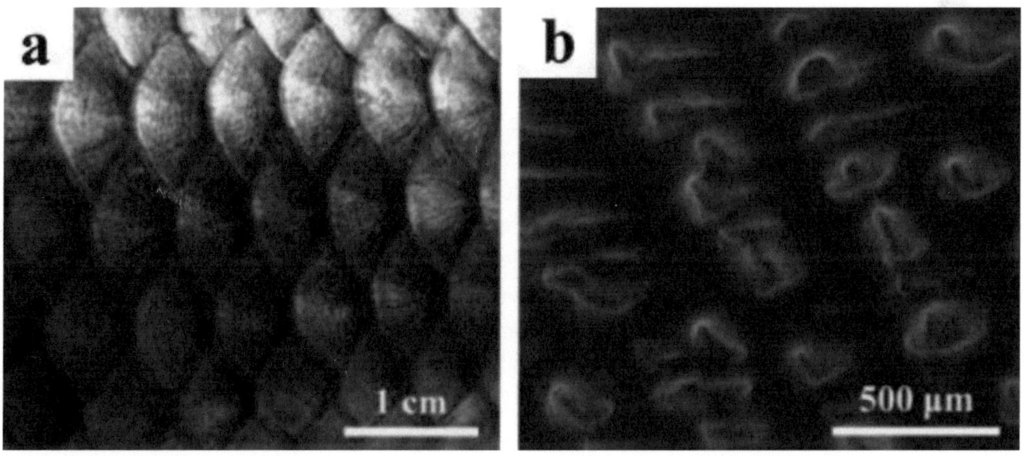

Figure 12. Surface structures of fish scales: (a) Optical image of the fish scales. (b) SEM image of fish scales. Source: Ref. [64], Copyright 2009; reproduced with permission from Wiley-VCH Verlag GmbH & Co. KGaA.

in contact with something like oil droplets in water, their fine-scale hierarchical structures can secure water molecules and then create an oil/water/solid interface. Further illustration of the underwater self-cleaning surface can be found in shark skin, which is protected by rather small separate tooth-like dermal denticles ribbed with longitudinal grooves. When these grooved scales are aligned parallel to the local water's flow direction, they can significantly reduce the creation of vortices over the smooth skin surface, thus enchasing water movement and flow efficiency [65]. To sum up, both, the microstructures and nanostructures, and the chemical properties of biological surfaces are capable of deterring contaminant matter from the surface and may offer an innovative direction for the construction of bioinspired and biomimetic self-cleaning membrane surfaces.

2. Synthesis of bioinspired and biomimetic membranes

Nature has always found a way to evolve common materials with functions that stand out as desirable. In fact, nature's sophisticated methods of selection have inspired advanced research directions in membrane materials and production. Biological structures, functions, formations, and compositions tend to take their forms on multiple scales ranging from molecular to microscale, macroscale, and nanoscale and in a manner that is strategically hierarchical and makes up a range of key functional elements. This exciting bioinspired and biomimetic trait has been especially appealing when it comes to designing and producing new synthetic membranes with superior structures, formations, functions, and compositions. The concise overviews of the six types of bioinspired and biomimetic membranes and their corresponding natural prototypes are covered in this review and are outlined in **Table 1**.

2.1. Fabrication of biomimetic and bioinspired membranes based on compositions of natural prototypes

2.1.1. Based on zwitterion and glycosyl

2.1.1.1. Fabrication of membranes via surface zwitterionization

Zwitterions are based on compounds that have an equal number of negatively and positively charged groups and, as a result, show an apparent neutral state. Research shows that there are biological zwitterionic phospholipids on the external lipid layer of the cell membrane. They are there to enhance biocompatibility with the surrounding tissues and stop the adhesion of exterior matters in biological fluids [4]. When it comes to bioinspired and biomimetic membranes, a variety of zwitterionic compounds have been used for the process of membrane surface zwitterionization. Because of the enhanced zwitterionic head group's fouling resistant qualities in cell membranes, the aim of surface zwitterionization is to stop foulants from attaching themselves to the membrane's surface. Research has shown that a number of typical zwitterionic moieties have been effectively introduced onto the surface of the membrane. A robust hydration of the zwitterionic moieties can create a sturdy hydration layer on the membrane's surfaces with the aid of electrostatic interactions that provide membranes with good fouling resistant abilities

Classifications	Natural prototypes	Biomimetic and bioinspired membranes
Based on composition	*Zwitterion and glycosyl*: the functional groups on the outside of cell membrane which renders antifouling properties	Antifouling membranes with functionalized surfaces resembling the composition of cell membrane through surface zwitterionization and glycosylation
Based on structure	*Biological channel*: the transmembrane proteins or protein assemblies which provide the fastest and specific transport channels for ions and small molecules *via* passive transport	Nanoporous membranes with ordered transport channels for ions and small molecules through incorporating biological channel proteins and/or artificial nanochannels
Based on formation	*Biomineralization*: the formation process of biominerals in organisms through precise hierarchical assembly of nanoscale building blocks under regulation of biomolecules	Organic-inorganic hybrid membranes with inorganic nanoparticles formed within polymeric matrix through the *in situ* mineralization reaction of inorganic precursors under the inducing and modulating of organics
	Bioadhesion: the high-strength conglutination of organisms (especially marine organisms) onto solid surfaces under mild condition and aqueous environment through the combination of multiple interactions	Composite membranes with high interfacial strength between different layers or different moieties through incorporating biomimetic adhesion strategy to form multiple interactions on interfaces
	Self-assembly: the spontaneous organization of molecules under thermodynamic equilibrium conditions into structurally well-defined arrangements based on numerous specific and nonspecific intermolecular/intramolecular interactions	Nanoporous membranes with ordered channels through self-assembly of block copolymers; nanoporous membranes with hydrophilic surface through self-assembly and spontaneous segregation of amphiphilic copolymer (surface segregation)
Based on function	*Self-cleaning*: the capacity of some biological surfaces to clear dirt away and keep themselves clean due to their superhydrophobic and nonwetting attributes	Self-cleaning membranes with superhydrophobic or superhydrophilic/oleophobicity surfaces through incorporating low surface energy moieties or high hydration energy moieties

Table 1. Introduction of natural prototypes and the corresponding biomimetic and bioinspired membranes [65].

and superior hydrophilicity [66]. The process of grafting zwitterionic moieties onto the surface of the membrane allows for an effectual method of realizing surface zwitterionization using covalent bonding. This method has received a lot of attention because of its potential applications. Different types of chemical reactions were engaged so as to fixate the zwitterionic moieties onto membrane's surface after it was formed. Graft polymerization offers an appealing route for membrane surface modification processes because of the monomer species' diverse range. The high-energy radiation-initiated graft polymerization process has gained considerable attention as one of the conventional methods for grafting functional polymer brushes from membrane surfaces and with the aid of which radiation-grafted zwitterionic brushes can be obtained using straightforward control. With the aid of the UV-irradiated technique and plasma pre-treatment, surface zwitterionization applied using graft polymerization of the zwitterionic monomer on the hydrophobic surface of poly(vinylidene fluoride) (PVDF) microfiltration (MF) membrane, polypropylene MF or nonwoven fabric membrane [67–70], polytetrafluoroethylene (PTFE) MF membrane, and polyethersulfone (PES) ultrafiltration (UF) membranes and polysulfone (PSf) UF membranes [71–74]. It should be noted that the high-energy types of excitation could likewise create unwanted branched, or cross-linked brush structure, as well as photo degradation of the substrate membrane [75]. Alternatively, chemically initiated graft polymerization is moderate and does not need special equipment. Zwitterionic SBMA and CBMA monomers were grafted using the surface of PVDF membranes and physisorbed free radical grafting methods that rely on azo-bis-isobutyrylnitrile (AIBN) as the initiator [76, 77]. In addition, the process of grafting the zwitterionic MPC and MPDSAH monomers from hydroxyl-containing membrane surface was similarly conducted with the aid of ceric ammonium nitrate, which was applied as a redox initiator in an aqueous medium [78, 79]. Numerous challenges still exist for the uniform zwitterionic brushes and high-grafting densities because of the steric effect of the monomers that have been already grafted. In the last decade, surface-initiated controlled radical polymerization, such as the surface-initiated reversible addition-fragmentation chain transfer polymerization (SI-RAFT) and surface-initiated atom-transfer radical polymerization (SI-ATRP), has been frequently used to create well-defined zwitterionic brushes on the membranes' surfaces. The process of surface zwitterionization based on surface-initiated controlled radical polymerization was applied extensively, and a mixture of various anionic and cationic pairs ($N^+(CH_3)_2/SO_3^-$, $N^+(CH_3)_2/COO^-$, $N^+(CH_3)_2/PO_4^-$) has been created so as to ensure the overall charge neutrality and high-membrane hydrophilicity [80]. In recent projects, the application of click chemistry, or the generation of products that follows nature's examples, for specific surface modification processes has offered a new route for membrane surface zwitterionization. This innovative direction features good control, high yield, and mild reaction dynamic. The surface attachment of long-chain and short-chain zwitterionic moieties has been obtained with the help of azide-alkyne cycloaddition reactions and surface-initiated thiolene coupling chemistry [81, 82]. The physical adsorbing and blending of zwitterionic copolymers with membrane-forming polymers are easier approaches to surface zwitterionization. Although zwitterionic brushes possess the high water affinity qualities, a number of amphiphilic zwitterionic copolymers were first synthesized and then implemented so as to augment the overall stability of zwitterionic brushes with the mediation of hydrophobic interaction occurring between the membrane's matrix and hydrophobic chains. While the membrane preparation process by in situ blending was occurring, the amphiphilic zwitterionic copolymers may stimulate surface separation of zwitterionic brushes onto the membrane's surface with hydrophobic chains fastened in the membrane's matrix [83]. Throughout the membrane's process of modification, the amphiphilic zwitterionic copolymers may be adsorbed on the membrane's surfaces and with hydrophobic chains fastened to them [84]. Once the expansion of surface modification methods has begun, new exciting chemical reactions and techniques are being used for the construction of composite zwitterionic membrane surfaces. These techniques include oxidative polymerization of zwitterionic amino acid 3,4-dihydroxy-l-phenylalanine (DOPA) and initiated chemical vapor deposition of zwitterionic polymers, chemical crosslinking of zwitterionic colloid particles, and interfacial polymerization of zwitterionic amide monomer [85]. The process of membrane surface zwitterionization may likewise result from membranes that include pyridine or N,N-dimethylamino-2-ethylmethacrylate (DMAEMA) moieties, featuring tertiary amine reactive sites [86].

2.1.1.2. Fabrication of membranes using surface glycosylation

A sufficiently hydrated glycocalyx (glycoprotein-polysaccharide) is located on the outside of the cell membrane and helps to manage specific interactions, like cell-cell recognition, as well as stop unwelcome nonspecific protein adhesion through a mixture of hydrogen bond indicated hydration and steric repulsion effects. When it comes to bioinspired and biomimetic membranes, certain glycolmonomers or glycopolymers are implemented as biomimetic materials for the process of membrane surface glycosylation. Due to the fouling resistant quality of glycocalyx on the membrane's surfaces and the glycoside cluster effect, the aims of surface glycosylation are the identification of proteins or the deterrence of nonspecific interactions between proteins and membrane surfaces through the production of extended hydroxyl group rich chains enclosed by molecules of water.

2.1.2. Shortcomings and challenges

While in-depth research has been conducted on glycosylation and membrane surface zwitterionization, the majority of the tests were limited to the lab scale. First of all, the scale-up of complex polymer modification and synthesis strategies is problematized by reaction conditions' precision control, including the control of temperature, residence-time, velocity, and catalyst distribution in the reactor. Second, zwitterionic moieties are frequently too costly to implement in larger quantities. In addition, glucosyl moieties may be susceptible to microbial degradation during the repeated and long-term application. Finally, the comprehensive knowledge of the membrane's structural evolution under varying condition has not been fully investigated. However, glycosylation and membrane surface zwitterionization offer the most encouraging directions in environmental, engineering, and biotechnical applications of innovative membrane technologies [87].

2.2. Fabrication of bioinspired and biomimetic membranes based on structures of natural prototypes

2.2.1. Biological channel structure

There is a substantial number of channels formed by proteins, as well as protein assemblies, within the cell's membrane and that effectively contribute to the transmembrane transport of water, nutrients, and ions [6, 87]. The rapid and relatively controllable transport of water, ions, and other nutrients through biological channels ensures the success of their essential movements within the organisms [6, 9, 88, 89]. Currently, the process of simulating the biological channels' structure of in cell membranes with the aim of producing artificial membranes, offering high performance and a range of useful functions, has been of enormous technological and scientific interest.

2.2.1.1. Fabrication of membranes with incorporated biological channel proteins

The direct way of producing biomimetic channels is to simulate the structure and composite of the cell membrane. This entails integrating biological channel proteins into the bilayer lipid membranes (BLM), as it is a basic model of the phosphor lipid bilayer for cell membrane [88, 89]. However, the key problem with BLM is the low stability potential. To solve this drawback, the supported BLM on different porous substrates can be implemented [88, 89]. Some of the most popular substrates used include gold, as well as silicon, glass, other metallic thin layers, polymers, and Si3N4 [90]. When compared to the organic substrates, the inorganic porous substrates offer a higher number of advantages with regard to the chemical, mechanical, thermal stability, and lifetime properties [91–94]. Moreover, the block copolymers' self-assembly provides another method for creating a bilayer that can function as an alternative to BLM. This is mostly due to its controllability, greater stability, and the capacity to stop the direct contact of protein to solid substrate, since this can inactivate and immobilize proteins [95]. These biomimetic membranes can be produced using a variety of approaches such as Langmuir-Blodgett/Langmuir-Schaefer monolayer transfer methods, spin-coating, and vesicle rupture. Within these methods, the vesicles rupture approach is one of the easiest to apply and most frequently implement [95]. **Figure 13** outlines the schematic process of producing biomimetic membrane using vesicles rupture method. In the first step, the vesicle incorporated channel proteins are produced using the film rehydration approach (**Figure 13a**). Next, the solution of vesicles is made and then released directly onto the substrates (**Figure 13b** and **c**). Finally, the vesicles proceed to rupture with the aid of covalent interaction or interfacial adsorption, as a result forming the planar bilayer membranes (**Figure 13d**) [92, 94]. To facilitate the suitable interactions with solid substrates, polymers forming the bilayers need to be used in a way that does not alter their self-assembly functionality and structure [92]. The substrates likewise must be functionalized if they are to remain chemically active. Various triblock copolymers end-functionalized with methacrylate, disulfide, and acrylate groups were created so as to react with silanization-modified substrate, gold-coated substrate, and amine using covalent interaction [92]. Gold is frequently selected as the surface modifier for

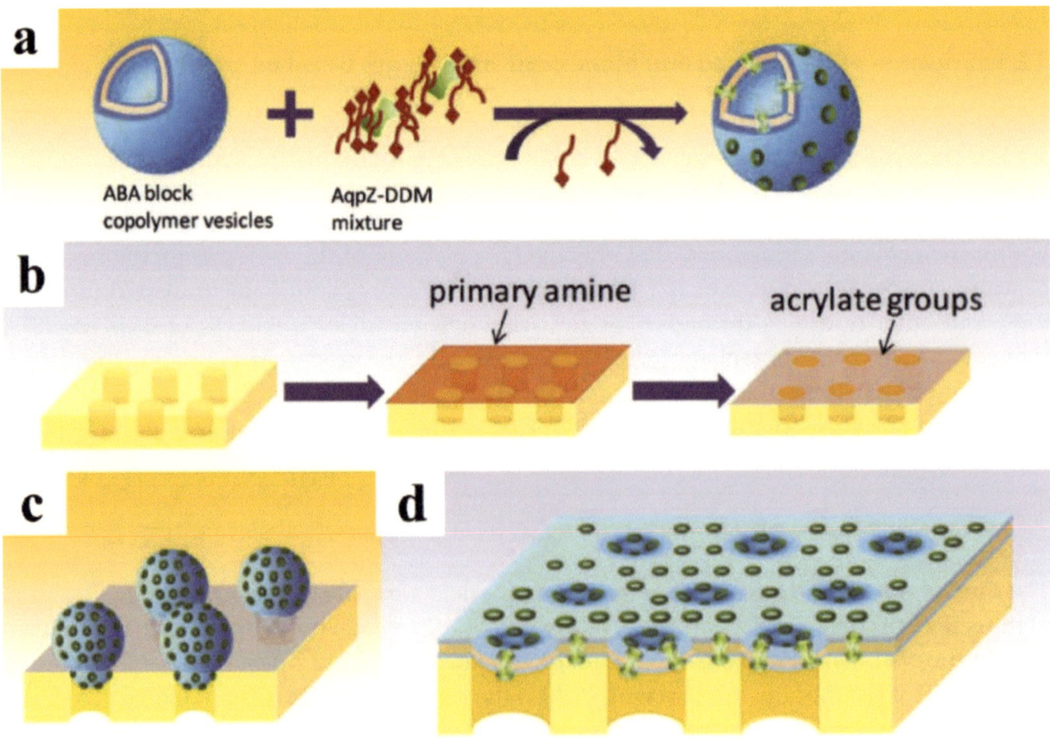

Figure 13. Schematic diagram of pore-spanning membrane design and synthesis. Source: Ref. [94], Copyright 2012; reproduced with permission from Wiley-VCH Verlag GmbH & Co. KGaA.

substrates, since it is not cytotoxic, stable, and very active, the latter allowing it to react with polymers and offer new reaction sites for membrane further modifications. **Figure 13b** shows polycarbonate tracked-etched membranes covered with a gold layer so as to attain the subsequent chemisorption of cysteamine monolayer and the later conversion to acrylate. The improved overall stability of biomimetic membranes may be achieved using the process of forming covalent interactions between the acrylate groups on the substrate and the methacrylate groups on triblock copolymer.

Recently, layer-by-layer (LbL) self-assembly and interfacial polymerization were used to develop strong and defect-free AQP-containing membranes that lend themselves to easier scaled up [96, 97]. First, the AQP-containing proteoliposomes were created and then embedded into the membrane's matrix. This helped to create a compatible and stable environment for AQP. These research studies generated valuable new methods for fabrication of biological channel proteins-containing membranes offering improved efficiency. The primary change from the earlier experimental studies is that in this case, AQPs functioned as the dispersed phase within the membrane rather than infiltrate the entire membrane.

2.2.1.2. Membrane fabrication by method of constructing artificial nanopores/nanochannels

Artificially created nanopores/nanochannels featuring functional groups may behave as equivalents of biological channel proteins, due to their great flexibility in terms of shape and size, high stability, chemically and mechanically robust properties, and the various tunable surface qualities [98]. Membranes that possess artificially created nanopores/nanochannels can be produced using bottom-up and top-down routes. What these routes entail is the creation of engineered solid-state nanopores/nanochannels on nonporous substrates using micromachining and then producing nanopores/nanochannels with the aid of self-organization of molecules and atoms, respective of the directions [98]. Specifically, the top-down route is primarily based on electron beam, ion-track-etching, laser, and electrochemical etching technologies, using which the nanopores/nanochannels of varying sizes and shapes on organic and inorganic substrates can be produced [99]. The nanopores/nanochannels developed made with bottom-up route incorporate hexagonally packed cylindrical block copolymer, carbon nanotube (CNT) by chemical vapor deposition (CVD), organic nanotubes by self-assembly, anodic aluminum oxide (AAO) and titania nanotube (TNT) by anodic oxidation, and other using respective methods [100]. If compared with the top-down route, the bottom-up route can help develop membranes with higher pore/channel density potential, a highly beneficial characteristic for molecular separations, as well as other research areas that require a greater channel array area [88, 89]. For example, the AAO porous template may feature a pore/channel density of 10^{15} m^{-2}, while the TNT membrane has a density of 5–10 × 10^{13} m^{-2} pore/channel, a rate that is greater than the natural cells' ion-channel density of nearly 10^{12} m^{-2} [88, 89]. The nanopore/nanochannel entails the channel or pore with a diameter value in the range of 1–100 nm, a number that is bigger than the sizes of most molecules and ions. As a result, the process of entrance or inner surface functionalization is required in order to lower the operational nanopore/nanochannel size or act as the "gate" ion channels located in cell membranes, effectually helping to achieve selective permeation ultimate. Furthermore, for the use of nanoporous membranes in bio-recognition and energy conversion, the process of inner modification is frequently needed to recognize and immobilize bio-molecules. A common method used for this is the immobilization of functional molecules on the nanopores/nanochannels' interior surface with the help of diversified chemical covalent reactions [101]. For example, gold nanopores/nanochannels can be modified with molecules carrying S-S or SH groups so as to create S-Au bonds, while the oxide surface can be altered using a range of silane derivatives [102]. Alternative approaches to the modification of nanopores/nanochannels are plasma modification, electro-static self-assembly, and the deposition of metals using ion sputtering deposition, electron beam evaporator, or electroless deposition [103]. **Figure 14** outlines an instance of inner surface–modified nanochannel with pH-responsive and employing the chemical covalent type of reaction. In this case, the cylindrical nanochannels with a 15 nm diameter on poly (ethylene terephthalate) (PET) membrane were first produced with ion-tracked technology. Next, the nanochannels were modified with 4,4′ azobis (4-cyanopentanoic acid), as a surface-confined polymerization initiator, and a 4-vinyl pyridine as the monomer for forming pH-responsive polymer brushes [98]. These types of brushes can alter their form from the charged hydrophilic state, collapsed, swollen, and neutral hydrophobic state, when an environmental pH alternates between 2 and 10.

Synergistic coassembly of block copolymers (BCPs) and nanotube subunits (cyclic peptide, 8CP) was used to produce thin membranes that include subnanometer organic [104]. First, polymers were tethered onto 8CP so as to augment solubility and help mediate interactions between one part of BCP and 8CP, as shown in **Figure 15**. Once blended with BCPs, the 8CP-polymer conjugates were restricted in the BCP cylindrical microdomains, which have an affinity with polymers, and then formed into nanotubes in the nanoscopic domains once heated by the hydrogen bonding occurring between amino acid residues located on neighboring

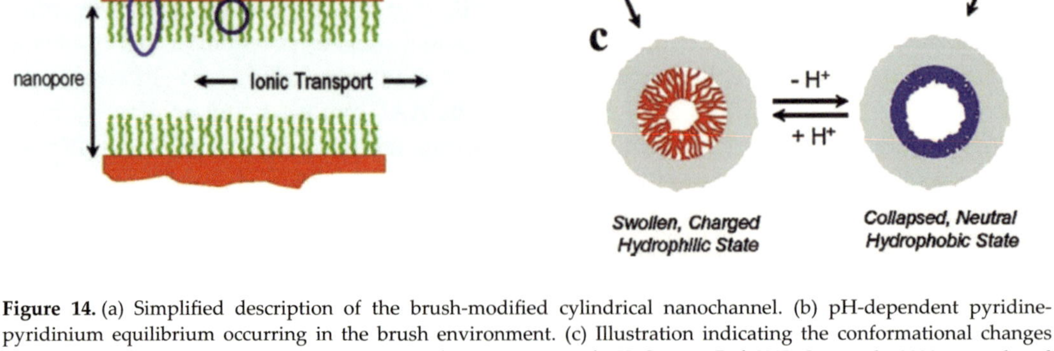

Figure 14. (a) Simplified description of the brush-modified cylindrical nanochannel. (b) pH-dependent pyridine-pyridinium equilibrium occurring in the brush environment. (c) Illustration indicating the conformational changes happening in the brush layer upon variations in the environmental pH. Source: Ref. [98], Copyright 2009; reproduced with permission from the American Chemical Society.

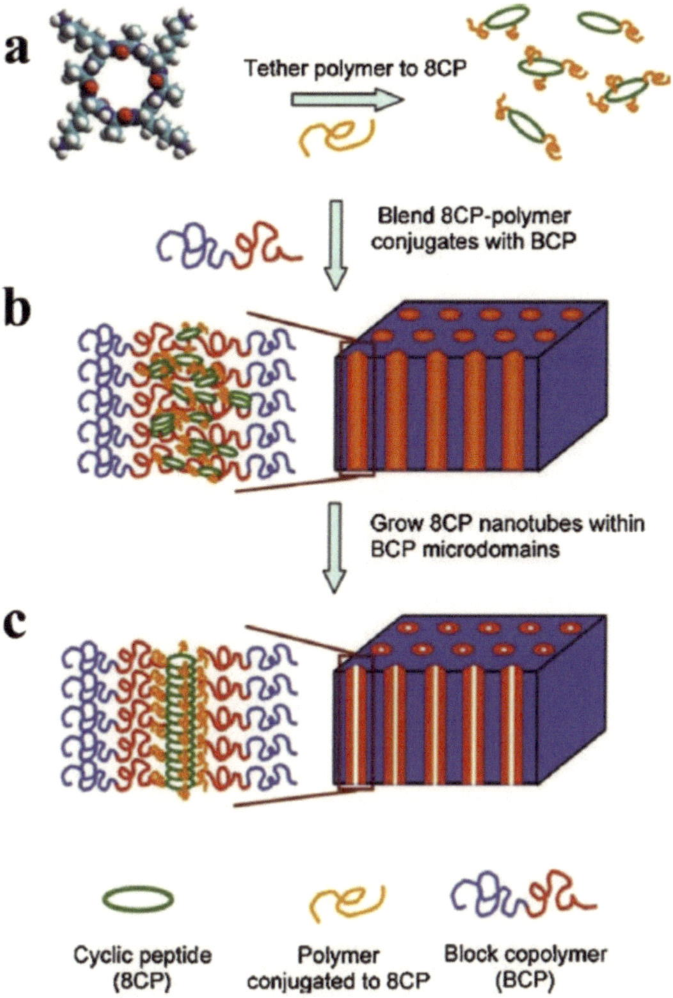

Figure 15. Schematic illustration of the process generating subnanometerporous films using direct coassembly. Source: Ref. [104], Copyright 2011; reproduced with permission from the American Chemical Society.

peptides. Finally, the membranes with sub-nanometer channels, that are oriented in a normal way toward the surface, were successfully created. The shape and size of the nanotubes be tailored through the process of changing the nanotube subunits' molecular structure beyond the restrictions imposed by block copolymer self-assembly. As a result, selective and swift molecular transport can be obtained. Within the artificial nanopore/nanochannel types, the CNT stands out since it functions as an alternative to water channels and biological ion channels due its propensity for narrow diameter, inherent smoothness of the inner surface, and hydrophobicity [105]. Molecular dynamics (MD) simulations have been used to research the transport mechanisms of water and ions in CNT, as well as the possible uses of CNT in membrane processes and applications [106]. Research suggests that water molecules indicate single-file transport in CNT because of the creation of a robust hydrogen bond chain, similar to the water transport detected in AQP [107]. As a consequence, the CNT's water transport rate is analogous to the one occurring in AQP. To gain higher selectivity values, CNTs are frequently modified at the entrance using organic groups that can help achieve a lower diameter and a more selective ion interaction [108]. While the hands-on applications of CNT-containing membranes, implementing nanochannels in CNT, is rather limited, the possibilities it promises has attracted attention and incited new research directions. Because of its enhanced controllability, high channel density, and superior mechanical robustness, the biomimetic membranes featuring artificial nanopores/nanochannels can be highly relevant to processes that require advanced size-selective separations.

2.2.1.3. Fabrication of membranes via incorporation of artificial nanopores/nanochannels and biological channel proteins

When it comes to this type of biomimetic membranes, the biological channels can guarantee the advantages of intelligence and an atomically precise structure that resembles living cells. Alternatively, the artificial nanopores/nanochannels offer qualities like durability, size, shape control, and robustness. The ion channel protein Gramicidin-A, in the track-etched nanopores with a diameter of 15 nm, was filled on the polycarbonate thin film, while the ion diffusion coefficient of Na^+, K^+, Ca^{2+}, and Mg^{2+} ions was measured in order to calculate the nanoporous membrane's selectivity and permeability values [87]. The Gramicidin-A's adsorption into the nanopores was preferred by the surface hydrophilic treatment featuring ethanol and contributed to a greater affinity of Gramicidin-A toward hydrophobic pores rather than toward the hydrophilic surface. It should be noted that although the effective ion diffusion coefficients were amplified after Gramicidin-A inclusion, the increase in values was not as substantial as was expected. This can be due to the fact that the nanopores were not completely filled up with Gramicidin-A, and as a result, the ions likewise diffused within the "free" electrolyte inside the nanopores. Thus, additional experimental research is necessary so as to achieve cases where the entire nanopores are filled out.

2.2.2. Challenges and shortcomings

The overall performance of a biomimetic nanoporous membrane primarily relies on the membrane's channel density, and membrane integrity [91]. Extensive research initiatives have been established with the specific purpose of developing bioinspired and biomimetic membranes that include biological channel proteins. There are still multiple challenges that problematize practical applications of bioinspired and biomimetic membranes: (1) The first challenge is that the membrane channel density is difficult to control [91], determine, and limit, since it needs to make the self-assembly structure unaffected [93]; (2) the second problem is that the channel protein activity needs to be maintained and that limits the preparation process environment [94]; (3) third, it is challenging to make a defect-free bi-layer in large-scale type of production [94]; (4) the fourth challenge is that the costs of production become excessively high because of how complicated the process of extracting proteins can be. Nanoporous membranes featuring artificial nanopores/nanochannels can offer an exiting range of separation applications due to their superior stability; however, they also have to confront a number of common challenges. When it comes to membranes using top-down routes, the process of the homogeneous modification of interior surfaces through the nanoscale channels, as well as the large-scale modification, is hard to perform successfully on consistent basis. Furthermore, their applications are limited by their lower channel density and the need to use costly equipment. Of all the membranes employing bottom-up routes, the CNT-containing membranes have gained the most attention due to their theoretically superior water permeability. However, the fabrication of large-scale membranes with low selectivity and aligned CNTs is extremely difficult, and this limits their advancement from the theoretical research stages into practical application testing.

2.3. Fabrication of biomimetic and bioinspired membranes based on formations of natural prototypes

2.3.1. Based on biomineralization

Production of bioinspired and biomimetic membranes based on biomineralization is a method that stimulates the creation of inorganic nanoparticles in the polymeric matrix using mineralization reaction that resembles biomineralization that occurs under somewhat milder circumstances. During the last several decades, organic-inorganic hybrid membranes have gained a lot of interest and became widely applicable, since they offer the benefits of stability and rigidity of inorganic moiety, together with the improved adaptability and efficient membrane-forming property in polymeric moiety [109]. At the same time, organic-inorganic hybrid membranes allowed for new properties due to their hybrid structures. The most direct way of producing a hybrid membrane is the process of physical blending of inorganic nanoparticle and polymer. This process is relatively easy to undertake and regulate. One of the drawbacks is the creation of nonselective voids due to collection of inorganic nanoparticles and their lack of proper compatibility with polymeric matrices [110]. The in-situ sol-gel process is an alternative method of producing these types of membranes and may be capable of overcoming this limitation. During the in-situ sol-gel process, the polycondensation and hydrolysis of the inorganic precursors happen under the catalysis of an acid or a base that creates inorganic nanoparticles in a polymeric casting solution [111]. If compared with the physical blending, inorganic nanoparticles can disperse more homogeneously and offer improved compatibility with the polymeric matrix. The sol-gel method likewise has a number of inherent problems, such severe conditions like strong acidic or alkaline environment, and a relatively low controllability [112]. Biomineralization process mixes organic materials with inorganic as it produces materials with hierarchically sophisticated structures and improved physicochemical qualities at normal pressure and temperature values and an almost neutral pH value in an aqueous environment featuring straightforward chemical compositions [15, 113]. These types of materials are much better than many of the artificially synthesized materials because of the critical control that can be exercised over their size, shape, structure, and assembly of constituent parts [15]. The process of biomineralization in nature offers a uniquely rich source of inspiration, when it comes to the design and production of hybrid membranes.

2.3.1.1. Biomimetic mineralization

The process of biomimetic mineralization mimics the biomineralization method during the material-synthesizing using organic inducer to incite the creation of inorganic nanoparticles from inorganic precursor and as a result producing materials with distinct properties and microstructures [20]. In this case, the inorganic precursor can take the form of metal alkoxide or metal salt. The organic inducer can take the shape of macromolecules, or smaller sized molecules, and have the necessary functional groups and would help activate the inorganic precursor reaction. The macromolecules can be in the form of the amino group for silica and titania or phosphate, sulfate, and carboxylate groups for calcium phosphate and calcium carbonate [114]. For example, the inducers frequently implemented for the creation of silica are macromolecules, such as protein and small molecules comprised of amines and amino acids [115]. The in situ biomimetic mineralization process is an appealing method for the production of hybrid membranes. It prevents the inhomogeneous filler distribution and filler agglomeration that has happened during the physical blending approach, which suffers from harsh conditions like alkaline or strong acidic environment or poor controllability that can occur during in-situ sol-gel method [112]. In order to produce hybrid membranes using the in-situ biomimetic mineralization, two methods have been created. The first method requires the addition of organic inducer and inorganic precursor into the solution that includes membrane-forming polymer. This allows the mineralization process to transpire at the same time as the membrane formation. The second method requires the immersion of the membrane with inducers into precursor-containing solution. For both these approaches, the inorganic precursors interact with organic inducers through metal-organic chelation or electrostatic attraction. As a consequence, the inorganic precursors become enriched in the microdomains near organic inducers, and this in turn creates the necessary conditions and locations for the mineralization reaction to occur and then homogeneously generate inorganic nanoparticles. Recent research has applied both approaches, and varying types of membrane-forming polymers, inorganic precursors, and organic inducers produce diversified hybrid membranes [112].

2.3.1.2. Membrane fabrication using biomimetic mineralization during membrane formation

The process of mixing raw materials is a relatively easy method for producing hybrid membranes using in-situ biomimetic mineralization. In this case, the gelatin-silica hybrid membranes were created by dissolving sodium silicate and gelatin in water, followed by the solidification of the casting solution [112]. As part of this solution, the positively charged amino groups on gelatin molecules absorbed silicic acid oligomers created by sodium silicate through electrostatic attractions that augmented the local oligomer concentration and quickened the polycondensation progression. As a consequence, silica nanoparticles featuring a diameter smaller than 100 nm were created homogeneously in the gelatin matrix. Chitosan (CS) was used as an inducer and was meant to regulate the production of CdS nanoparticles, since it offers superior metal ion adsorption potential [116]. Once the $CdCl_2$ solution was combined with chitosan, the CS-Cd^{2+} complexes were produced using the adsorption and chelation of the hydroxyl and amino groups on chitosan with Cd^{2+} ions [116]. Once the adsorption and chelation balance were obtained, the fresh sulfocarbamide solution was gradually dropped. After that, the S^{2-} ions were discharged from the sulfocarbamide and then prompted a reaction with the Cd^{2+} ions in CS/Cd^{2+} complexes so as to generate chitosan/nano-CdS (CS/n-CdS). During this process of membrane production, chitosan or gelatin showcased a minimum of four key roles, that is, creating an especially thin membrane scaffold, inciting in-situ production of inorganic nanoparticles, restricting the growth of inorganic nanoparticles to the polymeric network, and lowering the particles' accumulation. There are two frequently applied methods for biomimetic mineralization in the case of membrane-forming polymers that do not have mineralization-inducing groups. Specifically, these methods entail adding other organic inducers into the casting solution or grafting functional groups onto the polymers. Although the former approach may appear easier, the organic inducer has to be selected correctly. A problem could occur if the inducer's catalytic activity was too elevated, implying that the inorganic nanoparticles are forming too rapidly, and the nanoparticles will grow and collect in a shorter time period, thus precipitating before the actual membrane casting. Furthermore, the inducers added must be compatible with the membrane-forming polymers within the required range of compositions. During the process of creating silica-containing hybrid membranes, amino group as well as analogous cationic groups can be crucial. In this case, the quaternized modification was applied to poly(vinyl alcohol) (PVA) and poly(2,6-dimethyl-1,4-phenylene oxide) (BPPO) [117]. The polymers' quaternary ammonium groups incited the production of silica using a variety of silica sources, while the network created by polymers and silica during the reaction caused the hybrid membranes to be more compact. Moreover, the addition of other organic inducers to the membrane casting solution allowed to form a silica-containing hybrid membrane [118]. While it is a commonly applied membrane-forming polymer, the PVA cannot incite silica formation. Alternatively, gelatin is a popular silification inducer that can be compatible with PVA at low content values. As a consequence, gelatin was selected as the inducer and then added into the PVA solution with the 9/1 (PVA/gelatin) mass ratio. In the next stage, membrane casting solution was created by mixing the PVA-gelatin solution with the precursor silicate solution. Finally, the silica nanoparticles were produced homogeneously within the network of PVA chains.

The process of functionalizing polymers with proper negatively charged groups offers a method for inciting as well as controlling the development of $CaCO_3$ nanoparticles. For instance, hyperbranched polyglycidol (hb-PG) functionalized using a variety of groups, such as phosphate monoester, sulfate, and carboxylate groups, was implemented during the preparation of $CaCO_3$ hybrid membranes with the spray technique. The application of this method indicated that the functional group type had a substantial effect on the structure and morphology of $CaCO_3$. **Figure 16** shows how phosphate-ester-functionalized hb-PG, sulfate, and carboxylate helped to form calcite composite, vaterite composite, and vaterite-calcite composite, respectively.

Biomimetic mineralization is a water-demanding process because of the implicit need for water during the reactions and water solubility of inducers. As a result, hybrid membranes cannot be developed with the methods mentioned above for water-insoluble polymers, since they require organic solvents in order to dissolve polymers. As a solution for this requirement, the process of water in oil W/O reverse microemulsion was developed. This process is based on the addition of surfactant and the tracing of water inorganic casting solution [119]. In this instance, the water-soluble inducer contacts the oil-soluble inorganic precursor at the interface of two phases and then encourages the hydrolysis-condensation reaction. Once this occurs, the silica nanoparticles are produced in this narrowed space, as outlined in **Figure 17**, and the creation of hydrophobic/oleophilic polymer-silica hybrid membrane is completed.

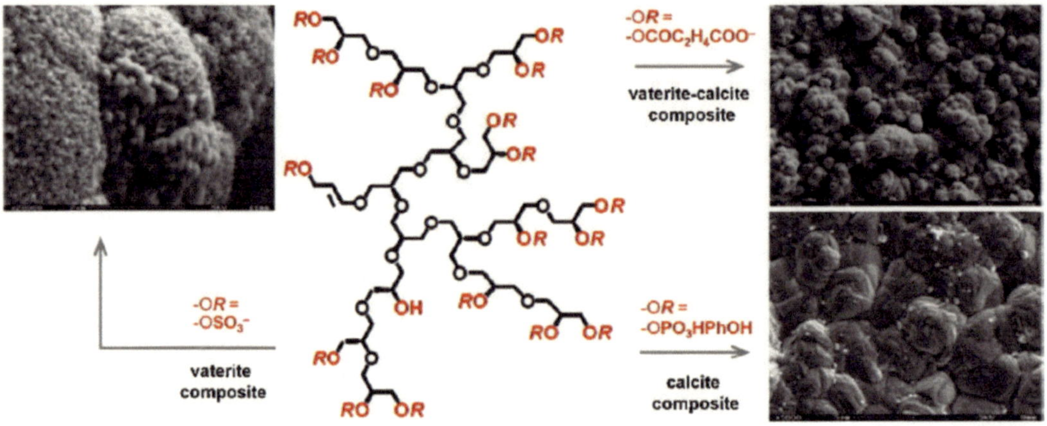

Figure 16. The molecular structure of hb-PG and SEM micrographs of $CaCO_3$ hybrid membranes formed in the presence of differently functionalized hb-PGs. Source: Ref. [114], Copyright 2012; reproduced with permission from Elsevier Ltd.

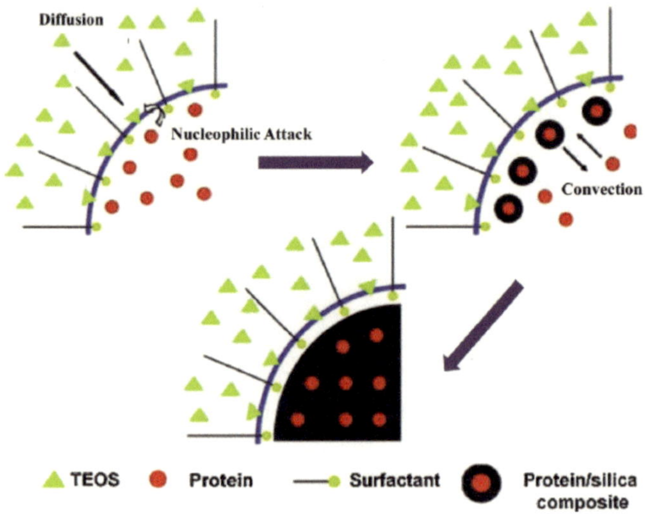

Figure 17. The formation mechanism of silica mediated by macromolecule inducer in reverse microemulsion. Source: Ref. [119], Copyright 2012; reproduced with permission from Elsevier Ltd.

The development of W/O microemulsion was a critical step toward a successful biomimetic mineralization process in hydrophobic/oleophilic polymer solution. In this process, the water in microemulsion has the capacity to dissolve inducers. Water likewise is essential for the hydrolysis reaction of silica precursors, and the water/oil interface guarantees the presence of reaction sites where the mineralization process can occur.

2.3.1.3. Membrane fabrication using biomimetic mineralization after membrane formation

The process if immersing the membrane with inducers into a precursor-containing solution offers a post-treatment method for developing hybrid membranes using the in-situ biomimetic mineralization. These types of inducers may exist either on the surface or within the membrane's matrix and cause a variety of inorganic nanoparticles' distributions. Recent research suggests that smaller sized molecular inducers, such as amino acids, are seldom applied because of their weaker interactions with membranes and smaller size, since that makes them susceptible to leaking in aqueous solutions [120]. Whenever the inducers are adsorbed solely on the membrane's surface, they can have contact with inorganic precursors once the membrane is immersed into the solution and as a result lead to the creation of inorganic layer on the surface [120]. Microcapsule type membranes were produced with the aid of this approach. In this instance, the sacrificial templates were distributed in protamine aqueous solution for the duration of several minutes, after which they were suspended in titanium-source or silica-source solutions and once the residual protamine was washed away [120]. During this process, the titania or inorganic silica layer was created on the external surface. Whenever the inducers occur within the membrane's matrix and when the membrane-forming polymers have the mineralization-inducing ones, functional groups are mixed and then situated within the membrane, and the inorganic nanoparticles can develop in the membrane's matrix, once it is immersed in the precursor-containing solution [121]. Notably, the process of mineralization happens only if the precursors are diffused into the membrane's matrix and then interact with the inducers, since this ensures that both reaction and diffusion occur simultaneously. The distribution of inorganic nanoparticles is directly connected to the membrane matrix's structure, as well as the rate at which mineralization reaction happens. In most instances, the amount of inorganic components within the membrane slowly lowers as one moves away from the external surface and toward the interior. In a research report, CS membrane was immersed into a simulated body fluid (SBF) for the duration of 3 weeks so as to try and produce hydroxyapatite (HA) [122]. During this experiment, the cationic groups in the CS membrane helped the adsorption of $(PO_4)^{3-}$ and the subsequent nucleation. The inducer protamine was secured within the confined spaces created by cross-linked PVA molecular chains. During the immersion of the PVA-protamine membrane into the precursor-containing solution, the inorganic precursor first diffused into the membrane's matrix and then generated silica nanoparticles through the templating and catalysis of protamine (**Figure 18**). The silica nanoparticles' sizes could be easily regulated through the controlled alteration of the precursor solution's concentration and pH values. The production of silica nanoparticles may be manipulated by adjusting the membrane matrix structure and altering the annealing temperature so as to control the bulk polymer network and cross-linking of PVA (**Figure 19**). To sum up, biomimetic mineralization method offers an innovative as well as applicable approach for the development of hybrid membranes with homogeneous dispersion, advantageous interfacial interactions under mild conditions, and nanoscale filler sizing. As more research emerges on the mineralization mechanism of biominerals, the process of biomimetic mineralization will gain more ground when it comes to production of diversified hybrid membranes.

2.3.2. *Based on bioadhesion*

In addition to the separation performance, stability is a key parameter when it comes to the practical aspects of producing a functioning membrane. For those composite type membranes that are made of two distinct layers, the varying surface properties of the two layers

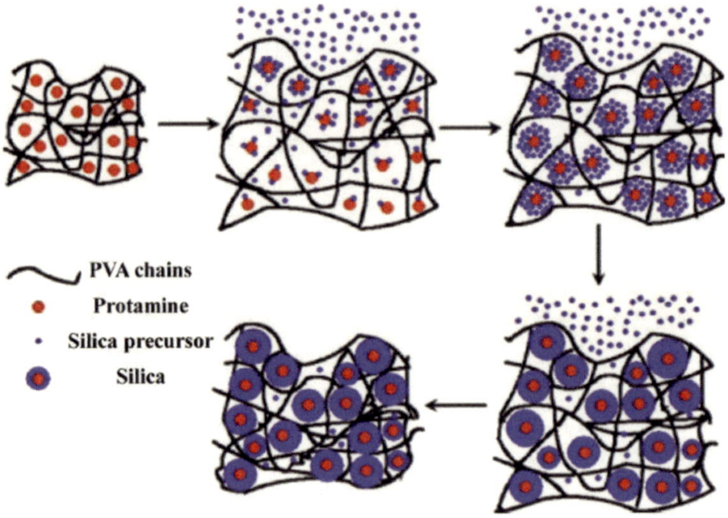

Figure 18. The formation process of silica nanoparticles within PVA matrix. Source: Ref. [122], Copyright 2010; reproduced with permission from Elsevier Ltd.

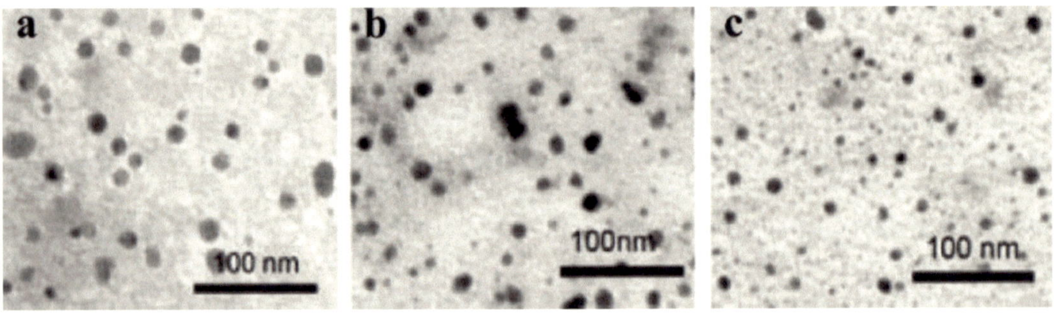

Figure 19. Transmission electron microscopy (TEM) images of silica in the nanohybrid skin layer after it is annealed at (a) 293 K, (b) 333 K, and (c) 373 K. Source: Ref. [122], Copyright 2010; reproduced with permission from Elsevier Ltd.

can cause adverse interface compatibility as well as weakened interfacial interaction between the layers. Whenever the swelling amounts of the two layers are different, a significant stress will appear at the interface. This stress can force the two layers to peel off relatively easily, if the stress surpasses the interfacial interaction. Improving interfacial interaction and interface compatibility between the two layers is an effective and straightforward method for obtaining high stability in composite membranes [123]. When it comes to the surface-functionalized membranes, preserving functional groups during long-time operations is a crucial prerequisite. For membranes with weak interactions between molecular chains and flexible molecular chains, the membrane structure can decline if it is in contact with solvent, water, or other plasticizers during use, since this interaction can significantly decrease selectivity. Enhancing the membrane's cohesive energy will help improve the membrane's stability and maintain its structure. Bioadhesives have served as an inspiration because of their controllable adhesive/cohesive capacity, greater strength, and broader applicability. In fact, biomimetic adhesion strategies that use bioadhesives, or their analogs like biomimetic adhesives, have been implemented so as to better deal with the challenges listed above.

2.3.2.1. Fabrication of membranes via incorporation of bioadhesives

The addition of bioadhesives as an intermediate layer during the composite membrane production is a relatively easy and efficient way to improve the interfacial interaction between the two layers [124]. Furthermore, bioadhesives that have been derived from natural sources, including dextrin, shellac, and gelatin, are complying with the basic requirement listed as part of the environmental protection. The bioadhesive carbopol (CP) was used for the first time as an intermediate layer for connecting the polyacrylonitrile (PAN) support layer and the CS separation layer. Specifically, CP is a mucoadhesive polymer that features many of the carboxylic groups (COOH) that partially dissociate when in water and offer high viscosity and flexible structure at low concentration values. The schematic depiction of the interface interaction for CS/CP/PAN composite membrane is shown in **Figure 20**. In addition to the carboxy group (COOH) of CP, the hydroxyl group (OH), the amino group (NH_2) of CS, and the van der Waals force, the cyano group (CN) of PAN may generate a multiplicity electrostatic interactions or hydrogen bonds. Once the CP layer is incorporated, the peak peeling strength value was found to be four times greater than that of the CS/PAN membrane. Furthermore, the absolute values of interfacial energy for CP/PAN and CS/CP interfaces were greater than those present in CS/PAN interface based on molecular dynamic MD simulation. The SEM images available in **Figure 21** showcase that the composite membrane features a three-layered structure made up of the support layer,

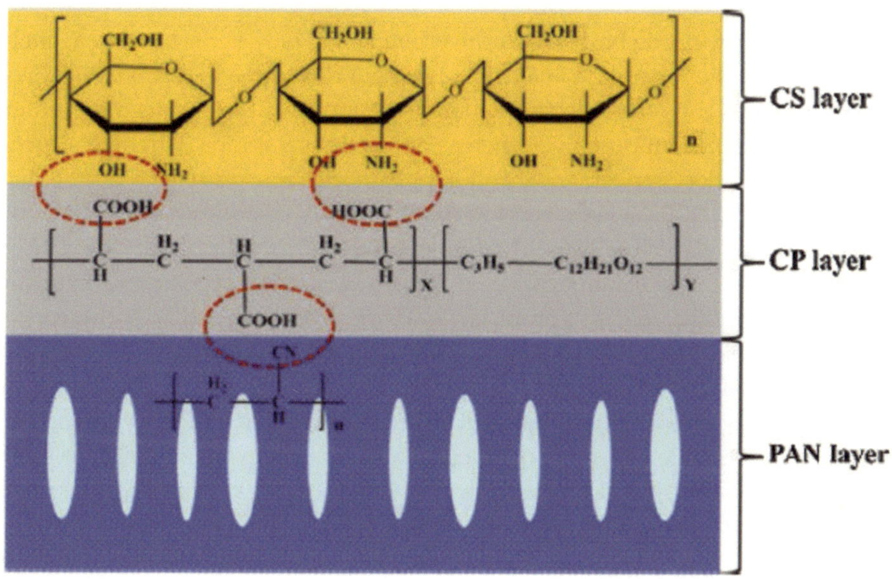

Figure 20. Schematic representation of the interfacial interaction in CS/CP/PAN composite membrane. Source: Ref. [124], Copyright 2010; reproduced with permission from Elsevier Ltd.

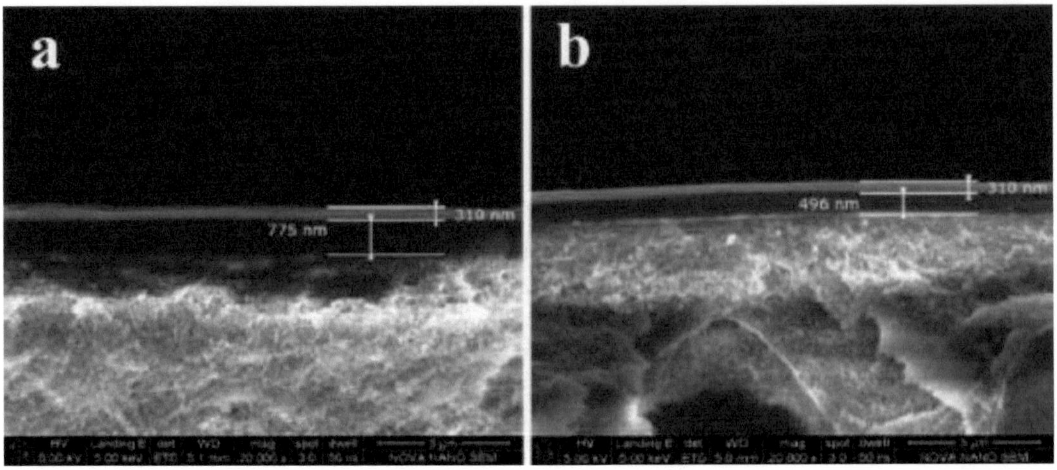

Figure 21. SEM images of cross-section: (a) GCCS(30)/CP(0.5)/PAN membrane, (b) GCCS(30)/CP(0.05)/PAN membrane. Source: Ref. [124], Copyright 2010; reproduced with permission from Elsevier Ltd.

intermediate layer, and separation layer. The fact that there is an intermediate layer has a number of effects on the composite membranes' overall properties and structure. In particular, the additional layer can augment the mass transfer resistance for permeating molecules. The interactions occurring between the intermediate layer and the other layers can affect the stability and structure of interfaces. Moreover, the intermediate layer can function as a defensive coating when it generates a more compatible surface and in turn allows for the casting of polymer solution with low concentration, thus helping to produce a much thinner separation layer.

Cases discussed above focus on bioadhesives that functioned only as the binding agent between the support layer and the separation layer. An argument can be made that if a bioadhesive can generate a thin membrane with selective separation functions, that act as a separation layer while bound tightly to the support layer, then a composite membrane with high structural stability, desirable separation performance, and simple fabrication procedure can be developed [125]. In this case, the bioadhesive that acts as the separation layer needs to possess dual functions of separation and adhesion. These dual functions have varying demands for its chemical and physical properties. The bioadhesive must have a number of specific characteristics so as to provide durable binding to the support layer. These properties can be summed up as follows: (1) numerous polar groups, like OH and COOH; (2) electro negativity; (3) larger molecular weight; (4) flexible chain; and (5)relatively moderate surface tension [126]. Bioadhesive likewise needs to offer advantageous free volume distribution, selective adsorption for one of the permeating molecules, and suitable molecular chain rigidity that can help obtain higher selectivity and permeability. A bioadhesive hyaluronic acid, or a type of acidic polysaccharide, was used for the separation layer of the composite membrane and intended for dehydration of organic solvents because of its excellent chain flexibility, higher molecular weight, elevated negative charge density, favorable membrane-forming properties, and strong affinity to water. MD simulations as well as experimental inquiries were conducted with the aim to corroborate the strong interfacial interaction and promising interface compatibility of this as-prepared composite membrane.

2.3.2.2. Fabrication of membranes via incorporation of biomimetic adhesives

In addition to the bioadhesives obtained from organisms, biomimetic adhesives featuring comparable functional groups and structure can be used as possible alternatives, if the matching bioadhesives are very costly and difficult to extract. As noted earlier, the cement secreted by the sandcastle worm and the adhesive proteins found in mussel byssus contain DOPA that plays an essential role in the bioadhesion process [26]. Dopamine, as an analog of DOPA, has almost identical properties and structural arrangement. Dopamine and DOPA are able to conduct self-polymerization and oxidation under mild conditions in an aqueous environment so as to generate an exceptionally thin coating with favorable biocompatibility, robust interface binding force with diverse substrates, and higher hydrophilicity potential, similar to the operational characteristics of adhesive proteins found in marine organisms [26]. Such an adhesive capacity and enhanced structural stability of the as-prepared coating may be obtained using a range of chemical and physical interactions, for instance, metal chelation, hydrogen-bonding, covalent interaction, and $\pi-\pi$ interaction [34]. Polydopamine (PDA) was added onto various support layers, including PES, PTFE, PSf, and ceramic, before the creation of the separation layer. A number of interactions occurring between the intermediate PDA layer and the other two layers facilitated an enhanced interfacial compatibility between the two contrasting layers, as well as the improved membrane structure's stability for the long-time operation. In a recent experimental setting, PDA layer was used to form a more suitable surface for the interfacial polymerization reaction by actively manipulating the surface roughness, support layer's pore structure, and hydrophilicity. In addition to functioning as the intermediate layer, poly (DOPA)/PDA was likewise used as the membrane's skin layer, like the surface modification coating [127] or the composite membrane's separation layer [34]. If compared to the frequently employed methods, the deposition of poly (DOPA)/PDA is more efficient and green-conscious, as well as advantageous in terms of durability. As illustrated in **Figure 22**, the composite membrane was produced

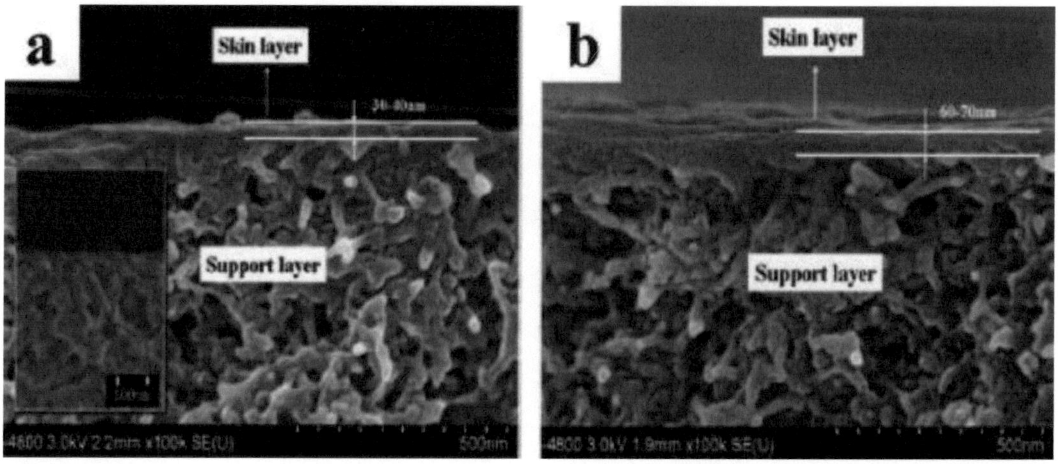

Figure 22. SEM image of the cross-section area of the PDA/PSf composite membranes: (a) single coating (inset: the uncoated PS membrane), (b) double coating. Source: Ref. [34], Copyright 2009; reproduced with permission from the American Chemical Society.

using defect-free and ultrathin PDA separation layer and by immersing the support layer into dopamine aqueous solutions, thus allowing the self-polymerization reaction to happen on the surface [34]. The separation layer's structure and thickness may be controlled through alteration of coating time, number of coats, and the pH and concentration values of dopamine solution. Up until now, the poly (DOPA)/PDA coating has been implemented in membranes that are made out of different materials and feature a variety of pore sizes [127]. In every experimental case, the membrane's hydrophilicity gained visible improvement. However, when it comes to the surface roughness, the values changed depending on the pore sizes found on the membrane's surface. Research also indicated that poly(DOPA)/PDA-coated layer is comprised of accumulated nanoparticles [127]. In the case of MF membrane, pore sizes are bigger than in the case of poly (DOPA)/PDA nanoparticles, which in turn suggests that they are made inside the pores and lead to surface smoothing [128]. In membranes with similar or smaller pore sizes, including reverse osmosis (RO), UF, and nanofiltration (NF) membranes, the process of pore blocking occurs as well as dominates at beginning, thus causing an escalation in roughness [129]. When compared to dopamine, DOPA offers advantageous quality in the form of zwitterionic, as it can help construct a surface with higher hydrophilicity potential.

The poly(DOPA)/PDA derivatives with DOPA/dopamine grafted with other molecules embody an innovative surface modification method that can improve stability, diversity, and operation. The anchoring abilities of mussel adhesive proteins and cell membrane's fouling resistance were integrated through the fabrication of doubly biomimetic copolymer as antifouling coating, which contains both catechol groups and phosphorylcholine (PC) side groups. **Figure 23** showcases that the doubly biomimetic copolymer may be successfully adsorbed onto a range of substrates using the robust anchoring force created by catechol groups,

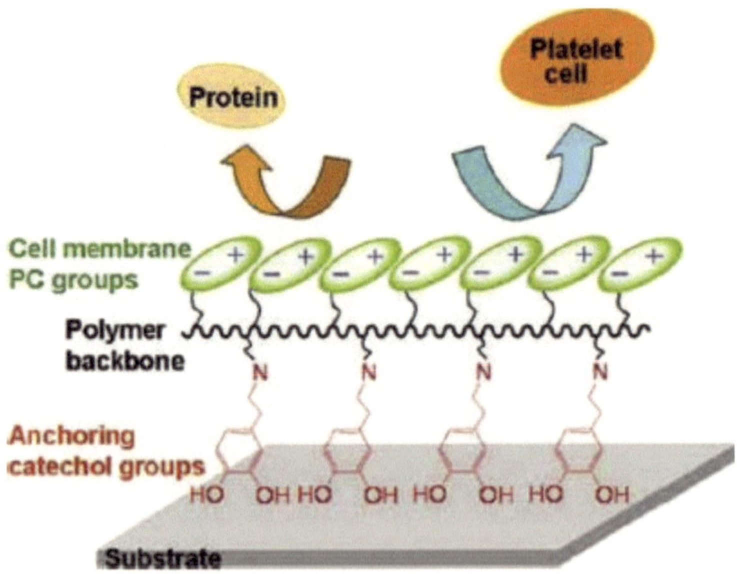

Figure 23. Schematic illustration of the structure and fouling resistance of the doubly coating of biomimetic copolymers. Source: Ref. [130], Copyright 2012; reproduced with permission from Wiley-VCH Verlag GmbH & Co. KGaA.

and the PC groups are oriented toward the external side and are creating the antifouling surface resembling cell membrane. As a result, the antifouling surfaces can be generated on different devices and materials with the help of dip-coating in the doubly biomimetic polymer solution. In addition to the enhanced adhesive capacity, Poly(DOPA)/PDA's other advantage is its high reactivity, as it offers reaction sites that help perform additional modifications for the membrane's surface.

Moreover, polyethylene (PE) porous membranes were modified with PDA coating and then immobilized with heparin and bovine serum albumin (BSA), respectively, using covalent bonds in aqueous environment with the aim of gaining improved biocompatibility and higher hydrophilicity. The schematic of the PDA deposition on the PE porous membranes and the subsequent heparin immobilization are shown in **Figure 24**. High numbers of o-benzoquinonyl groups, occurring on the PDA layer's surface after the oxidation and self-polymerization of dopamine, had reacted with the amino/imino groups on heparin, once the membrane was immersed into a heparin solution. As a consequence, the deposition of poly (DOPA)/PDA and their derivatives offers an approach with added versatility and long-time durability that can modify the membrane's surface and help to integrate diverse functions, the latter being particularly valuable for membranes suffering from chemical inertness.

To help increase cohesive energy, as well as membrane's structural stability, dopamine was added into the membrane's matrix as a potential modifier. The polymerization and oxidation of dopamine can happen prior to, during, and after the process of membrane production [132]. A variety of oxidizing agents, such as iron ions, sodium periodate, and oxygen, have been applied in order to encourage the reaction. Multiple interactions between the membrane's matrix and PDA can help make the membrane significantly more stable. Moreover, the adhesive and cohesive balance of PDA, together with the produced membrane structure, may be efficiently regulated through the process of fluctuating the oxidation condition, for example, the ratio of dopamine to oxidizing agent if the production of PDA was during or after membrane's creation, as illustrated in **Figure 25**.

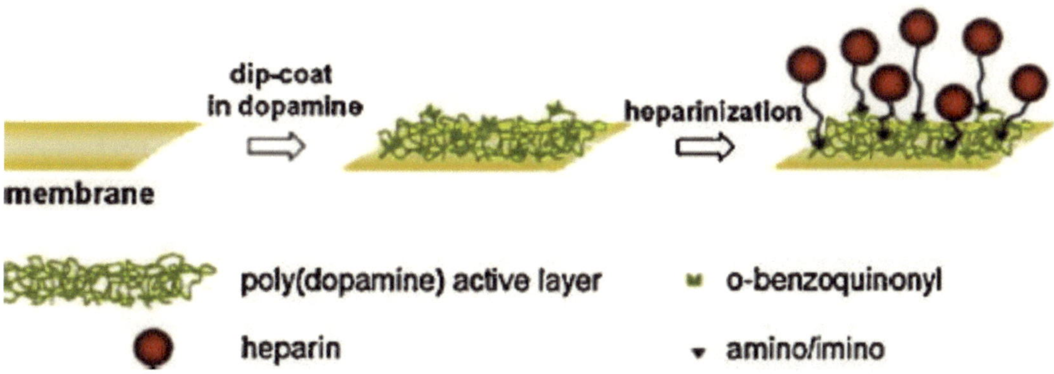

Figure 24. The schematic of the PDA deposition on PE porous membranes and subsequent heparin immobilization. Source: Ref. [131], Copyright 2010; reproduced with permission from Elsevier Ltd.

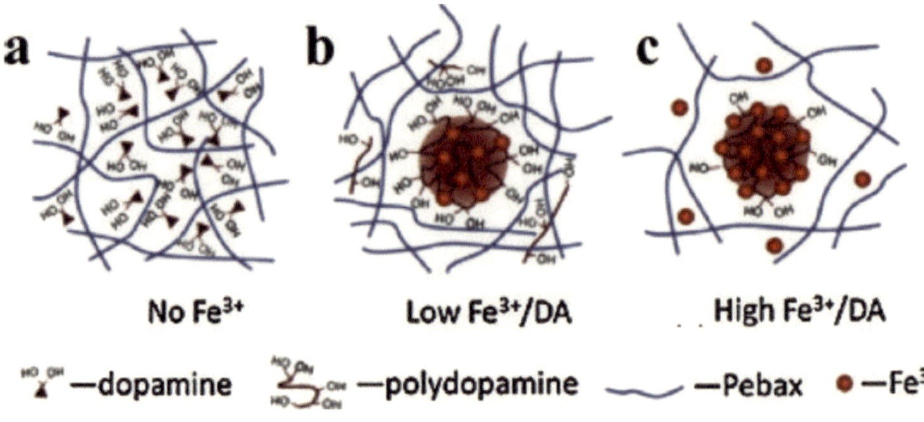

Figure 25. Schematic illustration of the possible nanoscale structures of hybrid membranes with different Fe^{3+}/DA. (a) DA monomers bearing abundant phenyl groups indicate high adhesion ability but weak cohesive ability. (b) Low Fe^{3+}/DA leads to aggregated $Fe3^{+}$-DA complexes with enhanced cohesive interaction and adequate adhesion ability. (c) High Fe^{3+}/DA leads to robust Fe^{3+}-DA nano aggregates with few available phenyl groups and poor adhesion ability. Source: Ref. [133], Copyright 2012; reproduced with permission from the Royal Society of Chemistry.

2.3.3. Based on self-assembly

Self-assembly can offer an efficient method for duplication of natural manufacturing processes from bioinspired and biomimetic pathways, as both of these share a key characteristic in the form of spontaneous organization, namely biomacropolymer and phospholipids self-assembly. These comparable structures and interaction mechanisms suggest that the self-assembly process offers a distinct nanoscale approach for regulating the membrane's chemistries and structures. The following section provides an outline of various self-assembly processes that are presently used in the production of ordering nanoporous membranes, as well as modification of polymer membranes.

2.3.3.1. Fabrication of membranes via block copolymer self-assembly

The synthetic block copolymers containing two or more thermodynamically conflicting blocks may experience microphase separation into aggregates of multiple morphologies with extremely ordered structures. Microdomain morphologies of diblock copolymers, including cylinders or spheres, are composed of a one phase in a matrix of another, in addition to lamellar and gyroids (**Figure 26**) [134]. Membranes exhibiting improved selectivity and higher flux can be produced using self-assembled block copolymers. Several dense type membranes based on self-assembly of block copolymer were produced with the aim of offering useful applications in pervaporation, fuel cells, and CO_2 membrane separation processes. However, the majority of researchers are turning to the development of nanoporous membranes offering properties such as narrow pore size distributions, tunable chemical and mechanical characteristics, higher porosity, and enhanced ordered and oriented nanopores. For instance, if the composition of block copolymers and the molecular weight happen within specific restrictions, then the spontaneous self-assembly progression can help facilitate ordered cylinders that are aligned perpendicularly with respect to the surfaces and successfully converted into properly ordered nanoporous membranes.

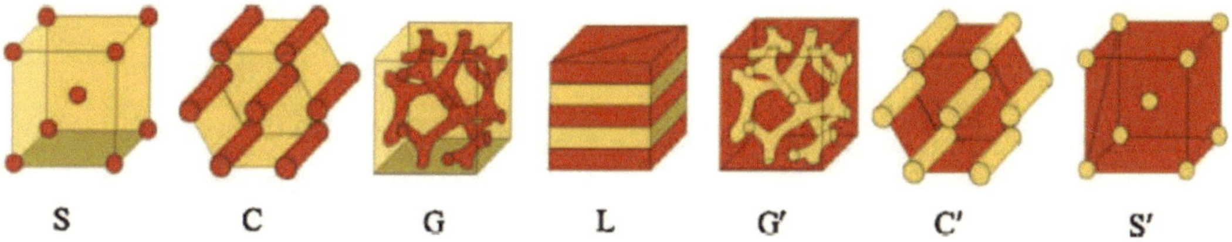

Figure 26. Diagram of the microdomain morphologies of diblock copolymers. Source: Ref. [134], Copyright 2005; reproduced with permission from the Materials Research Society.

Innovative research work has been conducted with the aim of producing nanoporous membranes based on the self-assembly of PS block copolymers with poly (ethylene oxide) (PEO) or hydrophilic poly (methyl methacrylate) (PMMA) blocks. These two approaches were established specifically in order to achieve nanoporous membranes. The first approach relies on the elimination of minor PMMA component that can lead to the creation of cylindrical microdomains oriented in a normal manner with respect to the membrane's surface [135]. The method's representative extremely ordered nanoporous thin films developed using self-assembled PEO-b-PMMA-b-PS were created through initial solvent annealing that is followed by ultraviolet (UV) irradiation that degrades the PMMA block [136]. The terminal PEO block and the central PMMA block degradability ensure the long-range order within the overall system. Nanoporous type membranes, featuring narrow pore size distribution, have the capacity to provide improved selectivity values as well as enhanced filtration flux potential. The second approach is based on the removal of homopolymer from the block copolymer/homopolymer blends where the homopolymer is more constricted to the cylindrical microdomains' center [136]. Moreover, a double-layered nanoporous membrane was created using a combination of PS-b-PMMA together with cylindrical microdomains of homopolymer PMMA. In this case, the film was first constructed on top of the sacrificial silicon oxide layer, after which it was released into the HF solution and then relocated onto the PS membrane's surface. Finally, it was treated by selectively eliminating the PMMA homopolymer from the cylindrical PMMA microdomains using acetic acid (**Figure 27**). As a result of this experimental run, an 80-nm thick membrane was produced with cylindrical 15-nm diameter pores for virus filtration applications.

Polylactide (PLA) is an innovative type of degradable blocks, as well as a multipurpose moiety for developing efficiently ordered nanoporous block copolymer membranes. An approach for making monodisperse nanoporous membranes was designed based on the block polymer PS-b-PLA self-assembly [138]. In this instance, a cautious regulation of the copolymer film's solvent evaporation rate can facilitate a perpendicular orientation. Furthermore, the exposure of the composite membrane to a dilute aqueous base can selectively etch the PLA block, thus fabricating a porous structure. The nanoporous membranes were likewise created based on cylinder-forming triblock copolymer polystyrene-b-poly (dimethylacrylamide)-b-polylactide (PS-b-PDMA-b-PLA) and PS-b-PI-b-PLA and the etching of the PLA block. An effective new method was shown to generate strong bicontinuous nanoporous block copolymer self-assembled membranes using the process of ring-opening metathesis polymerization of norbornene-functional PS-b-PLA and dicyclopentadiene (DCPD) additive (polymerization induced phase separation), which was then followed by the selective elimination of PLA block [139, 140].

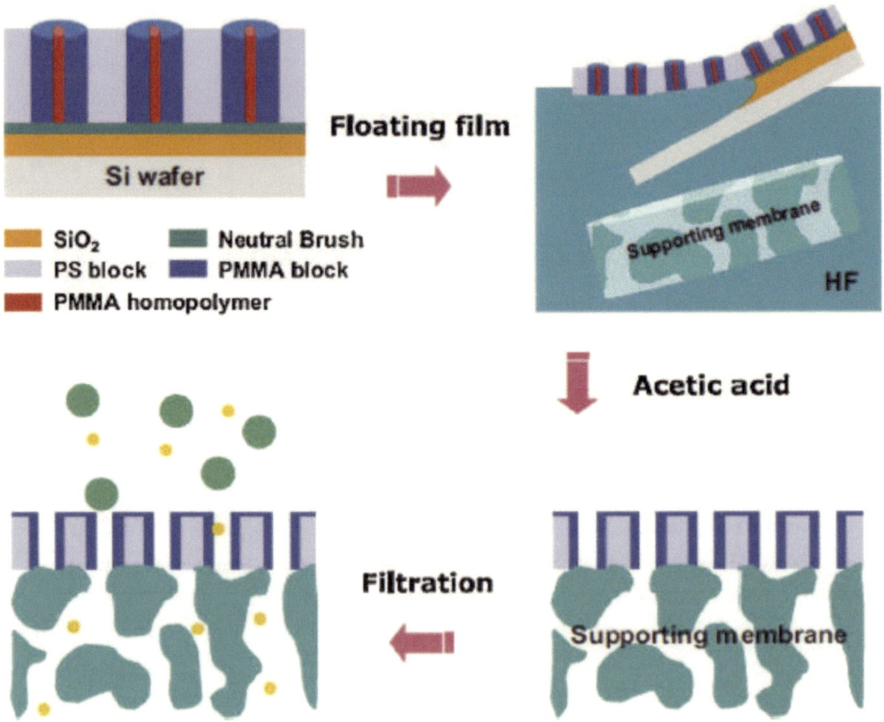

Figure 27. Schematic depiction of the procedure for the production of asymmetric nanoporous membranes through the removal of homopolymer from block copolymer/homopolymer blend films. Source: Ref. [137], Copyright 2006; reproduced with permission from Wiley-VCH Verlag GmbH & Co. KGaA.

The application of this method has resulted in the cross-linked nanoporous membranes featuring narrow pore size distributions. Furthermore, PS-based block copolymer composites (PS-b-PLA and PS-b-PEO) were implemented for the production of ordered nanoporous membranes featuring hydrophilic pore surfaces. The pegylated pore type surfaces were created through the process of degradative elimination of the PLA block from the self-assembled PLA/PEO microdomains. The bicontinuous gyroid morphology together with the hexagonally packed cylindrical morphology was implemented based on the specific annealing circumstances [141]. The PE-based block copolymer composites can likewise be applied during the process of nanoporous membranes' production with hydrophilic pore surfaces and using crystallization-induced self-assembly followed by the PLA removal. Block copolymer composites of PLA-b-PE-b-PLA and poly (2-(2-methoxyethoxy) ethyl methacrylate)-b-polyethylene-poly(2-(2-methoxyethoxy) ethyl methacrylate) (PMe(OE)xMA-b-PE-b-PMe(OE)xMA) were responsible for producing a disorderly bicontinuous structure with semicrystalline PE domains and a mixed PLA/PMe(OE)xMA domains. An adequately selective PLA etching from the PLA/PMe(OE)xMA domains using mild base treatment can effectively manufacture a nanoporous PE with pore walls covered with PMe(OE)xMA polymer chains (**Figure 28**) [141].

The uniquely self-assembled block copolymer featuring a cleavable covalent linking unit in the middle of the block copolymer has the capacity to successfully remove small component domains without the application of tough chemicals. An innovative method for producing

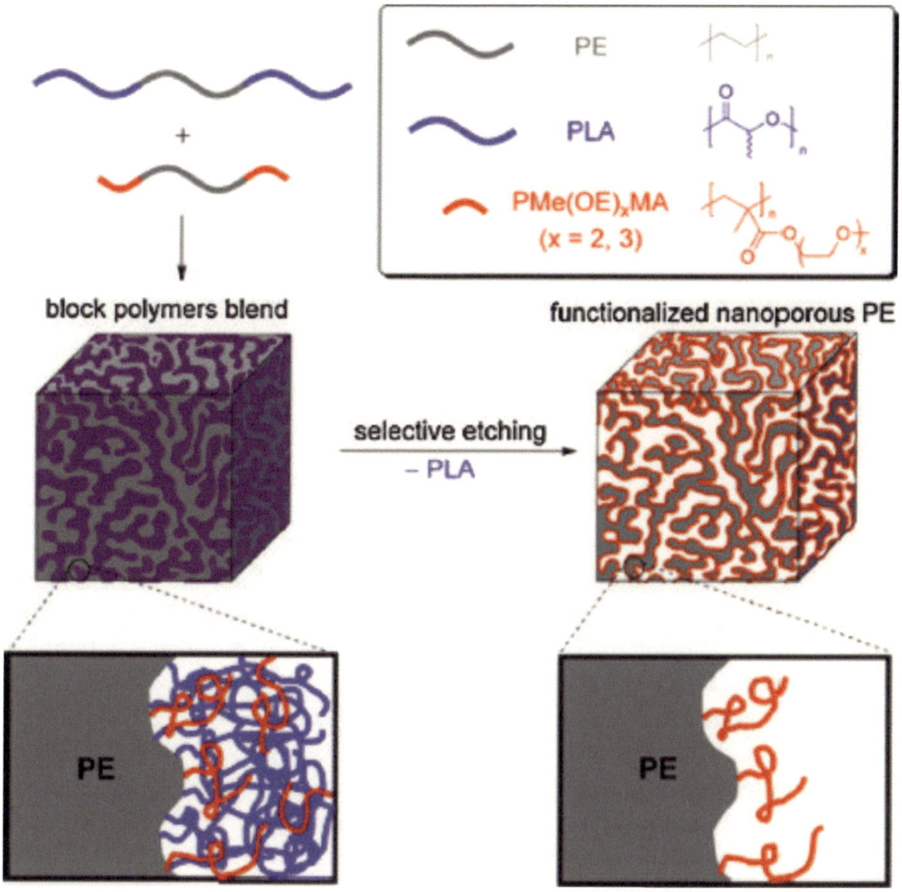

Figure 28. Preparation strategy of the nanoporous PE membrane with pore wall lined with PMe(OE)xMA by the PLA selective etching from the reactive block copolymer blends. Source: Ref. [142], Copyright 2012; reproduced with permission from the American Chemical Society.

nanoporous PS films was established with the help of a selectively photocleavable PS-b-PEO block copolymer (ONB-(PS-b-PEO)), where a photochemically sensitive orthonitrobenzyl (ONB) group was fitted in as a type of photocleavable linking unit [143]. In this case, the cylindrical PEO domains can be taken out following the UV light irradiation and the selective solvent rinse. This approach was likewise used to create nanoporous thin films based on PS-b-PEO block copolymer carrying a photo-cleavable o-nitrobenzyl ester junction. Furthermore, the nanoporous films from the connected poly(styrene-ss-ethylene oxide) (PS-ss-PEO) were shown through the redox cleavable disulfide bond [144]. Once the annealing in a benzene/water vapor environment had occurred, the PS-ss-PEO films had reoriented the PEO cylindrical microdomains in a manner normal with respect to the film's surface. Next, the PEO block could be effortlessly cleaved through the immersion of PS-ss-PEO thin films into a d, l-dithiothreitol-containing ethanol solution, thus forming nanoporous thin films (**Figure 29**). The films accumulated, due to the PS-b-PEO acquiring cleavable triphenylmethyl ether juncture between PEO and PS, can likewise generate nanopores through the process of selective PEO removal and using trifluoroacetic acid etching [146].

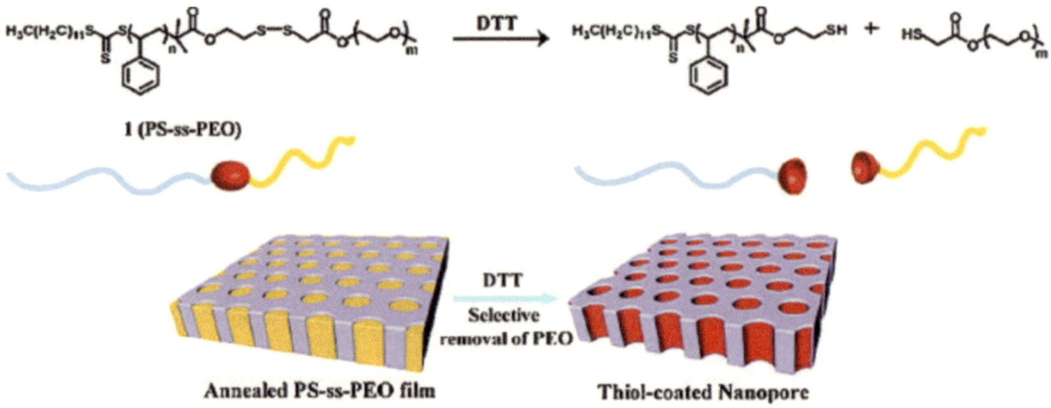

Figure 29. Structure of the PS-ss-PEO block polymer connected by a disulfide bond and schematic representation of the nanoporous thin film preparation. Source: Ref. [145], Copyright 2009; reproduced with permission from the American Chemical Society.

An approach developed with the help of block copolymer supramolecular assemblies together with hydrogen bond donors and acceptors was implemented for the production of ordered nanoporous membranes using the process of removal of minor component enriched nanodomains. Another method relied on hydrogen bonding between 3-pentadecyl phenol(PDP) and 4-vinylpyridine monomer units in order to form a comb-like molecular architecture and effectively alter the gyroid/cylinder morphology of polystyrene-b-poly(4-vinylpyridine) (PS-b-P4VP). The two-dimensional films fabricated using such supramolecular assemblies can produce nanoporous membranes through the removal of amphiphilic PDP domains, which occurs when washed with a selective type solvent (**Figure 30**) [147]. Moreover, PS-b-P4VP/PDP comb-like block copolymer systems was applied so as to acquire a lamellae-within-cylinders films with periodic and well-defined nanoporous structures. In order to obtain thin films with perpendicularly oriented hexagonally ordered cylinders of P4VP, the 2-(4′-hydroxy benzeneazo) benzoic acid (HABA) was implemented as hydrogen bond donors. Made out of cylindrical nanodomains and generated by P4VP-HABA associates surrounded by PS, the supramolecular assemblies were likewise produced using the PS-b-P4VP/HABA system [148]. As part of this process, the HABA molecules made hydrogen bonds with the P4VP units and then uniformly spread within the domain of P4VP (HABA). The HABA could be taken out with relative ease from the P4VP (HABA) domain, once it is rinsed in a selective solvent, which renders a systematic array of nanochannels. Furthermore, substances that can engage with the poly (vinylpyrrolidone) (PVP) block like dodecyl benzene sulfonic acid (DBSA), poly (methyl methacrylate)-dibenzo-18-crown-6-poly(methyl methacrylate) (PMCMA), PMMA, 1,5-dihydroxynaphthalene (DHN), 1-pyrenebutyric acid (PBA), and phenolic resin were similarly applied so as to produce ordered nanoporous films using the self-assembly of block copolymer supramolecules grounded in physical interactions [149].

Figure 31 outlines the two-step method established for the development of the nanoporous structure based on metallo-supramolecular block copolymers with amphiphilic blocks and connected together using metal-ligand complexes. This method's initial stage involves the

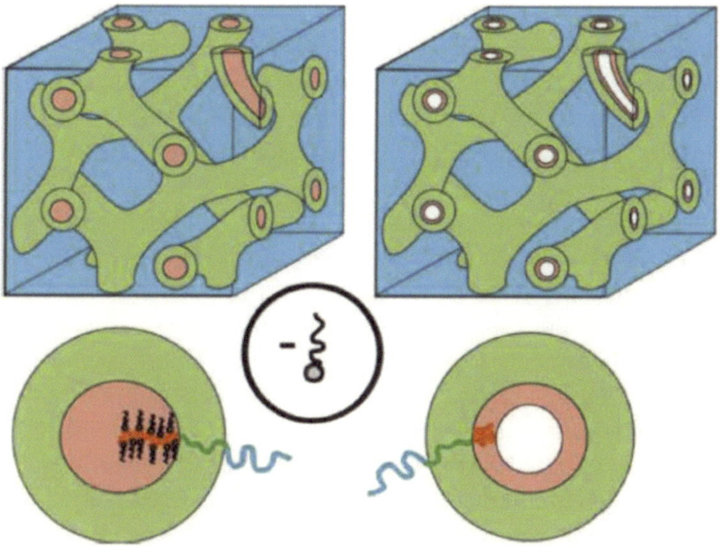

Figure 30. The schematics illustrate supramolecular self-assembly of PS-b-P4VP triblock copolymers. Source: Ref. [147], Copyright 2011; reproduced with permission from Wiley-VCH Verlag GmbH & Co. KGaA.

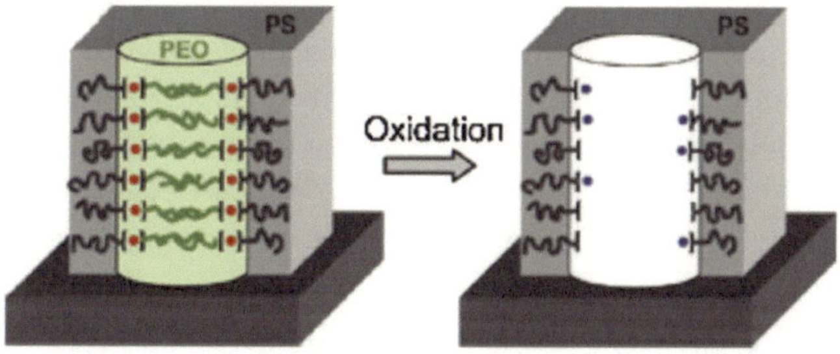

Figure 31. Schematic representation of the preparation of functionalized nanoporous thin films from metallo-supramolecular block copolymers. Source: Ref. [150], Copyright 2005; reproduced with permission from Wiley-VCH Verlag GmbH & Co. KGaA.

self-assembly of block copolymer, which then yields cylindrical microdomains oriented in a normal manner with respect to the substrate. The next stage in the process requires opening metal-ligand complex with the aid of redox chemistry that releases minor PEO block and generates nanopores. The metallo-supramolecular block copolymers show superior characteristics in so far as the supramolecular bond's potential reversibility bestows "smart materials" with controllable properties [151].

The overall difficulty of up-scaling, time-consuming preparation steps, and a lack of adequate long-range order are all serious challenges for the process of block copolymer-based membrane development. A new approach was designed for the fabrication of isoporous membranes with nanometer-sized pores based on the idea of joining the self-assembly of block copolymer PS-b-P4VP with the nonsolvent-induced phase separation (NIPS), as shown in

Figure 32 [153]. This solvent evaporation caused a concentration gradient within the block copolymer solution, specifically between the interface turned toward the air section and the interface turned toward the bottom. The microphase separation took place and then moved along the gradient within the greater concentration region of the surface, as a consequence directing the progressive growth of the cylindrical domains into the swollen layer area. As part of the process of phase separation, the nonsolvent water initially moved into the swollen P4VP blocks' cylindrical domains and then was switched with solvent. The solvent coming from the swollen PS matrix was primarily dispersed into the channels, since the interface area available for the solvent/nonsolvent exchange within the channels was significantly greater than the area available at the top surface. Research literature overviews likewise outline innovative methods for producing isoporous asymmetric membranes designer using solvent selectivity, supramolecular assembly of PS-b-P4VP block copolymer micelles, and complexation-directed supramolecular chemistry [154]. The supra molecular assembling methods of block copolymer micelles offered a flexible, nondestructive, and relatively low effort way for forming mesoporous block copolymer films featuring well-defined pore size values. The PI-b-PS-b-P4VP triblock copolymer-derived mesoporous films were produced with a joint application of controlled solvent evaporation, which directed the self-assembly of the terpolymer micelles to structurally shape the mesoporous selective layer, and of NIPS, which shaped the basic macroporous support structure. Once developed, the mesoporous films showed distinctive stimuli responsive permeation behavior.

A confined swelling-induced pore production approach has recently appeared as the latest method for the development of porous materials through the exposure of self-assembled block copolymers to solvents with highly selective minor phase. Some of the synergic benefits of this approach are its higher pore regularity, lack of weight loss, pore forming process reversibility, relative simplicity, and absence of chemical reactions. [152].

A new methodology of collective osmotic shock was formulated with respect to the swelling-induced expansion of the minor phase and the self-assembled block copolymer micelles [155]. At the core of this approach, a spherical block copolymer, or PS-b-PMMA, was applied so as to form materials receptive to the influences of collective osmotic shock (**Figure 33a**). In this

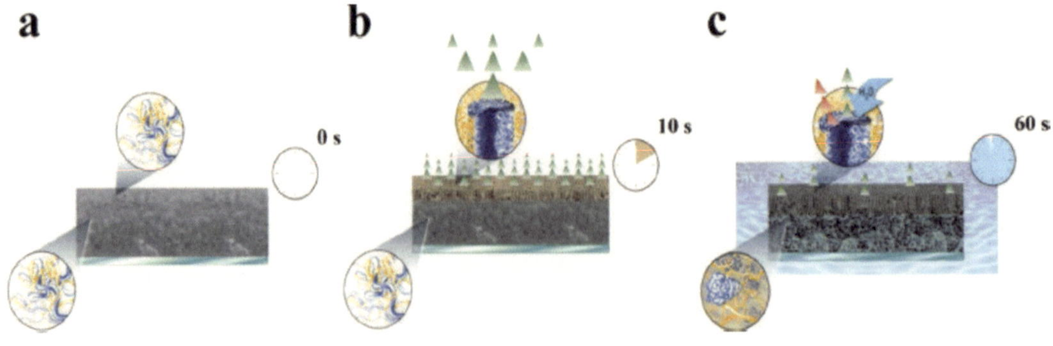

Figure 32. Schematic diagram of the asymmetric film formation process combining NIPS with the self-assembly of block copolymer PS-b-P4VP. Source: Ref. [152], Copyright 2007; reproduced with permission from the Nature Publishing Group.

dynamic, the PS-b-PMMA film was created using multiple layers of close-packed PMMA spherical cores that were carefully spaced out and enclosed within a PS matrix. The exposure to UV light that followed cross-linked the PS phase and dismantled the PMMA into smaller oligomers. Next, the film was submerged into acetic acid, which is a solvent for PMMA oligomers, and a substantially greater osmotic pressure was created within the PS matrix due to the solvation of degraded PMMA oligomers. This collective osmotic shock caused breakages among the spheres and formed a path for the complete release of PMMA oligomers. Synchronized and explosive fracture within the structured materials prompted the formation of nanoperforated multilayer constructions (**Figure 33b**) that can be applied in ultrafiltration as well as other diversified membrane processes [156].

This section's overview of the ordered porous membranes' production using self-assembly of block copolymers is quickly recapped in **Table 2**. This summary is a convenient reference guide for the production of ordered porous type membranes.

2.3.3.2. Fabrication of membranes via amphiphilic copolymer surface segregation

As an example of an in-situ method for modifying the membrane's surface, the surface separation of amphiphilic copolymers as part of the membrane surface development has shown distinct benefits, including the formation of effective brush layers on pore as well as membrane surfaces [186]. This self-assembly and surface segregation approach may be defined as a series of specific steps. First, the amphiphilic copolymers are mixed into the membrane's casting solution. During the following phase inversion process, the hydrophilic sections of the copolymers close to the interface are separated from the membrane's surface in a spontaneous manner, until the chemical potentials of the brush and bulk layers are balanced. Meanwhile, the hydrophobic sections are securely trapped within the membrane's matrix with the help of hydrophobic interaction [187]. Until recently, the majority of porous membranes were created using surface segregation of amphiphilic copolymers together with the commercially applied membrane production method called the wet phase inversion. In particular, PEO-based comb polymer and methyl methacrylate (MMA) were applied as the surface-segregating additives that would enhance PVDF (polyvinylidene fluoride) membrane's surface hydrophilicity potential.

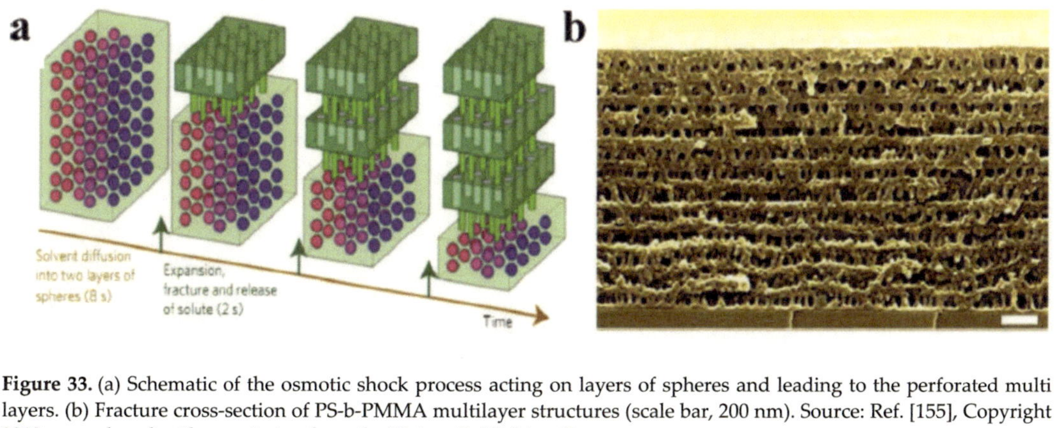

Figure 33. (a) Schematic of the osmotic shock process acting on layers of spheres and leading to the perforated multi layers. (b) Fracture cross-section of PS-b-PMMA multilayer structures (scale bar, 200 nm). Source: Ref. [155], Copyright 2012; reproduced with permission from the Nature Publishing Group.

Membranes	Assemblies and assembly approaches	Pore generation	References
PS-b-PI	PS-b-PI; coating PS-b-PI onto silicon substrates, followed by solvent evaporation	Degrading PI by O_3 and methanol rinsing	[157]
PS	PS-b-PMMA; coating PS-b-PMMA onto PS-r-PMMA neutral layer, followed by vacuum high temperature annealing and rapid quenching	Degrading PMMA by UV exposure and acetic acid rinsing	[137, 158, 159]
PS	PEO-b-PMMA-b-PS; coating PEO-b-PMMA-b-PS onto silicon substrates, followed by solvent annealing; PEO block-permitting long-range ordering	Degrading PMMA by UV exposure and acetic acid rinsing	[136, 160]
PS	(PS-r-BCB)-b-PMMA; coating (PS-r-BCB)-b-PMMA onto P(S-r-BCB-r-MMA) neutral layer, followed by thermal annealing and cross-linking at elevated temperatures	Degrading PMMA by UV exposure and acetic acid rinsing	[161]
PS	PS-b-PMMA; coating PS-b-PMMA onto glass substrates along with fast solvent evaporation	Degrading PMMA by UV exposure and acetic acid rinsing	[135, 162]
PS	PS-b-PMMA/PEO; coating PS-b-PMMA/PEO onto silicon substrates, followed by solvent annealing	Removing PMMA/PEO domains by UV exposure and acetic acid rinsing	[163]
PS-b-PEO	PS-b-PEO/PAA; coating PS-b-PEO/PAA onto porous supports along with fast solvent evaporation	Removing PAA by soaking in water	[164]
PS-b-PMMA	PS-b-PMMA/PMMA; coating PS-b-PMMA/PMMA onto PS-r-PMMA neutral layer, followed by vacuum high-temperature annealing and rapid quenching	Degrading PMMA by acetic acid rinsing	[165, 166]
PS	PS-b-PLA; coating PS-b-PLA porous support, followed by controlled solvent evaporation	Removing PLA by dilute aqueous base rinsing	[138]
PS-b-PI	PS-b-PI-b-PLA; coating PS-b-PI-b-PLA onto hexamethyldisilazane neutral layer or porous supports, followed by vacuum high-temperature annealing	Removing PLA by dilute aqueous base rinsing	[167]
PS-b-PDMA	PS-b-PDMA-b-PLA; molding PS-b-PDMA-b-PLA, followed by vacuum high-temperature annealing	Removing PLA by dilute aqueous base rinsing	[168]
PS	NPS-b-PLA/DCPD; cross-linking NPS-b-PLA/DCPD using the Grubbs catalyst, followed by controlled solvent evaporation	Removing PLA by dilute aqueous base rinsing	[139, 140]
PS/PS-b-PEO	PS-b-PEO/PS-b-PLA; controlled solvent evaporation followed by vacuum high-temperature annealing	Removing PLA by dilute base or concentrated HI solution rinsing	[141, 169]
PE	PLA-b-PE-b-PLA; melt molding, followed by cooling induced PE crystallization	Removing PLA by dilute aqueous base rinsing	[170]
PE/PMe(OE)$_x$MA-b-PE-b-PMe(OE)$_x$MA	PMe(OE)$_x$MA-b-PE-b-PMe(OE)$_x$MA/PLA-b-PE-b-PLA; melt molding, followed by cooling induced PE crystallization	Removing PLA by dilute aqueous base rinsing	[142]

Membranes	Assemblies and assembly approaches	Pore generation	References
PE	PS-b-PE; melt molding, followed by cooling induced PE crystallization	Removing PS by fuming nitric acid	[171]
PB	PB-b-PDMS; coating PB-b-PDMS onto glass substrates along with controlled solvent evaporation	Removing PDMS by tetra-n-butylammonium fluoride solution	[172]
PS	(ONB-(PS-b-PEO); coating (ONB-(PS-b-PEO) onto silicon substrates, followed by solvent annealing	Removing PEO by UV cleavage of ONB and methanol rinsing	[143]
PS	PS-ss-PEO with disulfide juncture; coating PS-ss-PEO onto silicon substrates, followed by solvent annealing	Removing PEO by DDT cleavage of disulfide juncture and ethanol rinsing	[145]
PS	PS-b-PEO with triphenylmethyl ether juncture; coating PS-b-PEO onto silicon substrates, followed by solvent annealing	Removing PEO by trifluoroacetic acid cleavage of triphenylmethyl ether juncture and methanol rinsing	[146]
PS	PS-b-PEO with o-nitrobenzyl juncture; coating PS-b-PEO onto silicon substrates, followed by solvent annealing	Removing PEO by UV cleavage of o-nitrobenzyl ester and methanol rinsing	[144]
PtBOS-b-PS-b-P4VP	PtBOS-b-PS-b-P4VP/PDP; coating PtBOS-b-PS-b-P4VP/PDP onto glass substrates, followed by solvent annealing	Removing PDP by ethanol rinsing	[147]
PS-b-P4VP	PS-b-P4VP/PDP; molding along with vacuum high-temperature annealing and rapid quenching	Removing PDP by ethanol rinsing	[148]
PS-b-P4VP	PS-b-P4VP/HABA; coating PS-b-P4VP/HABA onto silicon substrates, followed by solvent annealing	Removing HABA by methanol rinsing	[173]
PS-b-P4VP	PS-b-P4VP/HABA; casting PS-b-P4VP/HABA on porous support followed by nonsolvent induced phase inversion	Removing HABA by ethanol rinsing	[86]
PS-b-P4VP	PS-b-P4VP/PBA; coating PS-b-P4VP/PBA onto silicon substrates, followed by solvent annealing	Removing PBA by ethanol rinsing	[174]
PS-b-P4VP/DBSA	PS-b-P4VP/DBSA/PDP; coating PS-b-P4VP/DBSA/PDP PBA onto silicon substrates, followed by controlled solvent evaporation	P4VP/DBSA domains collapsing upon annealing	[149]
PS-b-P4VP/PMCMA	PS-b-P4VP/PMCMA; coating PS-b-P4VP/PMCMA onto silicon substrates, followed by controlled solvent evaporation	P4VP/PMCMA domains collapsing upon annealing	[175]
PS-b-P4VP	PS-b-P4VP/DHN; coating PS-b-P4VP/DHN onto silicon substrates, followed by controlled solvent evaporation	Removing DHN by methanol rinsing	[176]
Phenolic resin	PS-b-P4VP/phenolic resin; coating PS-b-P4VP/phenolic resin onto silicon substrates, followed by controlled solvent evaporation	Removing PS-b-P4VP by pyrolysis	[177]
PS	PS-[Ru^{2+}]-PEO; coating PS-[Ru^{2+}]-PEO onto silicon substrates, followed by solvent annealing	Removing PEO by oxidizing the Ru(II) into Ru(III)	[150, 178]

Membranes	Assemblies and assembly approaches	Pore generation	References
PS	PS-[Ni^{2+}]-PEO; coating PS-[Ni^{2+}]-PEO onto silicon substrates, followed by solvent annealing	Removing PEO by methanol rinsing	[179]
PS-b-P4VP	PS-b-P4VP; casting PS-b-P4VP onto glass substrates, followed by initial solvent evaporation and nonsolvent-induced phase inversion	Solvent/nonsolvent exchange	[153]
PS-b-PEO	PS-b-PEO; casting PS-b-PEO onto glass substrates, followed by initial solvent evaporation and nonsolvent-induced phase inversion	Solvent/nonsolvent exchange	[180]
PI-b-PS-b-P4VP	PS-b-PS-b-P4VP; casting PS-b-PS-b-P4VP onto glass substrates, followed by initial solvent evaporation and nonsolvent-induced phase inversion	Solvent/nonsolvent exchange	[181–183]
PS-b-P2VP	PS-b-P2VP; coating PS-b-P2VP onto silicon substrates, followed by controlled solvent evaporation	Shrinkage of P2VP chains after ethanol swelling	[184, 185]
PS	PS-b-PMMA; coating PS-b-PMMA onto silicon substrates, followed by high-temperature annealing	Degrading PMMA by UV exposure and acetic acid initiated collective osmotic shock	[155]

Table 2. Membrane fabrication using the block copolymer self-assembly.

The PEO side chains can connect to the membrane's surfaces because of their affinity to water. PEO side chains likewise display a long-lasting surface hydrophilicity property. PEO brushes, taken away from the surface during cleaning or operation, may be largely restored through additional segregation of the residual amphiphilic additives during the subsequent heat treatment or when others are driven by the developing gradient in the additives' chemical potential. In order to create porous membranes that follow the phase inversion method, amphiphilic copolymers were likewise applied as additives and include hyper-branched star polymers, comb-like copolymers, and block copolymers. Blended membranes had undergone testing in water at 60°C so as to assess the retentive stability of various amphiphilic polymers on the membrane's surface [188]. The testing showed a minor variation in water contact angles in blend membranes when they were continuously leached in hot water for the duration of 30 days, thus reflecting advanced membrane surface strength. Furthermore, a Pluronic block copolymer, poly (ethylene oxide)-b-poly(propylene oxide)-b-poly(ethylene oxide) (PEO-b-PPO-b-PEO), was applied as a surface-segregating additive for the creation of improved fouling resistant PES type membranes. The hydrophobic PPO segments in Pluronic block copolymers became securely fastened in the PES matrix and resulted in the covering of Pluronic block copolymers on PES. Alternatively, the hydrophilic PEO segments slowly drifted to the membrane's surface and lead to a membrane surface featuring improved stability and enhanced hydrophilicity (**Figure 34**) [190]. Moreover, high-performance PES/Pluronic membranes that show stable hydrophilic character and elevated flux values were developed using vapor-induced phase separation combined together with the nonsolvent-induced phase separation approach.

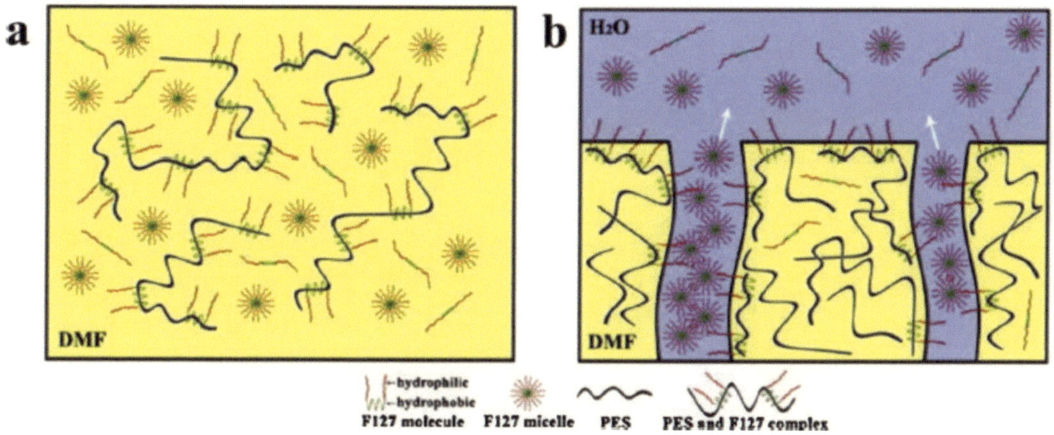

Figure 34. Illustration of dual roles of Pluronic F127 in the membrane formation process. (a) The self-assembly polymers lead to three forms of Pluronic F127 existing in a homogeneous casting solution. (b) Immersing the film in a water bath leads to phase separation and formation of ordered structure and pores within the membrane. Source: Ref. [189], Copyright 2008; reproduced with permission from Elsevier Ltd.

The design strategies for zwitterionic membrane surfaces were developed with the help of alternative amphiphilic zwitterionic ligands used as surface-segregating additives and included sulfobetaine copolymer, phosphorylcholine copolymer, and soybean phosphatidyl-choline [191]. As part of the phase-inversion procedure during membrane production, the surface segregation of zwitterionic segments was conducted spontaneously. This generated zwitterionic brushes on the membrane as well as pore's surface and lowered interfacial free energy value. Moreover, a forced surface segregation approach was applied to in-situ engineering process of a porous amphiphilic membrane's surface, with hydrophilic fouling resilient domains and lower surface free energy fouling release self-cleaning domains [192]. Lower surface energy segments, like silicone-segments or fluorine-containing segments, are not capable of spontaneous separation from the polymer-water interface, due to the unfavorable thermodynamics during NIPS process conducted through the free surface segregation [193]. As part of the NIPS procedure, hydrophilic segments were anticipated to separate at the membrane's surface coordinated by the amphiphilic copolymers' self-assembly. Alternatively, the covalently binding nonpolar hydrophobic sections were pulled onto the membrane's surfaces by hydrophilic segments using forced surface segregation (**Figure 35**). Due to the innate self-healing capacity of surface segregation methods, the long-term surface stability of the low surface energy sections located on the membrane surfaces was likewise anticipated.

2.3.4. Challenges and shortcomings

The prototypes inspired by nature considerably enhance the range of artificial material syntheses and their applications. The process of first extracting central principles at the core of natural material production and then imitating these processes is a rewarding method for replicating comparable physical and chemical structures. However, the complexity and

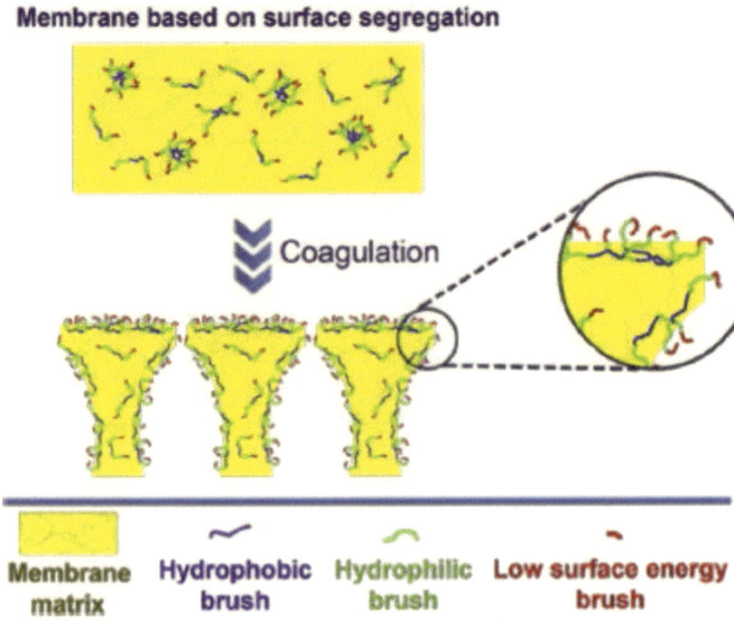

Figure 35. Forced surface segregation process during the membrane formation process.

precision involved in the formation of materials is difficult to imitate or grasp in its entirety. Some of the shortcomings and challenges faced by membrane-fabrication approaches based on simulations of natural prototypes are provided in **Table 3**.

2.4. Fabrication of bioinspired and biomimetic membranes based on functions of natural prototypes

2.4.1. Based on self-cleaning

The classification of self-cleaning surfaces is divided into hydrophobic and hydrophilic, as underwater oleophobic, type of surfaces. When it comes to the hydrophobic or oleophobic self-cleaning type surfaces, the interactions between the hydrophobic epicuticular waxes and multiscale geometrical surface structures are directly inspired by the surface characteristics of lotus leaves or other epidermis and plant leaves with similar properties. In fact, this correlation to the lotus leaves ensures that the hydrophobic self-cleaning type surfaces have an elevated water contact angle or a smaller sliding angle, which indicates low adhesion and superior hydrophobic or functionality. Alternatively, for the underwater oleophobic or hydrophilic self-cleaning surfaces, the correlation of elevated hydration energy moieties and physical heterogeneity, incited by the hydrated skin surfaces in marine organisms, suggests a greater underwater oil contact angle that helps to avert oil fouling [194]. Because of the distinct characteristics found in these self-cleaning surfaces, including nonwetting and anti-contamination, they may be applicable in a range of situations. Collectively, these new found properties signal a new era in self-cleaning membrane production and application.

Membrane fabrication methods	Challenges and shortcomings
Biomimetic mineralization	Controlled regulation of nanoparticle morphology and surface composition within polymer matrix In-depth analysis of mineralization reaction thermodynamics and kinetics with different inorganic precursors and organic inducers
Biomimetic adhesion	Unambiguous elucidation of formation mechanism and structure of PDA with convincing experimental evidences Long-term stability of PDA coating under extreme working environments
BCP self-assembly	Facile synthesis of well-defined block copolymers for rationally controlling the phase separation process Precise control of defect-free self-assembly process and pore size/morphology
Surface segregation	Synergistic control of the thermodynamics, kinetics, for selective surface segregation Manipulating multiple interactions for hierarchical structure creation

Table 3. Challenges and shortcomings of membrane fabrication methods that imitate the formations found in natural prototypes.

2.4.1.1. Fabrication of membranes via incorporation of low surface energy moieties

Lower surface energies and surface microscale and nanoscale geometrical structures are two of the primary dynamics when it comes to the efficacy of hydrophobic or oleophobic self-cleaning type membranes [195]. The methodologies applied when developing self-cleaning membrane surfaces may be grounded in two specific tactics. The first one requires building a rough surface using low surface energy materials, whereas the other is based on altering the rough surface with materials with lower surface energy values. The process of design and production of the bioinspired superhydrophobic membranes using electrospinning has become a popular research direction. In particular, electrospinning is an adaptable approach for making rough surfaces using low surface energy materials, that depend on this roughness, as hierarchically textured surfaces with microstructures or nanostructures, and added during the process of spinning. Roughness' length scale is credited to the smaller fiber diameters and hydrophobicity and is essential for the superhydrophobicity properties of fibrous membranes. Multiple methods have been noted for combining materials of lower surface energy together with higher surface roughness, including poly (3-phenyl-3,4-dihydro-2H-1,3-benzoxazine) blended with PAN, and electrospinning poly (styrene-b-dimethylsiloxane) block copolymers blended with homopolymer polystyrene (PS-b-PDMS/PS) [196]. A method based on in-situ was used to produce superhydrophobic fiber mats with the aid of electrospinning polystyrene that contains fluoroalkyl end-capped polymer additives [197]. Unrestricted surface segregation of these additives with respect to the polymer-air interface can generate fibers that have super-hydrophobic properties, are fluorine-rich, and show lower surface energy values. Although it is inspired by biological superhydrophobic surfaces offering hierarchical surface roughness characteristics on at least two different length scales, there is a need for a much finer scale structure so as to create a second level of roughness. A number of artificial superhydrophobic microporous and nanoporous fibrous membranes have been produced using an approach that builds a second level hierarchical surface based on nanohybrid systems. Nanomaterials, like Al_2O_3 nanoparticles, TiO_2 nanoparticles, SiO_2 nanoparticles, and graphene nanoflakes, collect within the polymeric fibers and as a result alter the surface chemistry and morphology, eventually allowing for superhydrophobicity with improved self-cleaning characteristics [198]. For example, an artificial composite fibrous membrane was created with the help of polyaniline (PANI) doped with azobenzenesulfonic acid blended with PS and using electrospinning (**Figure 36**). In this instance, a network of nanofibers with multiple sub-microspheres was spanning over the entire substrate connected with nanoknots, as well as nanoscale protuberances sheltering every sub-microsphere. The hierarchical roughness of microparticles and nanofibers has the capacity to improve the temperature-responsive wettability by switching between superhydrophobicity and superhydrophilicity activated by temperature.

Chemical vapor deposition process is a single-step solvent-free deposition method for surface modifications that may help add lower surface energy properties to nanoscale rough surfaces that can then generate improved hydrophobic self-cleaning membrane surfaces. Superoleophobic as well as super-hydrophobic self-cleaning nanocellulose aerogel type membranes were created with the aid of cellulose nanofibers that have been treated with fluorosilanes through CVD. The superoleophobic and superhydrophobic characteristics were due to the fluorinated fibrillar networks and aggregates with structures occurring at varying length scales. A noticeable improvement in fibrous membranes' hydrophobicity was observed, when CVD was combined with electrospinning [200]. Both, first level of roughness related to the fibers and second level of roughness related to the beads, were present in the poly caprolactone (PCL) fibrous membranes. The substantially lower surface energy in the coating layer produced using CVD allowed for a constant superhydrophobicity with a contact angle of 175°. This double-roughened highly hydrophobic fibrous type membrane was developed through the process of improving micrometer-scale electrospun fibers with nanometer-scale particles or pores [201]. The combination of chemical composition, roughened texture, and re-entrant surface curvature was likewise examined so as to design an oleophobic self-cleaning fabric membrane. Specifically, this membrane relies on the exceedingly low surface energy

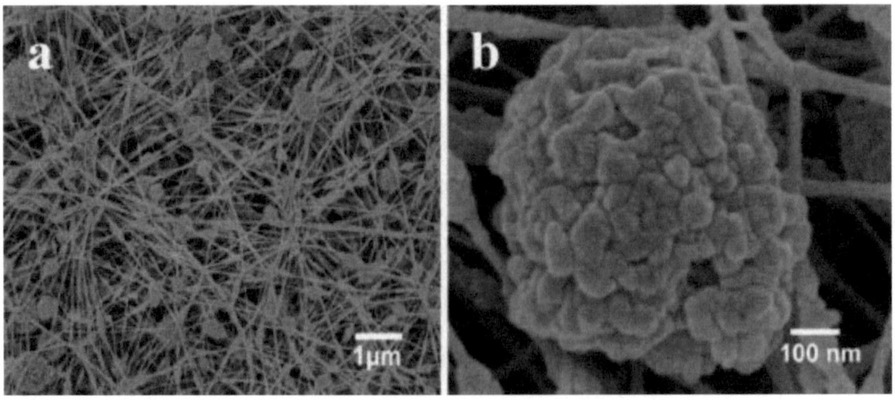

Figure 36. (a) SEM image of an electrospun PANI/PS composite fibrous membrane with lotus leaf–like structure. (b) Magnified view of a single sub-microsphere from (a). Source: Ref. [199], Copyright 2006; reproduced with permission from Wiley-VCH Verlag GmbH & Co. KGaA.

polyhedral oligomeric silsesquioxane (POSS) molecules, featuring a rigid silsesquioxane cage enclosed by per fluoro-alkyl groups (fluoro POSS) [201]. A number of experimental oleophobic membranes were created using a straightforward dip-coating and thermal annealing technique that applied a combination of fluoro POSS and PMMA, cross-linked poly (ethylene glycol) diacrylate (x-PEGDA), cross-linked PDMS, or poly (ethyl methacrylate) (PEMA) onto the textured substrates, like stainless steel wire meshes, that have re-entrant curvature on the rougher length scale [202]. For instance, a variety of fabric morphologies with multiple scales of roughness, high porosity, and "beads on a string" type morphology can be adjusted by changing the concentration of the fluoro POSS and PMMA blends [202]. Those surfaces that offer multiple scales of roughness allowed fiber membranes to obtain superhydrophobicity and oleophobicity at higher POSS concentration values and hydrophilicity oleophobicity at smaller POSS concentration values. There are, however, other methods available for the design and fabrication of self-cleaning membranes. For example, textile membranes covered with thiol-ligand nanocrystals, based on the interaction between the VIII and IB nanocrystals and n-octadecyl, can gain super-oleophilic and superhydrophobic qualities [203]. Furthermore, PVDF membranes, made out of linked spherical microparticles that have been uniformly dispersed on the surface, can be produced using an inert solvent-induced phase inversion that showcases superoleophilic as well as superhydrophobic potential [204]. Alternatively, the nanoparticle-polymer suspension coating was applied during the production of a self-cleaning stainless steel mesh membrane [204]. The synergistic effects of the micro/nanoscale hierarchical constructions produced with the help of SiO_2 nanoparticles and the hydrophilic-oleophobic groups of poly(diallyldimethylammonium chloride) (PDDA)-sodium per fluoro octanoate (PFO) allowed the spray-coated mesh membrane to successfully gain superoleophobic and superhydrophilic characteristics. A research study on this subject recently showed how amphiphilic self-cleaning membrane surfaces, that offer low surface energy characteristics and mixed domains of mosaic hydrophilic, were produced using surface grafting perfluoroalkyl molecules that instigated surface segregation in lower surface energy amphiphilic copolymers [205]. Constructed with fluorine-based polymers, the lower surface energy microdomains located on the membrane's surface were supposed to decrease the intermolecular interactions occurring between the membrane's surface and oil. The hydrophilic domains were intended to restrict water molecules and create a hydration layer that would become an oil, water, or solid interface for oleophobicity.

2.4.1.2. Fabrication of membranes via the incorporation of high hydration energy moieties

Research has shown that superhydrophilic surfaces submerged into water can likewise encourage self-cleaning behavior and oleophobicity. The elevated hydration conditions of hydrophilic moieties located on the membrane's surface have the capacity to restrict a relatively large ratio of water molecules by using hydrogen bond or electrostatic interaction, both of which essentially prevent the oil's entry onto the membrane's surface. These approaches to producing underwater hydrophilic or oleophobic self-cleaning type membranes place emphasis on incorporation of high hydration energy moieties onto the membrane's surfaces. On the other hand, the underwater self-cleaning superoleophobic membranes offer unusual microscale and nanoscale hierarchical structural organizations. Research collaborations have reported underwater superoleophobic membranes made out of polyacrylamide hydrogel-coated mesh membranes with microscale porous metal substrates and coarse nanostructured hydrogel coatings [206]. A thermal-responsive block copolymer PMMA-b-PNIPAAm was casted onto a steel mesh so as to create a membrane that can have two switchable states of wettability depending of temperature values (**Figure 37a**) [207]. A PMMA-b-PNIPAAm had undertaken a self-assembly process into a lamellar structure featuring PNIPAAm domains between the hard walls of PMMA on a nanometer scale. In this case, the alternating conformational modification of the PNIPAAm chain establishes the surface roughness at a value near lower critical solution temperature, while the collaboration between PMMA and PNIPAAm domains grants the film reversible switching between wettability conditions of hydrophobicity/oleophilicity and hydrophilicity/oleophobicity (**Figure 37b**). Moreover, underwater superoleophobic chitosan-coated meshes based on cross-linked chitosan network were successfully produced, and the overall stability potential of chitosan-coated meshes may be enhanced through the modification of the CS coating and its reduction, PVA addition, and full cross-linking [208].

Latest experimental research developments have reported the creation of underwater superoleophobic membranes based on PMAPS-g-PVDF and PAA-g-PVDF [207]. These ultralow oil-adhesion and superoleophobic properties of the PMAPS-g-PVDF membrane were caused by the enhanced surface energy and the hydrated conduct of the grafted zwitterionic PMAPS chains whenever in water. A prolonged conformation of hydrated PMAPS chains could incite the creation of a tightly bound hydration layer as well as encourage oil droplets to roll off from the membrane's surface [209]. The PAA-g-PVDF membranes' underwater superoleophobic wetting properties were impacted by the hydrophilic nature of PAA chains and the hierarchical micro/nanoscale structure. The micro/nanoscale spherical particles located on the

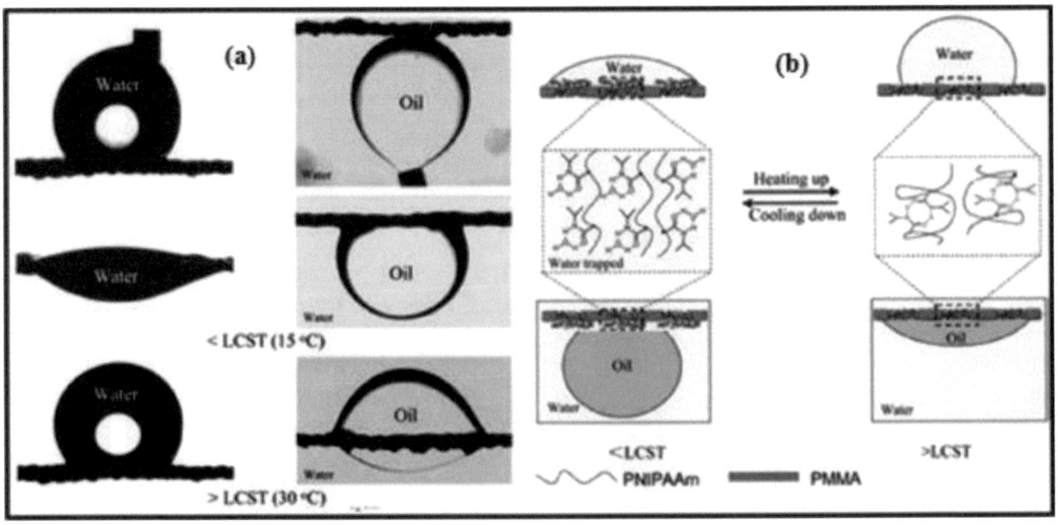

Figure 37. (a) Temperature-controlled water/oil wetting behavior on a block copolymer-coated mesh. (b) A schematic showing reversible conformational change of the PNIPAAm chain and the resultant surface roughness at different temperatures leading to two states of wettability [lower critical solution temperature (LCST)]. Source: Ref. [207], Copyright 2013; reproduced with permission from Wiley-VCH Verlag GmbH & Co. KGaA.

membrane's surface were produced using the PAA-g-PVDF micelle aggregates and during the application of the salt-induced phase-inversion method. Specifically, this occurred during the coagulation step, when the quick solvent exchange encouraged the NaCl's crystallization out from the water and the nascent small crystal seeds became accumulation points for aggregates around the PAA-g-PVDF micelles. Thus, this experimental approach was able to illustrate that an increase in roughness can improve antiwetting performance of underwater oils on the membrane's surfaces [210].

2.4.2. Challenges and shortcomings

Researchers that focus on the membrane surfaces' wetting performance have developed key sets of guidelines for the design of self-cleaning membranes. Admittedly, multiple challenges still have to be addressed in this area. When it comes to hydrophobic, or oleophobic, self-cleaning type membranes, the fluorinated moieties were used in the majority of cases in order to decrease surface energy values. The synthesis and application of fluorinated moieties could increase the likelihood of the ecosystem being contaminated by fluorine, which in turn can have a damaging effect on the living organic bodies and materials. As a result, environmentally conscious methods that can address self-cleaning are currently necessary. For the hydrophilic, or underwater oleophobic, self-cleaning type membrane, one of the critical concerns is the strength of the surface hydrophilic quality. Another aspect that has not been adequately investigated is the structural development of hydrophilic layer that experiences intense conditions like higher salinity, alkalinity/acidity, and temperature values. The future of membrane design must shift its focus toward combining strategies that take into consideration multiple interactions and that can significantly improve membrane hydrophilic stability and its applications.

Author details

Amira Abdelrasoul[1]*, Huu Doan[2] and Ali Lohi[2]

*Address all correspondence to: amira.abdelrasoul@usask.ca

1 Department of Chemical and Biological Engineering, University of Saskatchewan, Saskatoon, Saskatchewan, Canada

2 Department of Chemical Engineering, Ryerson University, Toronto, Ontario, Canada

References

[1] Abetz V, Brinkmann T, Dijkstra M, Ebert K, Fritsch D, Ohlrogge K, Paul D, Peinemann KV, Pereira-Nunes S, Scharnagl N, Schos-sig M. Developments in membrane research: From material via process design to industrial application. Advanced Engineering Materials. 2006;**8**:328-358

[2] Edidin M. Lipids on the frontier: A century of cell-membrane bilayers. Nature Reviews. Molecular Cell Biology. 2003;**4**:414-418

[3] Vermette P, Meagher L. Interactions of phospholipid and poly (ethylene glycol)-modified surfaces with biological systems: Relation to physicochemical properties and mechanisms. Colloids and Surfaces, B: Biointerfaces. 2003;**28**:153-198

[4] Andrew LL. Phosphorylcholine-based polymers and their use in the prevention of biofouling. Colloids and Surfaces, B: Biointerfaces. 2000;**18**:261-275

[5] Zardoya R. Phylogeny and evolution of the major intrinsic protein family. Biology of the Cell. 2005;**97**:397-414

[6] Sui H, Han BG, Lee JK, Walian P, Jap BK. Structural basis of water-specific transport through the AQP1 water channel. Nature. 2001;**414**:872-878

[7] Kozono D, Yasui M, King LS, Agre P. Aquaporin water channels: Atomic structure and molecular dynamics meet clinical medicine. The Journal of Clinical Investigation. 2002;**109**:1395-1399

[8] Coleman AW, Silva ED, Nouar F, Nierlich M, Navaza A. The structure of a self-assembled calixarene aqua-channel system. Chemical Communications. 2003;**7**:826-827

[9] Murata K, Mitsuoka K, Hirai T, Walz T, Agre P, Heymann JB, Engel A, Fujiyoshi Y. Structural determinants of water permeation through aquaporin-1. Nature. 2000;**407**: 599-605

[10] Dutzler R, Campbell EB, MacKinnon R. Gating the selectivity filter in ClC chloride channels. Science. 2003;**300**:108-112

[11] Yellen G. The voltage-gated potassium channels and their relatives. Nature. 2002;**419**:35-42

[12] Jiang Y, Lee A, Chen J, Cadene M, Chait BT, MacKinnon R. The open pore conformation of potassium channels. Nature. 2002;**417**:523-526

[13] Fyles TM. Synthetic ion channels in bilayer membranes. Chemical Society Reviews. 2007; **36**:335-347

[14] Gouaux E, MacKinnon R. Principles of selective ion transport in channels and pumps. Science. 2005;**310**:1461-1465

[15] Sommerdijk N, Colfen H. Lessons from nature-biomimetic approaches to minerals with complex structures. MRS Bulletin. 2010;**35**:116-121

[16] Nudelman F, Sommerdijk NA. Biomineralization as an inspiration for materials chemistry. Angewandte Chemie International Edition in English. 2012;**51**:6582-6596

[17] Matsunaga T, Sakaguchi T. Molecular mechanism of magnet formation in bacteria. Journal of Bioscience and Bioengineering. 2000;**90**:1-13

[18] Mann S. Biomineralization: Principles and Concepts in Bioinorganic Materials Chemistry. New York: Oxford University Press; 2001. 210 p

[19] Gower LB. Biomimetic model systems for investigating the amorphous precursor pathway and its role in biomineralization. Chemical Reviews. 2008;**108**:4551-4627

[20] AW X, Ma Y, Cölfen H. Biomimetic mineralization. Journal of Materials Chemistry. 2007; **17**:415-449

[21] Meldrum FC, Cölfen H. Controlling mineral morphologies and structures in biological and synthetic systems. Chemical Reviews. 2008;**108**:4332-4432

[22] Tang Z, Kotov NA, Magonov S, Ozturk B. Nano structured artificial nacre. Nature Materials. 2003;**2**:413-418

[23] Roth KM, Zhou Y, Yang W, Morse DE. Bifunctional small molecules are biomimetic catalysts for silica synthesis at neutral pH. Journal of the American Chemical Society. 2004;**127**:325-330

[24] Dickerson MB, Sandhage KH, Naik RR. Protein- and peptide-directed syntheses of inorganic materials. Chemical Reviews. 2008;**108**:4935-4978

[25] Wilker JJ. The iron-fortified adhesive system of marine mussels. Angewandte Chemie International Edition in English. 2010;**49**:8076-8078

[26] Lee H, Dellatore SM, Miller WM, Messersmith PB. Mussel-inspired surface chemistry for multi functional coatings. Science. 2007;**318**:426-430

[27] Waite JH, Andersen NH, Jewhurst S, Sun C. Mussel adhesion: Finding the tricks worth mimicking. The Journal of Adhesion. 2005;**81**:297-317

[28] Sun CJ, Srivastava A, Reifert JR, Waite JH. Halogenated DOPA in a marine adhesive protein. The Journal of Adhesion. 2009;**85**:126

[29] Stewart RJ, Wang CS, Shao H. Complex coacervates as a foundation for synthetic underwater adhesives. Advances in Colloid and Interface Science. 2011;**167**:85-93

[30] Lee H, Scherer NF, Messersmith PB. Single-molecule mechanics of mussel adhesion. Proceedings of the National Academy of Sciences of the United States of America. 2006; **103**:12999-13003

[31] Lin Q, Gourdon D, Sun C, Holten-Andersen N, Anderson TH, Waite JH, Israelachvili JN. Adhesion mechanisms of the mussel foot proteins mfp-1 and mfp-3. Proceedings of the National Academy of Sciences of the United States of America. 2007;**104**:3782-3786

[32] Waite JH. Surface chemistry: Mussel power. Nature Materials. 2008;**7**:8-9

[33] Hong S, Na YS, Choi S, Song IT, Kim WY, Lee H. Non-covalent self-assembly and covalent polymerization co-contribute to poly-dopamine formation. Advanced Functional Materials. 2012;**22**:4711-4717

[34] Li B, Liu WP, Jiang ZY, Dong X, Wang BY, Zhong YR. Ultra thin and stable active layer of dense composite membrane enabled bypoly (dopamine). Langmuir. 2009;**25**:7368-7374

[35] Zhang S. Emerging biological materials through molecular self-assembly. Biotechnology Advances. 2002;**20**:321-339

[36] Seidel SR, Stang PJ. High-symmetry coordination cages via self-assembly. Accounts of Chemical Research. 2002;**35**:972-983

[37] Hanczyc MM, Szostak JW. Replicating vesicles as models of primitive cell growth and division. Current Opinion in Chemical Biology. 2004;**8**:660-664

[38] Stone DA, Korley LTJ. Bioinspired polymeric nanocomposites. Macromolecules. 2010; **43**:9217-9226

[39] Hagn F, Eisoldt L, Hardy JG, Vendrely C, Coles M, Scheibel T, KesslerH A. Conserved spider silk domain acts as a molecular switch that controls fibre assembly. Nature. 2010;**465**:239-242

[40] Askarieh G, Hedhammar M, Nordling K, Saenz A, Casals C, Rising A, Johansson J, Knight SD. Self-assembly of spider silk proteins is controlled by a pH-sensitive relay. Nature. 2010;**465**:236-238

[41] Rising A, Hjälm G, Engström W, Johansson J. N-terminal nonrepetitive domain common to dragline, flagella form, and cylindri form spider silk proteins. Biomacromolecules. 2006;**7**:3120-3124

[42] Heim M, Keerl D, Scheibel T. Spider silk: From soluble protein to extraordinary fiber. Angewandte Chemie International Edition in English. 2009;**48**:3584-3596

[43] Du C, Falini G, Fermani S, Abbott C, Moradian-Oldak J. Supramolecular assembly of amelogenin nano spheres into birefringent microribbons. Science. 2005;**307**:1450-1454

[44] Li C, Kaplan DL. Biomimetic composites via molecular scale self-assembly and biomineralization. Current Opinion in Solid State & Materials Science. 2003;**7**:265-271

[45] Silverman HG, Roberto FF. Understanding marine mussel adhesion. Marine Biotechnology. 2007;**9**:661-681

[46] Zhang S. Molecular self-assembly: Another brick in the wall. Nature Nanotechnology. 2006; **1**:169-170

[47] Eadie L, Ghosh TK. Biomimicry in textiles: Past, present and potential. An overview. Journal of The Royal Society Interface. 2011;**8**:761-775

[48] Pechook S, Pokroy B. Self-assembling, bioinspired wax crystalline surfaces with time-dependent wettability. Advanced Functional Materials. 2012;**22**:745-750

[49] Liu K, Yao X, Jiang L. Recent developments in bio-inspired special wettability. Chemical Society Reviews. 2010;**39**:3240-3255

[50] Feng L, Li S, Li Y, Li H, Zhang L, Zhai J, Song Y, Liu B, Jiang L, Zhu D. Super-hydrophobic surfaces: From natural to artificial. Advanced Materials. 2002;**14**:1857-1860

[51] Bhushan B, Jung YC, Niemietz A, Koch K. Lotus-like biomimetic hierarchical structures developed by the self-assembly of tubular plant waxes. Langmuir. 2009;**25**:1659-1666

[52] Sun T, Feng L, Gao X, Jiang L. Bioinspired surfaces with special wettability. Accounts of Chemical Research. 2005;**38**:644-652

[53] Koch K, Bhushan B, Jung YC, Barthlott W. Fabrication of artificiallotus leaves and significance of hierarchical structure for super hydrophobicity and low adhesion. Soft Matter. 2009;**5**:1386-1393

[54] Koch K, Bhushan B, Barthlott W. Diversity of structure, morphology and wetting of plant surfaces. Soft Matter. 2008;**4**:1943-1963

[55] Zhu D, Li X, Zhang G, Zhang X, Zhang X, Wang T, Yang B. Mimicking the rice leaf-from ordered binary structures to anisotropic wettability. Langmuir. 2010;**26**:14276-14283

[56] Feng L, Zhang Y, Xi J, Zhu Y, Wang N, Xia F, Jiang L. Petal effect: A super hydrophobic state with high adhesive force. Langmuir. 2008;**24**:4114-4119

[57] Guo Z, Liu W. Biomimic from the super hydrophobic plant leaves in nature: Binary structure and unitary structure. Plant Science. 2007;**172**:1103-1112

[58] Bhushan B, Jung YC. Natural and biomimetic artificial surfaces for super hydrophobicity, self-cleaning, low adhesion, and drag reduction. Progress in Materials Science. 2011; **56**:1-108

[59] Weatherspoon MR, Cai Y, Crne M, Srinivasarao M, Sandhage KH. 3D rutile titania-based structures with morpho butterfly wing scalemorphologies. Angewandte Chemie International Edition in English. 2008;**47**:7921-7923

[60] ZZ G, Uetsuka H, Takahashi K, Nakajima R, Onishi H, Fujishima A, Sato O. Structural color and the lotus effect. Angewandte Chemie International Edition in English. 2003; **42**:894-897

[61] Zheng Y, Gao X, Jiang L. Directional adhesion of super hydrophobic butterfly wings. Soft Matter. 2007;**3**:178-182

[62] Hansen WR, Autumn K. Evidence for self-cleaning in gecko setae. Proceedings of the National Academy of Sciences of the United States of America. 2005;**102**:385-389

[63] Autumn K, Gravish N. Gecko adhesion: Evolutionary nano technology. Philosophical Transactions of the Royal Society A. 2008;**366**:1575-1590

[64] Liu M, Wang S, Wei Z, Song Y, Jiang L. Bioinspired design of a super oleophobic and low adhesive water/solid interface. Advanced Materials. 2009;**21**:665-669

[65] Bhushan B. Biomimetics: Lessons from nature – An overview. Philosophical Transactions of the Royal Society A. 2009;**367**:1445-1486

[66] Chen SF, Zheng J, Li LY, Jiang SY. Strong resistance of phosphorylcholine self-assembled monolayers to protein adsorption: Insights into non fouling properties of zwitterionic materials. Journal of the American Chemical Society. 2005;**127**:14473-14478

[67] Zhao J, Shi Q, Luan S, Song L, Yang H, Shi H, Jin J, Li X, Yin J, Stagnaro P. Improved biocompatibility and antifouling property of polypropylene non-woven fabric membrane by surface grafting zwitterionic polymer. Journal of Membrane Science. 2011;**369**:5-12

[68] Chen SH, Chang Y, Lee KR, Wei TC, Higuchi A, Ho FM, Tsou CC, Ho HT, Lai JY. Hemo compatible control of sulfobetaine-grafted polypropylene fibrous membranes in human whole blood via plasma-inducedsurface zwitterionization. Langmuir. 2012;**28**:17733-17742

[69] Zhao J, Song L, Shi Q, Luan S, Yin J. Antibacterial and hemo compatibility switchable polypropylene nonwoven fabric membrane surface. ACS Applied Materials & Interfaces. 2013;**5**:5260-5268

[70] Yang YF, Li Y, Li QL, Wan LS, Xu ZK. Surface hydrophilization of microporous polypropylene membrane by grafting zwitterionic polymer for anti biofouling. Journal of Membrane Science. 2010;**362**:255-264

[71] Jhong JF, Venault A, Hou CC, Chen SH, Wei TC. ZhengJ, Huang J, Chang Y. Surface zwitterionization of expanded poly (tetrafluoroethylene) membranes via atmospheric plasma-induced polymerization for enhanced skin wound healing. ACS Applied Materials & Interfaces. 2013;**5**:6732-6742

[72] Razi F, Sawada I, Ohmukai Y, Maruyama T, Matsuyama H. The improvement of antibiofouling efficiency of polyethersulfone membrane by functionalization with zwitterionic monomers. Journal of Membrane Science. 2012;**401-402**:292-299

[73] Susanto H, Ulbricht M. Photografted thin polymer hydrogel layers on PES ultrafiltration membranes: Characterization, stability, and influence on separation performance. Langmuir. 2007;**23**:7818-7830

[74] Yu H, Cao Y, Kang G, Liu J, Li M, Yuan Q. Enhancing antifouling property of polysulfone ultrafiltration membrane by grafting zwitterionic copolymer via UV-initiated polymerization. Journal of Membrane Science. 2009;**342**:6-13

[75] Ulbricht M, Yang H. Porous polypropylene membranes with different carboxyl polymer brush layers for reversible protein binding via surface-initiated graft copolymerization. Chemistry of Materials. 2005;**17**:2622-2631

[76] Zhou Q, Lei XP, Li JH, Yan BF, Zhang QQ. Antifouling, adsorption and reversible flux properties of zwitterionic grafted PVDF membrane prepared via physisorbed free radical polymerization. Desalination. 2014;**337**:6-15

[77] Li MZ, Li JH, Shao XS, Miao J, Wang JB, Zhang QQ, Xu XP. Grafting zwitterionic brush on the surface of PVDF membrane using physisorbed free radical grafting technique. Journal of Membrane Science. 2012;**405-406**:141-148

[78] Li Q, Bi QY, Zhou B, Wang XL. Zwitterionic sulfobetaine-grafted poly(vinylidene fluoride) membrane surface with stably anti-protein-fouling performance via a two-step surface polymerization. Applied Surface Science. 2012;**258**:4707-4717

[79] Liu Y, Zhang S, Wang G. The preparation of anti fouling ultrafiltration membrane by surface grafting zwitterionic polymer onto poly(arylene ether sulfone) containing hydroxyl groups membrane. Desalination. 2013;**316**:127-136

[80] Liu PS, Chen Q, SS W, Shen J, Lin SC. Surface modification of cellulose membranes with zwitterionic polymers for resistance to protein adsorption and platelet adhesion. Journal of Membrane Science. 2010;**350**:387-394

[81] HY Y, Kang Y, Liu Y, Mi B. Grafting poly zwitterions onto polyamide by click chemistry and nucleophilic substitution on nitrogen: A novel approach to enhance membrane fouling resistance. Journal of Membrane Science. 2014;**449**:50-57

[82] Huang J, Wang D, Lu Y, Li M, Xu W. Surface zwitterionically functionalized PVA-co-PE nano fiber materials by click chemistry. RSC Advances 2013;**3**:20922-20929

[83] Ye SH, Watanabe J, Iwasaki Y, Ishihara K. Anti fouling blood purification membrane composed of cellulose acetate and phospholipid polymer. Biomaterials. 2003;**24**:4143-4152

[84] Ye SH, Watanabe J, Ishihara K. Cellulose acetate hollow fiber membranes blended with phospholipid polymer and their performance for hemo purification. Journal of Biomaterials Science. Polymer Edition. 2004;**15**:981-1001

[85] Ji YL, An QF, Zhao Q, Sun WD, Lee KR, Chen HL, Gao CJ. Novel composite nanofiltration membranes containing zwitterions with high permeate flux and improved antifouling performance. Journal of Membrane Science. 2012;**390-391**:243-253

[86] Tripathi BP, Dubey NC, Choudhury S, Simon F, Stamm M. Antifouling and anti biofouling pH responsive block copolymer based membranes by selective surface modification. Journal of Materials Chemistry B. 2013;**1**:3397-3409

[87] Balme S, Janot JM, Berardo L, Henn F, Bonhenry D. KraszewskiS, Picaud F, Ramseyer C. New bioinspired membrane made of a biological ion channel confined into the cylindrical nanopore of a solid-state polymer. Nano Letters. 2011;**11**:712-716

[88] Zhu B, Li J, Xu D. Porous biomimetic membranes: Fabrication, properties and future applications. Physical Chemistry Chemical Physics. 2011;**13**:10584-10592

[89] Liu K, Jiang L. Bio-inspired self-cleaning surfaces. Annual Review of Materials Research. 2012;**42**:231-263

[90] Li X, Wang R, Tang C, Vararattanavech A, Zhao Y, Torres J, Fane T. Preparation of supported lipid membranes for aquaporin Z incorporation. Colloids and Surfaces, B: Biointerfaces. 2012;**94**:333-340

[91] Wang H, Chung TS, Tong YW, Meier W, Chen Z, Hong M, Jeyaseelan K, Armugam A. Preparation and characterization of pore-suspending biomimetic membranes embedded with aqua-porin Z on carboxylated polyethylene glycol polymer cushion. Soft Matter. 2011;**7**:7274-7280

[92] Duong PHH, Chung TS, Jeyaseelan K, Armugam A, Chen Z, YangJ HM. Planar biomimetic aquaporin-incorporated tri block copolymer membranes on porous alumina supports for nano filtration. Journal of Membrane Science. 2012;**409-410**:34-43

[93] Kumar M, Grzelakowski M, Zilles J, Clark M, Meier W. Highly permeable polymeric membranes based on the incorporation of the functional water channel protein aquaporin Z. Proceedings of the National Academy of Sciences of the United States of America. 2007;**104**:20719-20724

[94] Wang H, Chung TS, Tong YW, Jeyaseelan K, Armugam A, Chen Z, Hong M, Meier W. Highly permeable and selective pore-spanning biomimetic membrane embedded with aquaporin Z. Small. 2012;**8**:1185-1190

[95] Zhong PS, Chung TS, Jeyaseelan K, Armugam A. Aquaporin-embedded biomimetic membranes for nanofiltration. Journal of Membrane Science. 2012;**407-408**:27-33

[96] Sun G, Chung TS, Jeyaseelan K, Armugam A. A layer-by-layer self-assembly approach to developing an aquaporin-embedded mixed matrix membrane. RSC Advances. 2013;**3**:473-481

[97] Zhao Y, Qiu C, Li X, Vararattanavech A, Shen W, Torres J, Hélix-Nielsen C, Wang R, Hu X, Fane AG, Tang CY. Synthesis ofrobust and high-performance aquaporin-based biomimetic membranes by interfacial polymerization-membrane preparation and RO performance characterization. Journal of Membrane Science. 2012;**423-424**:422-428

[98] Yameen B, Ali M, Neumann R, Ensinger W, Knoll W, Azzaroni O. Synthetic proton-gated ion channels via single solid-statenano channels modified with responsive polymer brushes. Nano Letters. 2009;**9**:2788-2793

[99] Hou X, Guo W, Jiang L. Biomimetic smart nano pores and nanochannels. Chemical Society Reviews. 2011;**40**:2385

[100] Hinds BJ, Chopra N, Rantell T, Andrews R, Gavalas V, Bachas LG. Aligned multiwalled carbon nanotube membranes. Science. 2004;**303**:62-65

[101] Han C, Hou X, Zhang H, Guo W, Li H, Jiang L. Enantio selective recognition in biomimetic single artificial nano channels. Journal of the American Chemical Society. 2011; **133**:7644-7647

[102] Gyurcsányi RE. Chemically-modified nano pores for sensing. Trends in Analytical Chemistry. 2008;**27**:627-639

[103] Tian Y, Hou X, Jiang L. Biomimetic ionic rectifier systems: Asymmetric modification of single nano channels by ion sputtering technology. Journal of Electroanalytical Chemistry. 2011;**656**:231-236

[104] Xu T, Zhao N, Ren F, Hourani R, Lee MT, Shu JY, Mao S, Helms BA. Subnanometer porous thin films by the co-assembly of nano tube subunits and block copolymers. ACS Nano. 2011;**5**:1376-1384

[105] Fornasiero F, Park HG, Holt JK, Stadermann M, Grigoropoulos CP, Noy A, Bakajin O. Ion exclusion by sub-2-nm carbon nano tube pores. Proceedings of the National Academy of Sciences of the United States of America. 2008;**105**:17250-17255

[106] Kalra A, Garde S, Hummer G. Osmotic water transport through carbon nanotube membranes. Proceedings of the National Academy of Sciences of the United States of America. 2003;**100**:10175-10180

[107] Joseph S, Aluru NR. Why are carbon nanotubes fast transporters of water? Nano Letters. 2008;**8**:452-458

[108] Majumder M, Chopra N, Hinds BJ. Effect of tip functionalization on transport through vertically oriented carbon nanotube membranes. Journal of the American Chemical Society. 2005;**127**:9062-9070

[109] Peng FB, LY L, Sun HL, Wang YQ, Liu JQ, Jiang ZY. Hybrid organic–inorganic membrane: Solving the tradeoff between permeability and selectivity. Chemistry of Materials. 2005;**17**:6790-6796

[110] Li YF, He GW, Wang SF, SN Y, Pan FS, Wu H, Jiang ZY. Recent advances in the fabrication of advanced composite membranes. Journal of Materials Chemistry A. 2013; **1**:10058-10077

[111] Zhao J, Wang F, Pan FS, Zhang MX, Yang XY, Li P, Jiang ZY, Zhang P, Cao XZ, Wang BY. Enhanced pervaporation dehydration performance of ultrathin hybrid membrane by incorporating bioinspired multifunctional modifier and TiCl4 into chitosan. Journal of Membrane Science. 2013;**446**:395-404

[112] Pan FS, Cheng QL, Jia HP, Jiang ZY. Facile approach to polymer–inorganic nanocomposite membrane through a biomineralization-inspired process. Journal of Membrane Science. 2010;**357**:171-177

[113] Studart AR. Towards high-performance bioinspired composites. Advanced Materials. 2012;**24**:5024-5044

[114] Malinova K, Gunesch M, Montero Pancera S, Wengeler R, Rieger B, Volkmer D. Production of $CaCO_3$/hyper branched polyglycidol hybrid films using spray-coating technique. Journal of Colloid and Interface Science. 2012;**374**:61-69

[115] Klaysom C, Moon SH, Ladewig BP, GQM L, Wang LZ. The influence of inorganic filler particle size on composite ion-exchange membranes for desalination. Journal of Physical Chemistry C. 2011;**115**:15124-15132

[116] Copello GJ, Varela F, Vivot RM, Diaz LE. Immobilized chitos an asbiosorbent for the removal of Cd(II), Cr(III) and Cr(VI) from aqueous solutions. Bioresource Technology. 2008;**99**:6538-6544

[117] Xiong Y, Liu QL, Zhu AM, Huang SM, Zeng QH. Performance of organic–inorganic hybrid anion-exchange membranes for alkaline direct methanol fuel cells. Journal of Power Sources. 2009;**186**:328-333

[118] Cheng QL, Pan FS, Chen B, Jiang ZY. Preparation and dehumidification performance of composite membrane with PVA/gelatin–silicahy brid skin layer. Journal of Membrane Science. 2010;**363**:316-325

[119] Li B, Yu SN, Jiang ZY, Liu WP, Cao RJ, Wu H. Efficient desulfurization by polymer–inorganic nano composite membranes fabricated in reverse micro emulsion. Journal of Hazardous Materials. 2012;**211-212**:296-303

[120] Zhang YF, Wu H, Li J, Li L, Jiang YJ, Jiang Y, Jiang ZY. Protamine-templated biomimetic hybrid capsules: Efficient and stable carrier for enzyme encapsulation. Chemistry of Materials. 2007;**20**:1041-1048

[121] Baskar D, Balu R, Kumar TSS. Mineralization of pristine chitosan film through biomimetic process. International Journal of Biological Macromolecules. 2011;**49**:385-389

[122] Pan FS, Jia HP, Cheng QL, Jiang ZY. Bio-inspired fabrication of composite membranes with ultrathin polymer–silica nano hybrid skin layer. Journal of Membrane Science. 2010;**362**:119-126

[123] Zhao CH, Wu H, Li XS, Pan FS, Li YF, Zhao J, Jiang ZY, Zhang P, Cao XZ, Wang BY. High performance composite membranes with a poly-carbophil calcium transition layer for pervaporation dehydration of ethanol. Journal of Membrane Science. 2013;**429**:409-417

[124] Ma J, Zhang M, Wu H, Yin X, Chen J, Jiang ZY. Mussle-inspired fabrication of structurally stable chitosan/polyacrylonitrile composite membrane for pervaporation dehydration. Journal of Membrane Science. 2010;**348**:150-159

[125] Zhao J, Ma J, Chen J, Pan FS, Jiang ZY. Experimental and molecular simulation investigations on interfacial characteristics of gelatin/polyacrylonitrile composite pervaporation membrane. Chemical Engineering Journal. 2011;**178**:1-7

[126] Ma J, Zhang MH, Jiang ZY, Nie MC, Liu GX. Facile fabrication of structurally stable hyaluronic acid-based composite membranes inspired by bioadhesion. Journal of Membrane Science. 2010;**364**:290-297

[127] Cheng C, Li S, Zhao W, Wei Q, Nie S, Sun S, Zhao C. The hydrodynamic permeability and surface property of polyethersulfone ultrafiltration membranes with mussel-inspired polydop amine coatings. Journal of Membrane Science. 2012;**417-418**:228-236

[128] Xi ZY, Xu YY, Zhu LP, Wang Y, Zhu BK. A facile method of surface modification for hydrophobic polymer membranes based on the adhesive behavior of poly(DOPA) and poly(dopamine). Journal of Membrane Science. 2009;**327**:244-253

[129] McCloskey BD, Park HB, Ju H, Rowe BW, Miller DJ, Chun BJ, Kin K, Freeman BD. Influence of polydop amine deposition conditions on pure water flux and foulant adhesion resistance of reverse osmosis, ultrafiltration, and microfiltration membranes. Polymer. 2010;**51**:3472-3485

[130] Gong YK, Liu LP, Messersmith PB. Doubly biomimetic catecholic phosphorylcholine copolymer: A platform strategy for fabricating antifouling surfaces. Macromolecular Bioscience. 2012;**12**:979-985

[131] Jiang JH, Zhu LP, Li XL, YY X, Zhu BK. Surface modification of PE porous membranes based on the strong adhesion of poly-dopamine and covalent immobilization of heparin. Journal of Membrane Science. 2010;**364**:194-202

[132] Wu J, Zhang L, Wang Y, Long Y, Gao H, Zhang X, Zhao N, Cai Y, Xu J. Mussel-inspired chemistry for robust and surface-modifiable multilayer films. Langmuir. 2011;**27**:13684-13691

[133] Li YF, Wang SF, Wu H, Wang JT, Jiang ZY. Bioadhesion-inspired polymer–inorganic nanohybrid membranes with enhanced CO_2 capture properties. Journal of Materials Chemistry. 2012;**22**:19617-19620

[134] Hawker CJ, Russell TP. Block copolymer lithography: Merging"bottom-up" with "to p-down" processes. MRS Bulletin. 2005;**30**:952-966

[135] Joo W, Kim HJ, Kim JK. Broadband anti reflection coating covering from visible to near infrared wavelengths by using multi layered nanoporous block copolymer films. Langmuir. 2009;**26**:5110-5114

[136] Bang J, Kim SH, Drockenmuller E, Misner MJ, Russell TP, Hawker CJ. Defect-free nano-porous thin films from ABC triblock copolymers. Journal of the American Chemical Society. 2006;**128**:7622-7629

[137] Yang SY, Ryu I, Kim HY, Kim JK, Jang SK, Russell TP. Nano porous membranes with ultrahigh selectivity and flux for the filtration of viruses. Advanced Materials. 2006; **18**:709-712

[138] Phillip WA, O'Neill B, Rodwogin M, Hillmyer MA, Cussler EL. Self-assembled block copolymer thin films as water filtration membranes. ACS Applied Materials & Interfaces. 2010;**2**:847-853

[139] Chen L, Phillip WA, Cussler EL, Hillmyer MA. Robust nano porous membranes templated by a doubly reactive block copolymer. Journal of the American Chemical Society. 2007;**129**:13786-13787

[140] Amendt MA, Chen L, Hillmyer MA. Formation of nano structured poly(dicyclopentadiene) the rmosets using reactive block poly-mers. Macromolecules. 2010;**43**: 3924-3934

[141] Mao H, Arrechea PL, Bailey TS, Johnson BJS, Hillmyer MA. Control of pore hydrophilicity in ordered nanoporous polystyrene using an AB/AC block copolymer blending strategy. Faraday Discuss. 2005;**128**:149-162

[142] Kato T, Hillmyer MA. Functionalized nanoporous polyethylene derived from miscible block polymer blends. ACS Applied Materials & Interfaces. 2013;**5**:291-300

[143] Kang M, Moon B. Synthesis of photocleavable poly (styrene-block-ethylene oxide) and its self-assembly into nanoporous thin films. Macromolecules. 2008;**42**:455-458

[144] Schumers JM, Vlad A, Huynen I, Gohy JF, Fustin CA. Functionalized nanoporous thin films from photocleavable block copolymers. Macromolecular Rapid Communications. 2012;**33**:199-205

[145] Ryu JH, Park S, Kim B, Klaikherd A, Russell TP, Thayumanavan S. Highly ordered gold nanotubes using thiols at a cleavable block copolymer interface. Journal of the American Chemical Society. 2009;**131**:9870-9871

[146] Zhang M, Yang L, Yurt S, Misner MJ, Chen JT, Coughlin EB, Venkataraman D, Russell TP. Highly ordered nanoporous thin films from cleavable polystyrene-block-poly(ethylene oxide). Advanced Materials. 2007;**19**:1571-1576

[147] du Sart GG, Vukovic I, Vukovic Z, Polushkin E, Hiekkataipale P, Ruokolainen J, Loos K, ten Brinke G. Nanoporous network channels from self-assembled triblock copolymer supra molecules. Macromolecular Rapid Communications 2011;**32**:366-370

[148] Fahmi AW, Gutmann JS, Vogel R, Gindy N, Stamm M. Rheology pathway to macro scale ordered nano structures of polymericnano templates: Nanopores, nanosheets and nanofibers. Macromolecular Materials and Engineering. 2006;**291**:1061-1073

[149] Zhang P, Gao J, Li B, Han Y. Surface morphology evolution of a thinpolymeric supra-molecular film by tuning interactions. Macromolecular Rapid Communications. 2006; **27**:295-301

[150] Fustin CA, Lohmeijer BGG, Duwez AS, Jonas AM, Schubert US, Gohy JF. Nanoporous thin films from self-assembled metallo-supramolecular block copolymers. Advanced Materials. 2005;**17**:1162-1165

[151] Guillet P, Fustin CA, Wouters D, Hoeppener S, Schubert US, Gohy JF. Amphiphilic brushes from metallo-supramolecular block copoly-mers. Soft Matter. 2009;**5**:1460-1465

[152] Wang Y, Li F. An emerging pore-making strategy: Confined swelling-induced pore generation in block copolymer materials. Advanced Materials. 2011;**23**:2134-2148

[153] Peinemann KV, Abetz V, Simon PFWA. Symmetric super structureformed in a block copolymer via phase separation. Nature Materials. 2007;**6**:992-996

[154] Nunes SP, Behzad AR, Hooghan B, Sougrat R, Karunakaran M, Pradeep N, Vainio U, Peinemann KV. Switchable pH-responsive polymeric membranes prepared via block copolymer micelle assembly. ACS Nano. 2011;**5**:3516-3522

[155] Zavala-Rivera P, Channon K, Nguyen V, Sivaniah E, Kabra D, Friend RH, Nataraj SK, Al-Muhtaseb SA, Hexemer A, Calvo ME, Miguez H. Collective osmotic shock in ordered materials. Nature Materials. 2012;**11**:53-57

[156] Theato P, Ungar G. Nano porous ordered materials: Osmotically shocked. Nature Materials. 2012;**11**:16-17

[157] Lee JS, Hirao A, Nakahama S. Polymerization of monomers containing functional silyl groups. 5. Synthesis of new porous membranes with functional groups. Macromolecules. 1988;**21**:274-276

[158] Yang SY, Park J, Yoon J, Ree M, Jang SK, Kim JK. Virus filtration membranes prepared from nanoporous block copolymers with good dimensional stability under high pressures and excellent solvent resistance. Advanced Functional Materials. 2008; **18**:1371-1377

[159] Xu T, Stevens J, Villa JA, Goldbach JT, Guarini KW, Black CT, Hawker CJ, Russell TP. Block copolymer surface reconstruction: A reversible route to nanoporous films. Advanced Functional Materials. 2003;**13**:698-702

[160] Tang C, Bang J, Stein GE, Fredrickson GH, Hawker CJ, Kramer EJ, Sprung M, Wang J. Square packing and structural arrangement of ABC triblock copolymer spheres in thin films. Macromolecules. 2008;**41**:4328-4339

[161] Kim E, Shin C, Ahn H, Ryu DY, Bang J, Hawker CJ, Russell TP. Size control and registration of nano-structured thin films by cross-link able units. Soft Matter. 2008;**4**:475-479

[162] Joo W, Park MS, Kim JK. Block copolymer film with sponge-like nano porous structure for antireflection coating. Langmuir. 2006;**22**:7960-7963

[163] Park SC, Jung H, Fukukawa K, Campos LM, Lee K, Shin K. HawkerCJ, ha JS, bang J. Highly ordered nanoporous thin films by blending of PSt-b-PMMA block copolymers and PEO additives as structure directing agents. Journal of Polymer Science Part A: Polymer Chemistry. 2008;**46**:8041-8048

[164] Li X, Fustin CA, Lefevre N, Gohy JF, Feyter SD, Baerdemaeker JD, Egger W, Vankelecom IFJ. Ordered nanoporous membranes based on diblock copolymers with high chemical stability and tunable separation properties. Journal of Materials Chemistry. 2010; **20**:4333-4339

[165] Kim DH, Lau KHA, Joo W, Peng J, Jeong U, Hawker CJ, Kim JK, Russell TP, Knoll W. An optical waveguide study on the nano pore formationin block copolymer/homopolymer thin films by selective solvent swelling. The Journal of Physical Chemistry. B. 2006;**110**:15381-15388

[166] Jeong U, Ryu DY, Kim JK, Kim DH, Wu X, Russell TP. Precise control of nanopore size in thin film using mixtures of asymmetric block copolymer and homopolymer. Macromolecules. 2003;**36**:10126-10129

[167] Guo S, Rzayev J, Bailey TS, Zalusky AS, Olayo-Valles R, Hillmyer MA. Nanopore and nanobushing arrays from ABC tri block thin films containing two etchable blocks. Chemistry of Materials. 2006;**18**:1719-1721

[168] Phillip WA, Rzayev J, Hillmyer MA, Cussler EL. Gas and water liquid transport through nanoporous block copolymer membranes. Journal of Membrane Science. 2006; **286**:144-152

[169] Mao H, Hillmyer MA. Macroscopic samples of polystyrene with ordered three-dimensional nano channels. Soft Matter. 2006;**2**:57-59

[170] Pitet LM, Amendt MA, Hillmyer MA. Nanoporous linear polyethylene from a block polymer precursor. Journal of the American Chemical Society. 2010;**132**:8230-8231

[171] Uehara H, Kakiage M, Sekiya M, Yamagishi T, Yamanobe T, Nakajima K, Watanabe T, Nomura K, Hase K, Matsuda M. Novel design solving the conductivity vs. water-uptake

trade-off for polymer electrolyte membrane by bicontinuous crystalline/amorphous morphology of block copolymer. Macromolecules. 2009;**42**:7627-7630

[172] Li L, Szewczykowski P, Clausen LD, Hansen KM, Jonsson GE, Ndoni S. Ultrafiltration by gyroid nanoporous polymer membranes. Journal of Membrane Science. 2011;**384**: 126-135

[173] Luchnikov V, Kondyurin A, Formanek P, Lichte H, Stamm M. Moirépatterns in superimposed nanoporous thin films derived from block-copolymer assemblies. Nano Letters. 2007;**7**:3628-3632

[174] Kuila BK, Stamm M. Supramolecular complex of poly(styrene)-b-poly(4-vinylpyridine) and 1-pyrenebutyric acid in thin film. Macromolecular Symposia. 2011;**303**:85-94

[175] Gao J, Zhang P, Fu J, Li B, Han Y, Yu X, Pan C. Surfacemorphology evolution of poly(styrene-block-4-vinylpyridine) (PS-b-P4VP)(H+) and poly(methyl methacrylate)-dibenzo-18-crown-6-poly(methyl methacrylate) (PMCMA) supramolecular film. Polymer. 2007;**48**:2425-2433

[176] Laforgue A, Bazuin CG, Prud'homme RE. A study of the supramolecular approach in controlling di block copolymer nano patterning and nanoporosity on surfaces. Macromolecules. 2006;**39**:6473-6482

[177] Kosonen H, Valkama S, Nykänen A, Toivanen M, ten Brinke G,Ruokolainen J, Ikkala O. Functional porous structures based on the pyrolysis of cured templates of block copolymer and phenolic resin.Advanced Materials 2006;**18**:201-205

[178] Fustin CA, Guillet P, Misner MJ, Russell TP, Schubert US, Gohy JF. Self-assembly of metallo-supramolecular block copolymers in thinfilms. Journal of Polymer Science Part A: Polymer Chemistry. 2008;**46**:4719-4724

[179] Mugemana C, Gohy JF, Fustin CA. Functionalized nanoporous thinfilms from metallo-supramolecular diblock copolymers. Langmuir. 2012;**28**:3018-3023

[180] Karunakaran M, Nunes SP, Qiu X, Yu H, Peinemann KV. Iso-porous PS-b-PEO ultrafiltration membranes via self-assembly and water-induced phase separation. Journal of Membrane Science. 2014;**453**:471-477

[181] Phillip WA, Mika Dorin R, Werner J, Hoek EMV, Wiesner U, Elimelech M. Tuning structure and properties of graded tri-block ter polymer-based mesoporous and hybrid films. Nano Letters. 2011;**11**:2892-2900

[182] Dorin RM, Phillip WA, Sai H, Werner J, Elimelech M, Wiesner U. Designing block copolymer architectures for targeted membraneperformance. Polymer. 2014;**55**:347-353

[183] Hall AR, Scott A, Rotem D, Mehta KK, Bayley H, Dekker C. Hybrid pore formation by directed insertion of alpha-haemolysin into solid-state nano pores. Nature Nanotechnology. 2010;**5**:874-877

[184] Wang Y, He C, Xing W, Li F, Tong L, Chen Z, Liao X, Steinhart M. Nanoporous metal membranes with bicontinuous morphology from recyclable block-copolymer templates. Advanced Materials. 2010;**22**:2068-2072

[185] Wang Z, Yao X, Wang Y. Swelling-induced mesoporous block copolymer membranes with intrinsically active surfaces for size-selective separation. Journal of Materials Chemistry. 2012;**22**:20542-20548

[186] Shannon MA, Bohn PW, Elimelech M, Georgiadis JG, Marinas BJ, Mayes AM. Science and technology for water purification in the coming decades. Nature. 2008;**452**:301-310

[187] Ma XL, YL S, Sun Q, Wang YQ, Jiang ZY. Preparation of protein-adsorption-resistant polyether sulfone ultrafiltration membranes through surface segregation of amphiphilic comb copolymer. Journal of Membrane Science. 2007;**292**:116-124

[188] Zhao YH, Qian YL, Zhu BK, Xu YY. Modification of porous poly(vinylidene fluoride) membrane using amphiphilic polymers with different structures in phase inversion process. Journal of Membrane Science. 2008;**310**:567-576

[189] Zhao W, YL S, Li C, Shi Q, Ning X, Jiang ZY. Fabrication of antifouling polyether sulfone ultrafiltration membranes using Pluronic F127as both surface modifier and pore-forming agent. Journal of Membrane Science. 2008;**318**(12):405

[190] Wang YQ, Wang T, YL S, Peng FB, Wu H, Jiang ZY. Remarkable reduction of irreversible fouling and improvement of the permeation properties of poly(ether sulfone) ultrafiltration membranes by blending with Pluronic F127. Langmuir. 2005;**21**:11856-11862

[191] Li L, Schulte L, Clausen LD, Hansen KM, Jonsson GE, Ndoni S. Gyroid nanoporous membranes with tunable permeability. ACS Nano. 2011;**5**:7754-7766

[192] Chen WJ, Su YL, Peng JM, Zhao XT, Jiang ZY, Dong YN, Zhang Y, LiangYG LJZ. Efficient wastewater treatment by membranes through constructing tunable antifouling membrane surfaces. Environmental Science & Technology. 2011;**45**:6545-6552

[193] Chen WJ, YL S, Peng JM, Dong YN, Zhao XT, Jiang ZY. Engineering a robust, versatile amphiphilic membrane surface through forced surface segregation for ultra low flux-decline. Advanced Functional Materials. 2011;**21**:191-198

[194] Xue Z, Cao Y, Liu N, Feng L, Jiang L. Special wettable materials for oil/water separation. Journal of Materials Chemistry A. 2014;**2**:2445

[195] Wu W, Zhu Q, Qing F, Han CC. Water repellency on a fluorine-containing polyurethane surface: Toward understanding the surface self-cleaning effect. Langmuir. 2008; **25**:17-20

[196] Kao TH, Chen JK, Cheng CC, CI S, Chang FC. Low-surface-free-energy polybenzoxazine/polyacrylonitrile fibers for bionon fouling membrane. Polymer. 2013;**54**:258-268

[197] Hardman SJ, Muhamad-Sarih N, Riggs HJ, Thompson RL, Rigby J, WNA B, Hutchings LR. Electrospinning super hydrophobic fibers using surface segregating end-functionalized polymer additives. Macromolecules. 2011;**44**:6461-6470

[198] Shang Y, Si Y, Raza A, Yang L, Mao X, Ding B, Yu J. An in situ polymerization approach for the synthesis of super hydrophobic and super oleophilic nanofibrous membranes for oil–water separation. Nanoscale. 2012;**4**:7847-7854

[199] Zhu Y, Zhang J, Zheng Y, Huang Z, Feng L, Jiang L. Stable, super-hydrophobic, and conductive polyaniline/polystyrene films for corrosive environments. Advanced Functional Materials. 2006;**16**:568-574

[200] Ma M, Mao Y, Gupta M, Gleason KK, Rutledge GC. Super hydrophobic fabrics produced by electrospinning and chemical vapor deposition. Macromolecules. 2005;**38**:9742-9748

[201] Ma M, Gupta M, Li Z, Zhai L, Gleason KK, Cohen RE, Rubner MF, Rutledge GC. Decorated electrospun fibers exhibiting super hydrophobicity. Advanced Materials. 2007;**19**:255-259

[202] Tuteja A, Choi W, Ma ML, Mabry JM, Mazzella SA, Rutledge GC, McKinley GH, Cohen RE. Designing super oleophobic surfaces. Science. 2007;**318**:1618-1622

[203] Li J, Shi L, Chen Y, Zhang Y, Guo Z, Su B, Liu W. Stable super hydrophobic coatings from thiollig and nano crystals and their application in oil/water separation. Journal of Materials Chemistry. 2012;**22**:9774-9781

[204] Zhang W, Shi Z, Zhang F, Liu X, Jin J, Jiang L. Super hydrophobicand super oleophilic PVDF membranes for effective separation of water-in-oil emulsions with high flux. Advanced Materials. 2013;**25**:2071-2076

[205] Zhao XT, YL S, Chen WJ, Peng JM, Jiang ZY. Grafting perfluoro alkyl groups onto polyacrylonitrile membrane surface for improved fouling release property. Journal of Membrane Science. 2012;**415-416**:824-834

[206] Xue Z, Wang S, Lin L, Chen L, Liu M, Feng L, Jiang L. A novel super hydrophilic and underwater super oleophobic hydrogel coated mesh for oil/water separation. Advanced Materials. 2011;**23**:4270-4273

[207] Xue B, Gao L, Hou Y, Liu Z, Jiang L. Temperature controlled water/oil wettability of a surface fabricated by a block copolymer: Application as a dual water/oil on–off switch. Advanced Materials. 2013;**25**:273-277

[208] Zhang S, Lu F, Tao L, Liu N, Gao C, Feng L, Wei Y. Bio-inspired anti-oil-fouling chitosan-coated mesh for oil/water separation suitable for broad pH range and hyper-saline environments. ACS Applied Materials & Interfaces. 2013;**5**:11971-11976

[209] Zhu Y, Zhang F, Wang D, Pei XF, Zhang W, Jin J. A novel zwitterionic polyelectrolyte grafted PVDF membrane for thoroughly separating oil from water with ultra high efficiency. Journal of Materials Chemistry A. 2013;**1**:5758-5765

[210] Zhang W, Zhu Y, Liu X, Wang D, Li J, Jiang L, Jin J. Salt-induced fabrication of super hydrophilic and underwater super oleophobic PAA-g-PVDF membranes for effective separation of oil-in-water emulsions. Angewandte Chemie International Edition in English. 2014;**53**:856-860

Chapter 5

Aquaporin Biomimetic Membranes

Amira Abdelrasoul, Huu Doan and Ali Lohi

Additional information is available at the end of the chapter

http://dx.doi.org/10.5772/intechopen.71722

Abstract

Recent research looked at an array of aquaporin protein structures, their unique functions, and potential applications in the research and industrial sectors. This chapter focuses on the specific functional features of aquaporin biomimetic membranes to interrogate their permeability properties in relation to various biomimetic water-transporting membranes. This chapter discusses in detail functional characteristics of aquaporin, how to produce it, and the status of aquaporin development.

Keywords: aquaporin, function, permeability, protein, membrane design

1. Introduction

A series of recent reviews looked at an array of aquaporin protein structures, their unique functions, and potential applications in the research and industrial sectors [1–5]. This chapter focuses on the specific functional features of aquaporin biomimetic membranes to interrogate their permeability properties in relation to various biomimetic water-transporting membranes.

Aquaporin protein structures are a family of 24–30 kDa pores that form an essential type of membrane proteins. The process of red blood cell membrane protein purification (channel-forming key membrane protein of 28 kDa or CHIP28) [6] and consequent form of this protein in *Xenopus oocytes* [7] and liposomes [8] showcased a rather quick water diffusion process along the osmotic gradients. Since this process of purification, new research data had become available on this specific class of proteins, and a term aquaporins came to define it [9]. The primary aquaporin sequence features two repetitions where each one includes three transmembrane spanning α–helices (TM 1–3), as shown in **Figure 1**. Every repetition section includes a loop between TM2 and TM3 and an asparagine-proline-alanine (NPA) pattern signature.

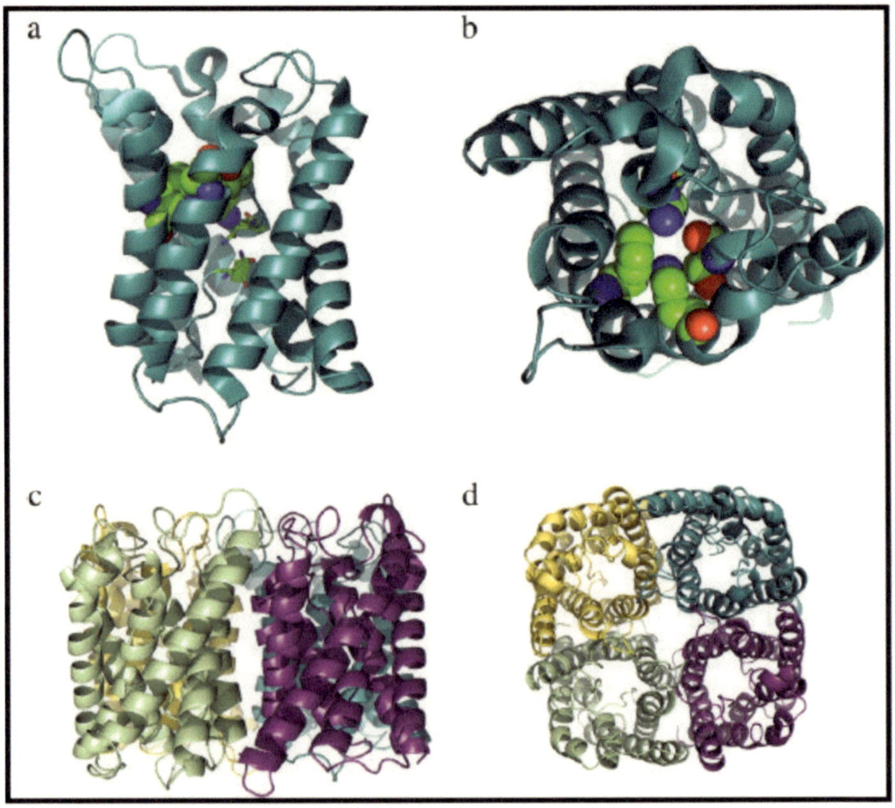

Figure 1. Aquaporin protein structure. (a): Side view of AqpZ monomer. Protein backbone (deep teal) with the two terminal asparagines from the NPA motifs shown using stick representation and the ar/R selectivity filter residues displayed in space fill representation. For stick and space fill representations, atoms have been color-coded as follows: carbon (green), oxygen (red), and nitrogen (blue). (b): Top view illustrating the selectivity filter (or constriction site) created by the four amino acids: F43, H174, R189, and T183. (c–d): Side and top view of the tetrameric AqpZ complex with the four monomers depicted in deep teal, violet purple, pale green, and yellow [1].

The aquaporin protein is assembled in a form of a hour-glass-shaped structure with six TM segments encircling a pore structure in the center and demarcated by the two opposing NPA motifs, as reflected in **Figure 1a** and **b**. The conserved aromatic/arginine (ar/R) region effectually describes the selectivity filter or constriction site, which is channel lumen's the narrowest part. The six TM AQP unit becomes a functional entity that acts as a pore, with the predominant unit assembly in biological membranes as a tetrameric arrangement [10], as indicated in **Figure 1c** and **d**. Due to their specific permeability properties, the mammalian homologs are categorized into two factions of aquaglyceroporins and aquaporins. The *Escherichia coli* model system can include both of these variants [11], the orthodox 'water only' channel AqpZ [12, 13], and the aquaglyceroporin GlpF, which is likewise permeable to glycerol [14]. While certain aquaporins may be categorized as solely water channels (e.g., AQP0, AQP4, and AqpZ), research that is being currently conducted indicates that a high number of aquaporins can have supplementary permeability properties [4].

Supplementing their already complex permeability profile parameters, multiple aquaporins showed a number of gating forms comparable to the opening and closing dynamics occurring in ion channels induced by external stimuli [15]. While numerous aspects of aquaporin gating and the process of regulating their permeability capacity remain unknown, the overall functionality of certain aquaporins has been confirmed as contingent on calmodulin [16, 17], phosphorylation, [18, 19] and pH [16, 20, 21].

2. Functional characterization of aquaporins

The properties of water's permeability and solute's rejection of single aquaporins cannot be measured very easily. Molecular dynamic simulations conducted with aquaporins suggest diffusional water permeabilities that correspond to the transport of 10^8–10^9 water molecules/s [22]. With regard to the quantity of the transported molecules, this data is about an order of magnitude greater than the parameters available in standard ion channel, with a single channel at a picoampere (pA) current level and a millisecond (ms) time scale corresponding to the transmembrane displacement of ~10^7 ions [23]. Although currents in the pA range can be calculated using the standard patch-clamp methods, the movement of 10^8–10^9 water molecules is not experimentally available due to the limitations of methods presently available to researchers. On the other hand, the macroscopic transport mediated with the aid of an ensemble of aquaporins can be measured. After this assessment is applied, the measured osmotic transport arising from a large (known) number of aquaporins can help estimate the single aquaporin permeability values. The two methods currently implemented for these calculations are as follows: *Xenopus oocyte* volume change, and light scattering from the proteoliposomes/proteopolymersomes.

In the *Xenopus oocyte* method expression, frog oocytes (~1 mm diameter) are cytoplasmically injected with mRNA that has been transcribed in vitro from a cDNA clone [24]. For aquaporins, in this instance, the resultant expression makes the oocyte membrane substantially more water permeable in comparison to the control oocytes [7]. During an osmotic challenge, the oocyte will alter its size (diameter value) and by implementing small osmotic gradients for brief period of time (e.g., 2.5 number of milliosmoles (mosM) for 5 s), the transport parameters (solute rejection and water permeability) can be calculated based on the initial rate of oocyte volume change rates in terms of shrinkage and swelling experimental runs [25].

Water permeability values of proteoliposomes/proteopolymersomes could likewise be assessed through the process of identifying the light scattering of the preparations within a stopped-flow apparatus setup, as indicated in **Figure 2a**. A suspension of aquaporin-containing vesicles of an original diameter approximately 200 nm briskly mixed with an identical volume of hyperosmolar solution featuring membrane impermeable solutes (e.g., sorbitol, sucrose, or mannitol) for proteoliposomes can create a dynamic where the subsequent transmembrane osmotic gradient will create water efflux from vesicles. As a result, the vesicle volume is lowered and may be measured with the help of an increase in the intensity values of scattered light. The rate constant, k, of the surge in normalized light intensity values is symptomatic of the water permeability coefficient and the water efflux rate constant. In this

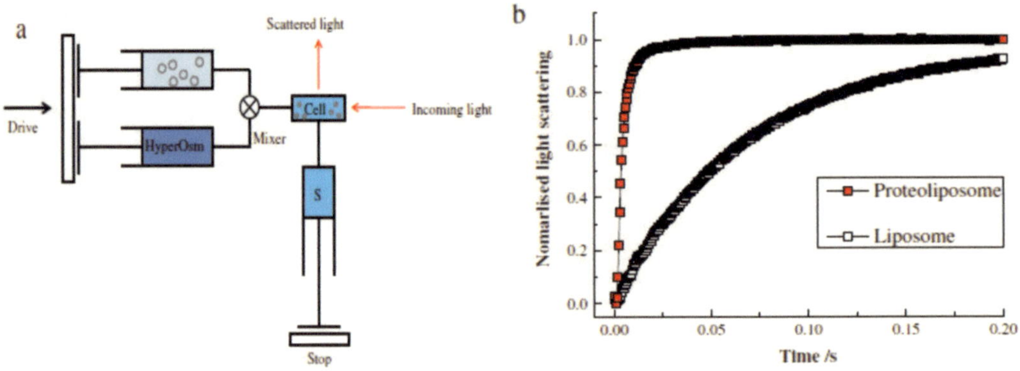

Figure 2. Stopped-flow characterization. (a) Schematics of stopped-flow measurement; (b): Typical stopped-flow values for lipid vesicles with (i.e., proteoliposomes) and without incorporated aquaporin (i.e., liposomes) [26].

instance, the light intensity values increase in an exponential manner with time, as indicated in **Figure 2b**. The response from the protein-free controls is connected to a single exponential, while the double-exponential function is applied for proteoliposomes/proteopolymersomes (vesicles) reflecting the dual pathways for water transport, either protein mediated or membrane mediated. Based on these processes, the k values can then be implemented for the calculation of the osmotic permeability P_f:

$$P_f = \frac{k}{\frac{S}{V_0} . V_w . \Delta osm}$$

where S/V_0 is the surface area of the initial volume ratio of the vesicle, V_w is the partial molar volume of water (18 cm^3/mol), and Δosm is the variance in osmolarity between the intravesicular and extravesicular aqueous solutions. Based on stopped-flow measurements, the water permeability of AqpZ is predicted to be in the range of $2–10 \times 10^{-14}$ cm^3/s [27–29] and showed a reasonable agreement with previously reported molecular dynamics simulation results ($3–30 \times 10^{-14}$ cm^3/s).

3. Production of aquaporins

Currently, the majority of recombinant aquaporins have only been created in lab-scale quantities for the purposes of screening, regulatory, functional, or structural research studies [30, 31]. The primary concern in protein production is that the membrane protein overexpression in vivo is problematized by their overly complex structure, hydrophobic transmembrane regions with host toxicity, and the low efficiency and time-consuming refolding steps necessary. Advances in high-expression systems and their potential applications could offer an insight into how a large-scale AQP production and implementation could be facilitated. These high-expression systems include *E. coli, Pichia pastoris, Saccharomyces cerevisiae,* and *baculovirus/* insect cell-based systems, and their recent review is available for additional information [32].

The *E. coli* expression-based methodologies that produce milligram quantities of protein have been effectively used to analyze the X-ray structure of the AqpZ and GlpF channels, AqpZ [13], GlpF [14], and of the archaeal aquaporin AqpM [33]. A high level of expression (200 mg/L) of the traditional aquaporin AqpZ was achieved in a recent study of the *E. coli* system, where a maltose binding protein (MBP) was used as a fusion partner protein, followed by a condition optimization process [34]. The *S. cerevisiae* system can be utilized for the production of large amounts of functional aquaporins [30, 35–39]. Alternatively, the methylotrophic yeast *P. pastoris* has been effectively used to produce a high number of distinct aquaporins. The potentially produced selection includes all 13 human aquaporins [40] and a wide range of active plant aquaporins [41–47]. Large-scale expressions of various functional recombinant aquaporins have been obtained with the aid of a baculovirus/insect cell system [48–55].

Research data indicate that there is a possibility for high-level membrane protein expression based on cell-free (CF) type of production. The essential requirement in this case is the process of synthesizing membrane proteins together with natural or synthetic lipids, as well as detergents, that can help solubilize the membrane's protein content. CF type of aquaporin production has been illustrated at analytical levels [31, 56–58], and recent tests showed high expression of properly folded AqpZ. Furthermore, plant aquaporin has been achieved with *E. coli* CF protocols and implementing various fusion vectors [59, 60]. Milligram of high potential AqpZ have been created in synthetic liposomes by using a CF approach [61]. Sutro Biopharma Inc. [62] showed the possibility of cost-effective cell-free protein synthesis in a 100-l reaction and the implicit advantages offered by CF systems that can act as an effective recombinant protein in industrial-scale production platforms.

The protein, that has been stabilized using a detergent, must be modified into its host biomimetic membrane; however, this creates challenges for industrial scaling and production. These challenges may be defined in terms of the detergent-stabilized intermediates, where the detergent cost and stability become primary concerns [63, 70]. Alternatively, the process of optimizing the interaction between membrane, protein (c.f. [64–66]), and yield may directly affect how much of functional protein content can be integrated into the final product (c.f. [67]).

4. Status of the aquaporin membrane development

Research conducted by Tang et al. suggests that membranes with excessively high salt rejection and permeability values can be created with the aid of aquaporin protein [68]. The measured water permeability values of AqpZ with proteoliposomes are used by researchers to argue that AqpZ-based biomimetic membranes can, in theory, obtain a membrane permeability as high as 167 $\mu m \cdot s^{-1} \cdot bar^{-1}$ (i.e., 601 $L \cdot m^{-2} \cdot h^{-1} \cdot bar^{-1}$, as shown in **Figure 3**). This value is more permeable by about two orders of magnitude when compared to the preexisting commercially available seawater RO membranes [69]. Even though there is a high level of membrane permeability present, there are considerable scaled-up concerns remaining. This is primarily due to the fact that the membranes are built using nanoscale elements, that is, aquaporins, and there are serious questions about how the biomimetic membranes can be scaled up and then stabilized to about 1 m^2 dimensions appropriate for industrial applications.

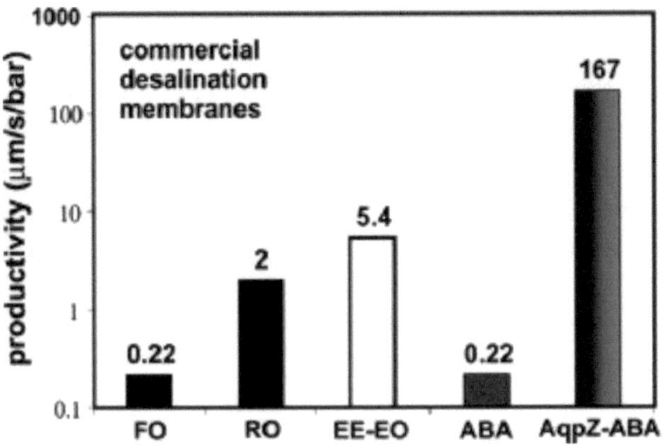

Figure 3. Comparison of water permeability of polymer vesicles with AqpZ (AqpZ-ABA) and without AqpZ (ABA) to those present in polymeric membranes [69].

Multiple design methods have been planned for the foreseeable future, as outlined in **Figure 4**. These strategies incorporate the creation of membranes based across multiple micron scale apertures, possibly as freestanding lipid or polymer membranes [44, 53, 68, 69], or alternatively as membranes stabilized using polymeric support materials [70]. Additional methods depend on nanoporous support materials on top of which the membranes are placed. For example, in this approach, there are several variations. The charged lipid vesicle can be deposited onto the commercially available nanofiltration membranes where the recipient surface is either a crosslinked polyamide or a sulfonated polysulfone negatively charged at pH7 [71]; there is an active rupture of aquaporin-containing polymersomes on top of the methacrylate functionalized cellulose acetate membranes [72], the method where detergent-stabilized tagged aquaporin are introduced into monolayers with nickel-chelating lipids [73]; and proteopolymersome deposition onto the surface of polycarbonate track-etched substrates covered with gold and then functionalized using photoactive acrylate groups. Research data suggest that [71–73] the implementation of spin coating and applied pressure improves vesicular coating/fusion on the substrate. As part of this process, surface charge and hydrophilicity values have been proven to play an essential role in shaping the overall quality of the supported lipid layers.

The data outlined in **Table 1** show the existing methods of creating aquaporin-based biomimetic membranes. In the majority of cases, most of the membranes discussed earlier feature a relatively low NaCl rejection value (or the rejection information is not available), a dynamic that prevents them from being used in desalination processes. Moreover, a large number of these membranes are not sufficiently stable for sophisticated industrial applications. In most cases, only small membrane areas are prepared, and the majority of methods depend on the implementation of specialized nanofabrication approaches, making the process exceedingly complicated and excessively expensive for scale up in forward osmosis (FO) and RO membrane fabrication. Ongoing innovation has allowed for the development of a new approach in the process of fabricating aquaporin-based biomimetic membranes. This innovative method relies on embedding aquaporin-containing proteopolymersomes or proteoliposomes into a

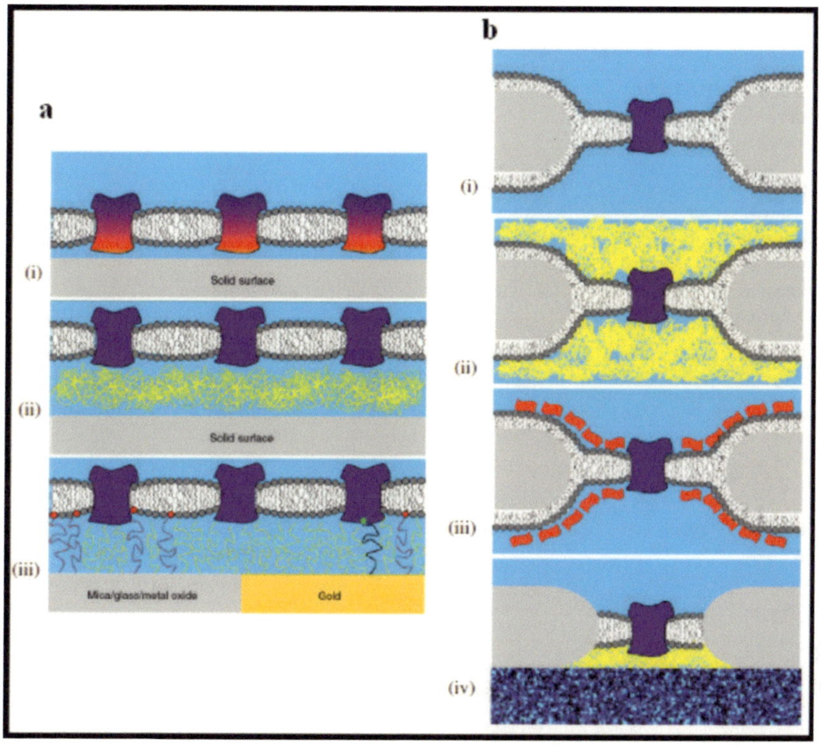

Figure 4. Summary of the existing biomimetic membrane designs. (a): Cross-sectional examples of solid-supported biomimetic membranes. (i) Direct deposition on a hydrophilic surface (light gray). This technique may introduce part of the integral membrane proteins (red protein shaded areas) embedded in the matrix formed from the self-assembly of lipids (dark gray molecules) too close to the surface, potentially inactivating (or even denaturing) the protein. (ii) Cushion-supported biomimetic membrane. (iii) Layers grafted covalently onto the support using spacers with silane groups and reacting with hydroxyl surfaces (light gray), or spacers with thiol groups bonding on gold surfaces (orange). Various hydrophilic spacers (e.g., poly(ethylene glycol) (PEG)) may be applied as cushion material. This cushion can be noncovalently interacting with the biomimetic membrane (yellow spacers) or covalently attached to lipids (red lipid headgroups) or proteins (green bonds), in the biomimetic membrane directly or through intermediates, for example, biotin-avidin complexes. (b): Cross-sectional illustrations of porous supported biomimetic membranes with an embedded protein (blue). (i) Freestanding membrane formed across a (micro or nano) porous support section. The membrane (solvent-free or solvent-containing) is formed in an aperture (light gray). (ii) Hydrogel-encapsulated biomimetic membrane. A hydrogel polymer meshwork (yellow) encapsulating the biomimetic membrane. (iii) A surface (S) layer-encapsulated membrane. The monomolecular layer of protein or glycoproteins (red) self-assembles into a two-dimensional lattice creating identical pores 2–8 nm in diameter. (iv) A cushioned membrane on a porous support [70, 71].

crosslinked polyamide matrix [69, 71]. In this approach, a microporous substrate was first soaked in an aqueous solution of m-phenylene-diamine (MPD) that likewise includes some aquaporin-containing vesicles, as shown in **Figure 4**. These soaked substrates were subsequently exposed to a tri-mesoyl chloride (TMC) solution to allow the formation of an interfacially polymerized polyamide rejection layer with vesicles dispersed in a thin rejection layer.

As part of this design, the aquaporin-containing vesicles offer superior water pathways through the polyamide layer and as a result substantially improve the membrane's water capacity for permeability. Furthermore, the crosslinked polyamide provides a scaffold that maintains the

Approach	WP ($L{\cdot}m^{-1}{\cdot}h^{-1}{\cdot}bar^{-1}$)	RNaCl (%)	Area (cm^2)	P_{max} (bar)	Upscaling issues	Remarks	Ref.
Charged lipid mixture vesicles depositions onto NF membranes	0.83	n.d.	3.5	10	Difficult to produce large defect-free membranes	No aquaporin incorporated	[11]
Vesicle fusion facilitated by hydraulic pressure on hydrophilic NF membranes coated with positively charged lipids	3.6 ± 0.2	35 ± 8	12.6	1	Difficult to produce large defect-free membranes	Low R_{NaCl}. Only suitable for NF. WP/RNaCl not tested. Not	[67]
Membranes across multiple micron scale apertures either as free-standing lipid or polymer membranes	n.d.	n.d.	4[a]	n.d.	Nanofabrication required Low robustness	WP/RNaCl not tested. Not suitable for RO.	[66, 67]
Membranes across multiple micron scale apertures and stabilized by hydrogel encapsulation	12–40	n.d.	3.5[a]	2	Nanofabrication required. High robustness	Characterized with gramicidin channels. No aquaporin incorporated.	[18]
Aquaporin containing polymersomes on methacrylate functionalized CA membranes	34.2 ± 6.9	32.9 ± 9.1	0.07	5	Medium robustness	Small area. High WP but low R_{NaCl}. Only suitable for NF.	[17]
Detergent-stabilized His-tagged aquaporin added to monolayers with nickel-chelating lipids	n.d.	n.d.	n.d.	n.d.	Complex fabrication. Low robustness	WP/RNaCl not tested. May not be suitable for desalination.	[51]
Proteopolymersome deposition onto gold-functionalized PC track-etched substrates	n.d.[b]	n.d.[b]	0.096	n.d.	Complex fabrication. Low robustness	Small area. Relatively high WP in FO. No RO data.	[35]
Interfacial polymerization method with embedded proteoliposomes	4 ± 0.4	96.3 ± 1.2	>200	14	Simple fabrication. High robustness	Combined high WP and RNaCl. Suitable for RO.	[8]

[a]Including membrane scaffold.

[b]RO tests were not performed. Based on FO tests, a WP of 16.4 $L{\cdot}m^{-2}{\cdot}h^{-1}$ and a salt flux of 6.6 $g{\cdot}m^{-2}{\cdot}h^{-1}$ were obtained for membranes prepared with a protein-to-polymer molar ratio of 1:100, with 0.3 M sucrose as draw and 200 ppm NaCl as feed.

Table 1. Existing approaches for the preparation of aquaporin-based biomimetic membranes. Performance data are presented as water permeability (WP) [$L{\cdot}m^{-2}{\cdot}h^{-1}{\cdot}bar^{-1}$], NaCl rejection (RNaCl) [%], membrane area (A) [cm^2], and maximal external pressure applied (PMax) [bar] when operated in RO. CA: cellulose acetate, PC: polycarbonate.

aquaporin-containing vesicles as well as defends them against the surrounding environmental conditions. This dynamic is anticipated to considerably increase the membrane's overall stability. This type of membrane exhibited a permeability value of ~4 $L{\cdot}m^{-2}{\cdot}h^{-1}{\cdot}bar^{-1}$, as shown in **Table 1**, while maintaining similar or even superior NaCl rejection values. This permeability is ~40% greater than the one occurring in commercial brackish water reverse osmosis membrane BW30 that was tested in the same conditions. Membranes featuring such a design have been extensively tested and showed enhanced stable flux and rejection performance capacity for the specified durations of weeks and months. A noticeable water improvement effect of aquaporins was likewise confirmed through a comparative analysis with membranes featuring vesicles containing inactive R189A AqpZ mutants [69–71]. This straightforward quality of this fabrication procedure ensures that this technique can be effortlessly scaled up for the manufacturing of large membrane areas crucial for industrial applications.

Author details

Amira Abdelrasoul[1]*, Huu Doan[2] and Ali Lohi[2]

*Address all correspondence to: amira.abdelrasoul@usask.ca

1 Department of Chemical and Biological Engineering, University of Saskatchewan, Saskatoon, Saskatchewan, Canada

2 Department of Chemical Engineering, Ryerson University, Toronto, Ontario, Canada

References

[1] Gonen T, Walz T. The structure of aquaporins. Quarterly Reviews of Biophysics. 2006; **39**:361-396

[2] Fu D, Lu M. The structural basis of water permeation and proton exclusion in aquaporins. Molecular Membrane Biology. 2007;**24**:366-374

[3] Ishibashi K. Aquaporin subfamily with unusual NPA boxes. Biochimica et Biophysica Acta. 2006;**1758**:989-993

[4] Verkman AS. More than just water channels: Unexpected cellular roles of aquaporins. Journal of Cell Science. 2005;**118**:3225-3232

[5] Nielsen S, Frokiaer J, Marples D, Kwon TH, Agre P, Knepper MA. Aquaporins in the kidney: From molecules to medicine. Physiological Reviews. 2002;**82**:205-244

[6] Denker BM, Smith BL, Kuhajda FP, Agre P. Identification, purification, and partial characterization of a novel Mr 28,000 integral membrane protein from erythrocytes and renal tubules. The Journal of Biological Chemistry. 1988;**263**:15634-15642

[7] Preston GM, Carroll TP, Guggino WB, Agre P. Appearance of water channels in Xenopus oocytes expressing red cell CHIP28 protein. Science. 1992;**256**:385-387

[8] Zeidel ML, Ambudkar SV, Smith BL, Agre P. Reconstitution of functional waterchannels in liposomes containing purified red cell CHIP28 protein. Biochemistry. 1992;**31**: 7436-7440

[9] Agre P, Sasaki S, Chrispeels MJ. Aquaporins: A family of water channel proteins. The American Journal of Physiology. 1993;**265**:F461

[10] Verbavatz JM, Brown D, Sabolic I, Valenti G, Ausiello DA, Van Hoek AN, Ma T, Verkman AS. Tetrameric assembly of CHIP28 water channels in liposomes and cell membranes: A freeze-fracture study. The Journal of Cell Biology. 1993;**123**:605-618

[11] Borgnia MJ, Agre P. Reconstitution and functional comparison of purified GlpF and AqpZ, the glycerol and water channels from *Escherichia coli*. Proceedings of the National Academy of Sciences of the United States of America. 2001;**98**:2888-2893

[12] Calamita G, Bishai WR, Preston GM, Guggino WB, Agre P. Molecular cloning and characterization of AqpZ, a water channel from *Escherichia coli*. The Journal of Biological Chemistry. 1995;**270**:29063-29066

[13] Savage DF, Egea PF, Robles-Colmenares Y, O'Connell JD 3rd, Stroud RM. Architecture and selectivity in aquaporins: 2.5 a X-ray structure of aquaporin Z. PLoS Biology. 2003;**1** E72

[14] Fu D, Libson A, Miercke LJ, Weitzman C, Nollert P, Krucinski J, Stroud RM. Structure of a glycerol-conducting channel and the basis for its selectivity. Science. 2000;**290**: 481-486

[15] Xin L, Su H, Nielsen CH, Tang C, Torres J, Mu Y. Water permeation dynamics of AqpZ: A tale of two states. Biochim. Biophys. Acta. 2011;**1808**:1581-1586

[16] Nemeth-Cahalan KL, Hall JE. pH and calcium regulate the water permeability of aquaporin 0. The Journal of Biological Chemistry. 2000;**275**:6777-6782

[17] Nemeth-Cahalan KL, Kalman K, Hall JE. Molecular basis of pH and Ca2+ regulation of aquaporin water permeability. The Journal of General Physiology. 2004;**123**:573-580

[18] Johansson I, Larsson C, Ek B, Kjellbom P. The major integral proteins of spinach leaf plasma membranes are putative aquaporins and are phosphorylated in response to Ca2+ and apoplastic water potential. Plant Cell. 1996;**8**:1181-1191

[19] Johansson I, Karlsson M, Shukla VK, Chrispeels MJ, Larsson C, Kjellbom P. Water transport activity of the plasma membrane aquaporin PM28A is regulated by phosphorylation. Plant Cell. 1988;**10**:451-459

[20] Zeuthen T, Klaerke DA. Transport of water and glycerol in aquaporin 3 is gated by H(+). The Journal of Biological Chemistry. 1999;**274**:21631-21636

[21] Yasui M, Hazama A, Kwon TH, Nielsen S, Guggino WB, Agre P. Rapid gating and anion permeability of an intracellular aquaporin. Nature. 1999;**402**:184-187

[22] Jensen MO, Mouritsen OG. Single-channel water permeabilities of *Escherichia coli* aquaporins AqpZ and GlpF. Biophysical Journal. 2006;**90**:2270-2284

[23] Hille B. Ionic Channels of Excitable Membranes. Sunderland: Sinauer; 2001

[24] Gurdon JB, Lane CD, Woodland HR, Marbaix G. Use of frog eggs and oocytes for the study of messenger RNA and its translation in living cells. Nature. 1971;**233**:177-182

[25] Meinild AK, Klaerke DA, Zeuthen T. Bidirectional water fluxes and specificity for small hydrophilic molecules in aquaporins 0-5. The Journal of Biological Chemistry. 1998; **273**:32446-32451

[26] Borgnia MJ, Kozono D, Calamita G, Maloney PC, Agre P. Functional reconstitution and characterization of AqpZ, the *E. coli* water channel protein. Journal of Molecular Biology. 1999;**291**:1169-1179

[27] Scheuring S, Ringler P, Borgnia M, Stahlberg H, DJM l, Agre P, Engel A. High resolution AFM topographs of the *Escherichia coli* water channel aquaporin Z. The EMBO Journal. 1999;**18**:7

[28] Mathai JC, Tristram-Nagle S, Nagle JF, Zeidel ML. Structural determinants of water permeability through the lipid membrane. The Journal of General Physiology. 2008;**131**:8

[29] Hovijitra NT, Wuu JJ, Peaker B, Swartz JR. Cell-free synthesis of functional aquaporin Z in synthetic liposomes. Biotechnol. Bioengineering. 2109;**4**:10

[30] Kaldenhoff R, Bertl A, Otto B, Moshelion M, Uehlein N. Characterization of plant aquaporins. Methods in Enzymology. 2007;**428**:505-531

[31] Shi LB, Skach WR, Ma T, Verkman AS. Distinct biogenesis mechanisms for the water channels MIWC and CHIP28 at the endoplasmic reticulum. Biochemistry. 1995;**34**:8250-8256

[32] Altamura N, Calamita G. Systems for production of proteins for biomimetic membrane devices. In: Hélix-Nielsen C, editor. Biomimetic Membranes for Sensor and Separation Applications. Dordrecht: Springer; 2012. p. 233-250

[33] Lee JK, Kozono D, Remis J, Kitagawa Y, Agre P, Stroud RM. Structural basis for conductance by the archaeal aquaporin AqpM at 1.68 A. Proceedings of the National Academy of Sciences of the United States of America. 2005;**102**:18932-18937

[34] Lian J, Ding S, Cai J, Zhang D, Xu Z, Wang X. Improving aquaporin Z expression in *Escherichia coli* by fusion partners and subsequent condition optimization. Applied Microbiology and Biotechnology. 2009;**82**:463-470

[35] Bienert GP, Moller AL, Kristiansen KA, Schulz A, Moller IM, Schjoerring JK, Jahn TP. Specific aquaporins facilitate the diffusion of hydrogen peroxide across membranes. The Journal of Biological Chemistry. 2007;**282**:1183-1192

[36] Fischer G, Kosinska-Eriksson U, Aponte-Santamaria C, Palmgren M, Geijer C, Hedfalk K, Hohmann S, de Groot BL, Neutze R, Lindkvist-Petersson K. Crystal structure of a yeast aquaporin at 1.15 angstrom reveals a novel gating mechanism. PLoS Biology 2009;7: e1000130

[37] Hedfalk K, Pettersson N, Oberg F, Hohmann S, Gordon E. Production, characterization and crystallization of the *Plasmodium falciparum* aquaporin. Protein Expression and Purification. 2008;**59**:69-78

[38] Laize V, Ripoche P, Tacnet F. Purification and functional reconstitution of the human CHIP28 water channel expressed in *Saccharomyces cerevisiae*. Protein Expression and Purification. 1997;**11**:284-288

[39] Pettersson N, Hagstrom J, Bill RM, Hohmann S. Expression of heterologous aquaporins for functional analysis in *Saccharomyces cerevisiae*. Current Genetics. 2006;**50**:247-255

[40] Oberg F, Ekvall M, Nyblom M, Backmark A, Neutze R, Hedfalk K. Insight into factors directing high production of eukaryotic membrane proteins; production of 13 human AQPs in *Pichia pastoris*. Molecular Membrane Biology. 2009;**26**:215-227

[41] Azad AK, Katsuhara M, Sawa Y, Ishikawa T, Shibata H. Characterization of four plasma membrane aquaporins in tulip petals: A putative homolog is regulated by phosphorylation. Plant Cell Physiol. 2008;**49**:1196-1208

[42] Azad AK, Sawa Y, Ishikawa T, Shibata H. Heterologous expression of tulip petal plasma membrane aquaporins in *Pichia pastoris* for water channel analysis. Applied and Environmental Microbiology. 2009;**75**:2792-2797

[43] Daniels MJ, Wood MR, Yeager M. In vivo functional assay of a recombinant aquaporin in *Pichia pastoris*. Applied and Environmental Microbiology. 2006;**72**:1507-1514

[44] Karlsson M, Fotiadis D, Sjovall S, Johansson I, Hedfalk K, Engel A, Kjellbom P. Reconstitution of water channel function of an aquaporin overexpressed and purified from *Pichia pastoris*. FEBS Letters. 2003;**537**:68-72

[45] Kukulski W, Schenk AD, Johanson U, Braun T, de Groot BL, Fotiadis D, Kjellbom P, Engel A. The 5A structure of heterologously expressed plant aquaporin SoPIP2;1. Journal of Molecular Biology 2005;**350**:611-616.

[46] Tornroth-Horsefield S, Wang K, Hedfalk Y, Johanson U, Karlsson M, Tajkhorshid E, Neutze R, Kjellbom P. Structural mechanism of plant aquaporin gating. Nature. 2006;**439**: 688-694

[47] Verdoucq L, Grondin A, Maurel C. Structure-function analysis of plant aquaporin AtPIP2;1 gating by divalent cations and protons. The Biochemical Journal. 2008;**415**:409-416

[48] Yang B, van Hoek AN, Verkman AS. Very high single channel water permeability of aquaporin-4 in baculovirus-infected insect cells and liposomes reconstituted with purified aquaporin-4. Biochemistry. 1997;**36**:7625-7632

[49] Swamy-Mruthinti S, Schey KL. Mass spectroscopic identification of in vitro glycated sites of MIP. Current Eye Research. 1997;**16**:936-941

[50] Werten PJ, Hasler L, Koenderink JB, Klaassen CH, de Grip WJ, Engel A, Deen PM. Large-scale purification of functional recombinant human aquaporin-2. FEBS Letters 2001;**504**: 200-205

[51] Drake KD, Schuette D, Chepelinsky AB, Crabbe MJ. Heterologous expression and topography of the main intrinsic protein (MIP) from rat lens. FEBS Letters 2002;**512**:191-198

[52] Drake KD, Schuette D, Chepelinsky AB, Jacob TJ, Crabbe MJ. pH-Dependent channel activity of heterologously-expressed main intrinsic protein (MIP) from rat lens. FEBS Letters. 2002;**512**:199-204

[53] Hiroaki Y, Tani K, Kamegawa A, Gyobu N, Nishikawa K, Suzuki H, Walz T, Sasaki S, Mitsuoka K, Kimura K, Mizoguchi A, Fujiyoshi Y. Implications of the aquaporin-4 structure on array formation and cell adhesion. Journal of Molecular Biology. 2006;**355**:628-639

[54] Yakata K, Hiroaki Y, Ishibashi K, Sohara E, Sasaki S, Mitsuoka K, Fujiyoshi Y. Aquaporin-11 containing a divergent NPA motif has normal water channel activity. Biochimica et Biophysica Acta. 2007;**1768**:688-693

[55] Hayakawa S, Mori M, Okuta A, Kamegawa A, Fujiyoshi Y, Yoshiyama Y, Mitsuoka K, Ishibashi K, Sasaki S, Hattori T, Kuwabara S. Neuromyelitis optica and anti-aquaporin-4 antibodies measured by an enzyme-linked immunosorbent assay. Journal of Neuroimmunology. 2008;**196**:181-187

[56] LaVallie ER, Lu Z, Diblasio-Smith EA, Collins-Racie LA, JM MC. Thioredoxin as a fusion partner for production of soluble recombinant proteins in *Escherichia coli*. Methods in Enzymology. 2000;**326**:322-340

[57] Lu Y, Turnbull IR, Bragin A, Carveth K, Verkman AS, Skach WR. Reorientation of aquaporin-1 topology during maturation in the endoplasmic reticulum. Molecular Biology of the Cell. 2000;**11**:2973-2985

[58] Paul DL, Goodenough DA. In vitro synthesis and membrane insertion of bovine MP26, an integral protein from lens fiber plasma membrane. The Journal of Cell Biology. 1983;**96**:633-638

[59] Xu Z, Lian J, Cai J. Efficient expression of aquaporin Z in *Escherichia coli* cell-free system using different fusion vectors. Protein and Peptide Letters. 2010;**17**:181-185

[60] Schwarz D, Junge F, Durst F, Frolich N, Schneider B, Reckel S, Sobhanifar S, Dotsch V, Bernhard F. Preparative scale expression of membrane proteins in *Escherichia coli*-based continuous exchange cell-free systems. Nature Protocols. 2007;**2**:2945-2957

[61] Hovijitra NT, Wuu JJ, Peaker B, Swartz JR. Cell-free synthesis of functional aquaporin Z in synthetic liposomes. Biotechnology and Bioengineering. 2009;**104**:40-49

[62] Zawada JF, Yin G, Steiner AR, Yang J, Naresh A, Roy SM, Gold DS, DS HHG, Murray CJ. Microscale to manufacturing scale-up of cell-free cytokine production—a new approach for shortening protein production development timelines. Biotechnology and Bioengineering. 2011;**108**:1570-1578

[63] Hansen JS, Vararattanavech A, Plasencia I, Greisen PJ, Bomholt J, Torres J, Emneus J, Helix-Nielsen C. Interaction between sodium dodecyl sulfate and membrane reconstituted aquaporins: A comparative study of spinach SoPIP2;1 and *E. coli* AqpZ. Biochimica et Biophysica Acta. 2011;**1808**:2600-2607

[64] Plasencia I, Survery S, Ibragimova S, Hansen JS, Kjellbom P, Helix-Nielsen C, Johanson U, Mouritsen OG. Structure and stability of the spinach aquaporin SoPIP2;1 in detergent micelles and lipid membranes. PLoS One. 2011;**6**:e14674

[65] Andersen OS, Nielsen C, Maer AM, Lundbæk JA, Goulian M, Koeppe RE. Ion channels as tools to monitor lipid bilayer–membrane protein interactions: gramicidin channels as molecular force transducers, Methods Enzymol. 1999;**294**:208-224

[66] Nielsen C, Goulian M, Andersen OS. Energetics of inclusion-induced bilayer deformations. Biophysical Journal. 1998;**74**:1966-1983

[67] Pszon-Bartosz K, Hansen JS, Stibius KB, Groth JS, Emneus J, Geschke O, Helix-Nielsen C. Assessing the efficacy of vesicle fusion with planar membrane arrays using a mitochondrial porin as reporter. Biochemical and Biophysical Research Communications. 2011;**406**:96-100

[68] Tang CY, Wang Z, Petrinic I, Fane AG, Hélix-Nielsen C. Biomimetic aquaporin membranes coming of age. Desalination. 2015;**368**:89-105

[69] Tang CY, Zhao Y, Wang R, Hélix-Nielsen C, Fane AG. Desalination by biomimetic aquaporin membranes: Review of status and prospects. Desalination. 2013;**308**:34-40

[70] Lander MR, Ibragimova S, Rein C, Vogel J, Stibius K, Geschke O, Perry M, Helix-Nielsen C. Biomimetic membrane arrays on cast hydrogel supports. Langmuir. 2011;**27**:7002-7007

[71] Kaufman Y, Grinberg S, Linder C, Heldman E, Gilron J, Freger V. Fusion of bolaamphiphile micelles: A method to prepare stable supported biomimetic membranes. Langmuir. 2013;**29**:1152-1161

[72] Zhong PS, Chung T-S, Jeyaseelan K, Armugam A. Aquaporin-embedded biomimetic membranes for nanofiltration. Journal of Membrane Science. 2012;**407**(408):27-33

[73] Zhang X, Tanner P, Graff A, Palivan CG, Meier W. Mimicking the cell membrane with block copolymer membranes. Journal of Polymer Science Part A: Polymer Chemistry. 2012;**50**:2293-2318

Chapter 6

Interactions between Aquaporin Proteins and Block Copolymer Matrixes

Amira Abdelrasoul, Huu Doan and Ali Lohi

Additional information is available at the end of the chapter

http://dx.doi.org/10.5772/intechopen.71723

Abstract

This chapter continues to further expand its focus on aquaporins (AQPs) by offering a general outline on how the AQPs block copolymers, and polymer support structures can interrelate and such connections can be comprehensively classified and defined. The first section of the overview will consider the relationship between block copolymers and AQPs. It will also examine the general membrane protein integration into block copolymers, since this can cause AQP-block copolymer complexes in vesicular (proteopolymersomes) as well as in planar forms. The majority of considerations taken into account during AQP incorporation come from the research conducted in relation to the process of incorporating other types of membrane proteins. This chapter includes an overview of the various characterization methodologies needed for the study of proteopolymersomes, as well as freeze-fracture transmission electron microscopy (FF-TEM), fluorescence correlation spectroscopy (FCS), small-angle X-ray scattering (SAXS), and stopped-flow light scattering (SFLS). The research data presented in this chapter emphasizes the fact that a successful process of membrane fabrication requires the integration of reconstituted AQPs into a suitable supporting matrix formation.

Keywords: aquaporin proteins, block copolymer, matrix, vesicular, membrane protein

1. Assessing aquaporin proteins and block copolymer matrixes interactions

Most research performed on membrane protein inclusion has been conducted primarily with lipids as the host matrix components (original proteoliposomes publication on the subject came out in 1971) [1]. Since then, polymer-based incorporation process has received increased attention and the earliest proteopolymersomes publication emerged in 2000 [2].

The initial work stages in this research area concentrated on the inclusion of membrane-spanning proteins, such as membrane-bound ion channels (ATPases) and bacteriorhodopsin incorporation into polymethyloxazoline-polydimethylsiloxane-polymethyloxazoline (PMOXA-PDMS-PMOXA) triblock copolymer bilayers occurring in planar [3] or vesicular forms [4–6]. It is quite fascinating that the membrane proteins may be functionally incorporated into polymeric bilayers (e.g., based on PMOXA-PDMS-PMOXA) that occur up to 10 times thicker than their lipidic counterparts [7]. Moreover, researchers have observed proteopolymersomes with protein density values that drastically surpassed those of proteoliposomes [8]. Research on these phenomena has helped to establish a theoretical approach for generalized membrane protein incorporation into the amphiphilic structures. This methodology is based on the notion that the efficiency of the membrane protein incorporation process relies on its hydrophobicity potential and its coupling capacity to the host membrane is directly connected to the hydrophobic mismatch parameters. In order to reduce this mismatch dynamic, the method calls for the host membrane to be deformed in such a way that it matches the hydrophobic length parameter of the membrane protein's transmembrane segment, in this case the hydrophobic length is 3–4 nm. A different manner of adaption would produce alternative results since the host membrane-induced membrane protein deformation is improbable due to the fact that membrane protein compressibility potential is generally one or two orders of magnitude greater than the one present in lipids [9, 10]. Researchers Srinivas and Discher argue that the application of coarse-grain simulations can ensure that the flexible hydrophobic chains would allow protein incorporation. Srinivas and Discher add that this may occur even in cases where there is a hydrophobic mismatch greater than 22% between hydrophobic interior of the chain region and membrane proteins [11, 12]. As a consequence, membrane proteins may be included with greater efficacy if the hydrophobic chains are sufficiently flexible [10]. Sine chains that are more flexible can possibly block the channel, a distinct lack of proteopolymersomes functionality can be perceived, even though the membrane protein was functionally incorporated [11]. Furthermore, elevated polydispersity may facilitate higher incorporation efficiency levels, since smaller chains can collect around the membrane protein and then offset the hydrophobic mismatch potential. Sufficiently positive incorporation data detected in the case of PMOXA-PDMS-PMOXA can thus likewise be credited to the much higher polydispersity index (PDI). Alternatively, in the setting such as natural lipid environment, the annual lipids surrounding the incorporated protein may be chosen partially due to the similarity with the lateral diffusion and protein surface [13]. The collective consequences of the hydrophobic mismatch are substantial for ATPases, ion channels [9], and co-transporter proteins. On the other hand, the effects of the mismatch are noticeably less for AQPs where they are reduced, since the protein structure itself is intrinsically more rigid [14].

Stoenescu and coworkers conducted the initial incorporation of AQPs in polymer bilayer in 2004 [15]. A research team by Stoenescu included an AQP0 originating from the mammalian eye lens directly into the polymersomes of three diverse block architectures (ABA, ABC, CBA, with A standing in for PMOXA, B for PDMS, and C for polyethylene oxide, PEO). This type of block configuration shapes how the orientation of AQP0 is being incorporated will occur. The data results obtained in this case suggest that ABA featured 50%

of included AQP0 with an orientation comparable to the one occurring in liposomes, CBA included only 20%, while ABC had 80%, as shown in the antibody labeling. In all these test examples, the process of incorporation was accomplished using the addition of AQP0 content into the detergent during the polymersome configuration, as well as with the application of size exclusion chromatography (SEC) for the removal of non-incorporated protein content [15].

Kumar and coworkers were able to produce the first evidence and samples of functional AQP incorporation in 2007. Specifically, Kumar integrated bacterial AqpZ from *E.coli* in PMOXA-PDMS-PMOXA polymersomes [7] and then verified their overall functional capacity within the stopped-flow light scattering (SFLS). As a relatively familiar permeability characterization methodology, SFLS permits the shrinkage of polymersomes because of the response to osmolarity changes that are monitored for the duration of the process using light scattering. The integration of AqpZ facilitated an osmotic response of proteopolymersomes that is 800 times greater than the one occurring with empty polymersomes. This test case likewise indicated that the activation energy, that is the barrier for water passing through the AqpZ, was analogous to the one present in AQP reconstituted in frog oocytes and proteoliposomes. During the testing, the molar protein-to-amphiphile ratio (mPAR) for ideal AqpZ performance within the triblock copolymer system was determined to be 1:50, a ratio that correspond to a 1:100 in a (diblock or lipid) bilayer system scenario [7]. The elevated density reconstitution parameter of AQP is also demonstrated by the creation of 2D AQP crystals that help collect structural (crystallographic) information about AQP, a process similar to the one applied to lipid based 2D AQP crystals [16]. In this type of process, a monolayer of nickel-functionalized polybutadiene-polyethylene oxide (PB-PEO) is collected at the water-interface, and includes the presence of aqueous solution, histidine-tagged AqpZ, PDMS-PMOXA-PDMS, and mixed micelles of detergent [17]. The property of nickel affinity to histidine further connects the AqpZ to the PB-PEO layer [18], effectually creating a dynamic of AqpZ high packing within the layer. Once the detergent is removed with the aid of biobeads and the PB-PEO is taken out with imidazole, the closely packed AqpZ PMOXA-PDMS-PMOXA crystals remained; however, the left over amount was not sufficient for retrieving key structural data [19, 20]. Data suggest that 2D crystals may in fact be useful when it comes to researching the effects of AQP on polymer self-assembly for general types of applications. The AQP0 has been shown to easily form 2D crystals because of its natural properties, as it occurs in stack formations within the eye lens [21]. The essential data findings collected during this experiment suggest that the AQP0 shapes the self-assembling behavior of both polymers in way that it is reciprocal to the hydrophilic volume ratio *f*. As the mPAR values are increasing, the interfacial curvature becomes lower and the polymersomes form into membrane sheets as well as a certain amount of crystals (see **Figures 1** and **2**). When it comes to PB-PEO, the construction of polymersomes happened only when AQP0 was added, while in the absence of AQP0 solely cylindrical structures were perceived. The greatest packing densities of functional AQPs within vesicular structures were noted at PB-PEO polymersomes featuring an mPAR of 1:15, a correlation that is much higher than the one that was obtained in the cases with frog oocytes and proteoliposomes. While not all of the AQP0 proteins were integrated, this sevenfold growth in osmotic

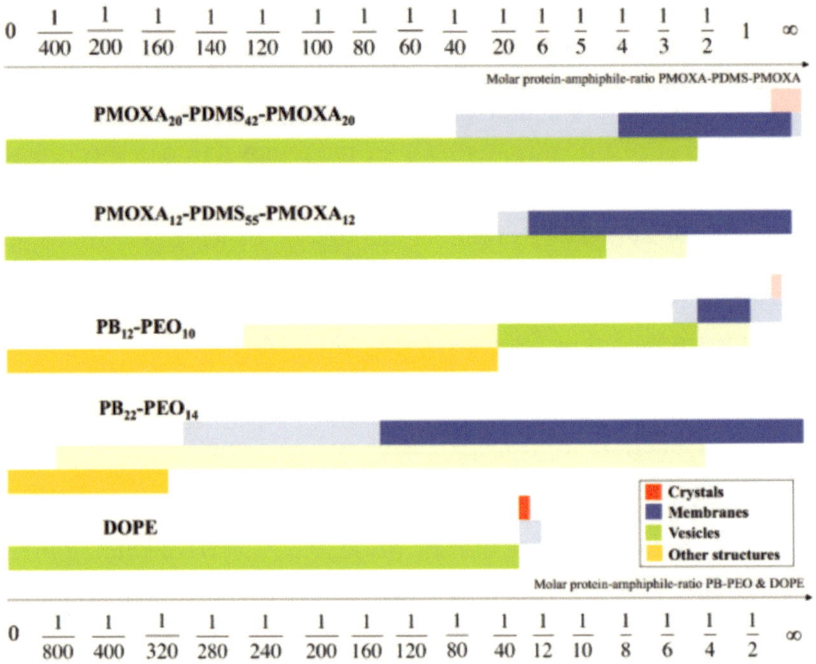

Figure 1. A schematic drawing of aggregate morphologies portrayed as a function of mPAR. PB12-PEO10 as it goes through four transitions. The morphologies shown in full color indicate the primary morphologies, where the pale colors signify the coexisting morphologies [24].

response values is relatively consistent with the high-packing density parameters and low permeability of AQP0 [22]. In this experimental run, the integration occurred through the process of mixing detergent-solubilized polymers with the detergent-solubilized AQP0, and then dialyzing the detergent out of the mixture [8, 23]. In this case, the vesicle's shape continued to show substantially greater densities at block copolymers, when correlated to standard types of lipids such as the 1,2-dioleoyl-sn-glycero-3-phosphoethanolamine (DOPE). The mPAR from the one-molecule bilaye- establishing ABA triblock copolymers was split by two, effectually allowing direct comparative analysis with the PB-PEO diblock copolymers and DOPE, both of which are acting as forming bilayers [24].

When it comes to the process of fabricating biomimetic membranes for a variety of applications, the original protein incorporation methodologies were from the period between 2009 and 2011 and were primarily based on the use of lipids [25, 26]. However, planar polymeric membranes have been effectively shown with the functional inclusion of gramicidin A [27]. Such research initiatives were first introduces by a Danish company called Aquaporin A/S. This company's innovative approaches to the process of biomimetic membrane fabrication will be examined in the later sections. This researched will be supplemented with an overview of the data generated by the research groups working at the National University of Singapore (NUS) and the Singapore Membrane Technology Center (SMTC) at Nanyang Technological University (NTU).

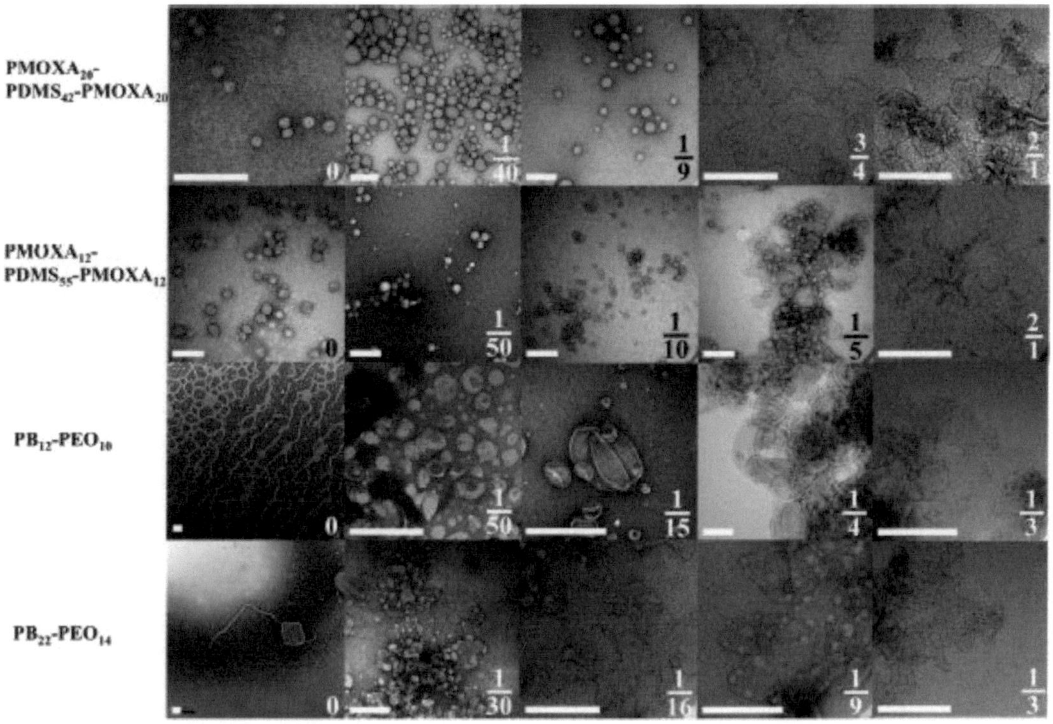

Figure 2. The TEM images of aggregate morphologies as a function of mPAR, where the PMOXA-PDMS-PMOXA copolymers self-assemble into vesicles, PB-PEO forms network- and sperm-like structures, however, only after incorporation of AQP0 vesicular structures are observed. Scale bar is 200 nm [23].

The overview in **Table 1** summarizes the critical research data on the experimental membrane protein and peptide, integration into block copolymer membranes. These data include information on a range of parameters, such as polymer chemistry and stochiometry, *PDI*, the incorporated membrane protein, the number-average molecular weight (M_n), *f*, the transport cargo (e.g., water for AQP) if there was functional incorporation, mPAR values, how polymer and membrane protein were mixed, and the shape of the polymer self-assembled structure. The date also show how the function incorporation values were measured and how M_n (which can be quantified using NMR) is related to M_w as $PDI = M_w / M_n$. This table does not include the incorporation studies that do not include block copolymer-protein interactions, such as cell-free expression systems [73–75], nanopores [76, 77], encapsulation in hydrophobic interior [6], hydrogel approaches [78, 79], and non-amphiphilic polymers [80]. Due to this restriction on the data, the table showcases the results that were made available by Wolfgang Meier and coworkers implementing PMOXA-PDMS-PMOXA triblock copolymers.

An overview in **Figure 3** outlines data on membrane protein integration into polymers in cases where M_n and *f* are known. Every black dot signifies a single polymer. The connected box summarizes the data on incorporated membrane protein family, polymer chemistry,

Polymer	M_n	PDI	f	Membrane protein	Transport Cargo	FI	mPAR	S	Incorporation method	Main functional incorporation measurement	References
$PMOXA_{13}$-$PDMS_{23}$-$PMOXA_{13}$	3.9				e^-	X	1:3300	V	MAq, biobeads, and SEC	Cargo → Reduction of MP → EPR signal	[28]
$PMOXA_{13}$-$PDMS_{23}$-$PMOXA_{13}$	4.7	NA	0.44	Alamethicin		X	1:590	P	MAq	Current change	[29]
$PMOXA_{13}$-$PDMS_{23}$-$PMOXA_{13}$	4.7	NA	0.44	Hemolysin		X	1:110,000,000	P	MAq	Current change	[29]
$PMOXA_{13}$-$PDMS_{23}$-$PMOXA_{13}$	4.7	NA	0.44	OmpG		X	1:33,000,000	P	MAq	Current change	[29]
$PMOXA_{20}$-$PDMS_{41}$-$PMOXA_{20}$	6.4	1.61	0.49	NtAQP1	CO_2	X	1:360	P	MOr	Cargo → Reaction inside vesicle → pH change	[30]
$PMOXA_{20}$-$PDMS_{41}$-$PMOXA_{20}$	6.4	1.61	0.49	NtPIP2:1	CO_2	X	1:360	P	MOr	Cargo → Reaction inside vesicle → pH change	[30]
$PMOXA_{20}$-$PDMS_{41}$-$PMOXA_{20}$	6.5	<1.2	0.51	AQP0	H_2O	ND	10:1–1:1	P	MAq and dialysis		[8]
$PMOXA_{20}$-$PDMS_{41}$-$PMOXA_{20}$	6.5	<1.2	0.51	AQP0	H_2O	ND	10:1–1:50	V	MAq and dialysis		[8]
$PMOXA_{20}$-$PDMS_{41}$-$PMOXA_{20}$	6.5	<1.2	0.51	AQP0	H_2O	—	1:2.5–0	V	MAq and dialysis	Vesicle size change	[8]
$PMOXA_{12}$-$PDMS_{54}$-$PMOXA_{12}$	6.0	1.01	0.2	AqpZ	H_2O	X	1:100–1:1600	V	MAq and biobeads	Vesicle size change	[31]

http://dx.doi.org/10.5772/intechopen.71723

Polymer	M_n	PDI	*f*	Membrane protein	Transport Cargo	FI	mPAR	S	Incorporation method	Main functional incorporation measurement	References
$PMOXA_{19}$-$PDMS_{74}$-$PMOXA_{19}$	8.7	1.46	0.23								
$PMOXA_{12}$-$PDMS_{54}$-$PMOXA_{12}$	6.0	1.01	0.3	AqpZ	H_2O	X	1:50–1:400	V	MAq and biobeads	Vesicle size change	[32, 33]
$PMOXA_{12}$-$PDMS_{54}$-$PMOXA_{12}$	6.0	1.01	0.3	Hemolysin		—	1:83,000,000	P	MAq	Current change	[29]
$PMOXA_{20}$-$PDMS_{54}$-$PMOXA_{20}$	7.4	NA	0.42	TsX	Nucleosides	X	1:450	V	MOr, SI, and SEC	Cargo → Encapsulated enzyme activity → Color change	[34]
$PMOXA_{8}$-$PDMS_{55}$-$PMOXA_{8}$	5.4	NA	0.22	AqpZ	H_2O	X	1:3500	V	PFR, biobeads, and SEC	Vesicle size change	[35]
$PMOXA_{12}$-$PDMS_{55}$-$PMOXA_{12}$	6.1	1.64	0.30	OmpF	ELF97	X	1:1200	V	MAq and SEC Cargo	Precipitation inside vesicle → Color change	[36]
$PMOXA_{12}$-$PDMS_{55}$-$PMOXA_{12}$	6.1	1.64	0.30	OmpF	Acridine orange	X	1:9,100,000	V	PPFR and SEC	Cargo release → Color change	[37]
$PMOXA_{12}$-$PDMS_{55}$-$PMOXA_{12}$	6.1	1.64	0.30	OmpF	Paraquat. Pyrocyanin	X	1:640	V	MAq and dialysis	No cargo → No detoxication of encapsulated enzyme → Cell death	[38, 39]
$PMOXA_{12}$-$PDMS_{55}$-$PMOXA_{12}$	6.1	1.64	0.30	AQP0	H_2O	ND	10:1–1:25	P	MAq and dialysis		[8]
$PMOXA_{12}$-$PDMS_{55}$-$PMOXA_{12}$	6.1	1.64	0.30	AQP0	H_2O	—	1:3–0	V	MAq and dialysis	Vesicle size change	[8]

Polymer	M_n	PDI	*f*	Membrane protein	Transport Cargo	FI	mPAR	S	Incorporation method	Main functional incorporation measurement	References
$PMOXA_{12}$-$PDMS_{55}$-$PMOXA_{12}$	6.1	1.64	0.30	AqpZ	H_2O	ND	1:4	Cr,V	MAq and biobeads		[19]
$PMOXA_{7}$-$PDMS_{60}$-$PMOXA_{7}$	5.6	NA	0.19	Gramicidin A	Monovalent cations	X	1:81,000	P	MOr	Current change	[27]
$PMOXA_{8}$-$PDMS_{60}$-$PMOXA_{8}$	5.8	NA	0.21	ApqZ	H_2O	X	1:3800	V	PFR, biobeads, and SEC	Vesicle size change	[35]
$PMOXA_{13}$-$PDMS_{62}$-$PMOXA_{13}$	6.8	1.47	0.29	NADH reductase	e^-	X	1:1900	V	MAq, biobeads, and SEC	Cargo → Reduction of MP → EPR signal	[28]
$PMOXA_{15}$-$PDMS_{62}$-$PMOXA_{15}$	7.1	1.50	0.32	NADH reductase	e^-	X	1:1800	V	MAq, biobeads, and SEC	Cargo → Reduction of MP → EPR signal	[28]
$PMOXA_{12}$-$PDMS_{65}$-$PMOXA_{12}$	6.9	1.67	0.27	MloK1	Potassium	X	1:390	P	MAq and biobeads	Current change	[40]
$PMOXA_{15}$-$PDMS_{68}$-$PMOXA_{15}$	7.6	NA	0.30	LamB	Maltohexaose	X	NA	P	MAq	Current change at varying cargo concentrations	[41]
$PMOXA_{15}$-$PDMS_{68}$-$PMOXA_{15}$	7.6	NA	0.30	OmpF	Actylthiocholine	X	1:10,000	V	PFR	Cargo → Encapsulated enzyme activity → Color change	[41]
$PMOXA_{15}$-$PDMS_{68}$-$PMOXA_{15}$	7.6	1.20	0.30	ApqZ	H_2O	X	1:10–1:1000	V	PFR and biobeads	Vesicle size change	[42]
$PMOXA_{15}$-$PDMS_{68}$-$PMOXA_{15}$	7.6	1.20	0.30	Hemolysin			1:66,000,000	V	MAq	Current change	[29]

http://dx.doi.org/10.5772/intechopen.71723

Polymer	M_n	PDI	*f*	Membrane protein	Transport Cargo	FI	mPAR	S	Incorporation method	Main functional incorporation measurement	References
$PMOXA_{21}$-$PDMS_{69}$-$PMOXA_{21}$	8.7	2.00	0.37	NADH reductase	e^-	X	1:1500	V	MAq, biobeads, and SEC	Cargo → Reduction of MP → EPR signal	[28]
$PMOXA_{16}$-$PDMS_{72}$-$PMOXA_{16}$	8.0	1.17	0.30	OmpF	Enone	X	1:220	V	PPFR and dialysis	Cargo → Encapsulated enzyme activity → Color change	[43]
PMOXA-PDMS-PMOXA	8.8	NA	NA	OmpF	ELF97	X	1:50	V	MAq and SEC	Cargo → Precipitation inside vesicle → Color change	[44]
$PMOXA_{32}$-$PDMS_{72}$-$PMOXA_{32}$	10.7	1.83	0.47	OmpF	7-ADCA. PGME	X	NA	V	PFR and dialysis	Cargo → Encapsulated enzyme activity → Bacterial death	[45]
$PMOXA_{11}$-$PDMS_{73}$-$PMOXA_{11}$	7.2	1.70	0.22	LamB	DNA	X	1:390	V	MOr, SI, and SEC	Fluorescence – labeled cargo	[46]
$PMOXA_{11}$-$PDMS_{73}$-$PMOXA_{11}$	7.2	1.70	0.22	OmpF	Nucleosides	X	1:10–1:100	V	PPFR and SEC	Cargo → Encapsulated enzyme activity → Color change	[47]
$PMOXA_{11}$-$PDMS_{73}$-$PMOXA_{11}$	7.2	1.70	0.22	TsX	Nucleosides	X	1:10–1:100	V	PPFR and SEC	Cargo → Encapsulated enzyme activity → Color change	[47]
$PMOXA_{11}$-$PDMS_{73}$-$PMOXA_{11}$	7.2	1.70	0.22	LamB	DNA	X	NA	P	MAq		[46]
Lipids											
$PMOXA_{21}$-$PDMS_{73}$-$PMOXA_{21}$	9.0	1.70	0.36	Alamethicin	Calcium	X	1:24	V	MAq	Cargo precipitation inside vesicle	[48, 49]

Polymer	M_n	PDI	*f*	Membrane protein	Transport Cargo	FI	mPAR	S	Incorporation method	Main functional incorporation measurement	References
$PMOXA_{21}$-$PDMS_{73}$-$PMOXA_{21}$	9.0	1.70	0.36	FhuA	Sulphorhodamine B	X	1:6,000,000	V	MOr, SI, and SEC	Cargo → Quenching inside vesicle → Color change	[50–52]
$PMOXA_{21}$-$PDMS_{73}$-$PMOXA_{21}$	9.0	1.70	0.36	FhuA	TMB	X	1:4500. 1:3,600,000	V	MAq/ and biobeads/MOr, SI, and SEC	Cargo → Encapsulated enzyme activity → Color change	[50, 51, 53]
$PMOXA_{21}$-$PDMS_{73}$-$PMOXA_{21}$	9.0	1.70	0.36	FhuA		ND	3000:1	P	MAq		[51]
$PMOXA_{21}$-$PDMS_{73}$-$PMOXA_{21}$	9.0	1.70	0.36	FhuA	NAD	—	NA	V	MAq	Cargo → Encapsulated enzyme activity → Absorbance change of cargo	[52]
$PMOXA_{21}$-$PDMS_{73}$-$PMOXA_{21}$	9.0	1.70	0.36	FhuA	DNA	—	NA	V	MOr, SI, and SEC	Fluorescence-labeled cargo	[52]
$PMOXA_{21}$-$PDMS_{73}$-$PMOXA_{21}$	9.0	1.70	0.36	LamB	Sugar	X	NA	P	MAq	Current change at varying cargo concentration	[54]
$PMOXA_{21}$-$PDMS_{73}$-$PMOXA_{21}$	9.0	1.70	0.36	OmpF	e^-	X	NA	P	MAq	Current change	[54]
$PMOXA_{21}$-$PDMS_{73}$-$PMOXA_{21}$	9.0	1.70	0.36	OmpF	Ampicillin	X	1:1000	V	MOr and SEC	Cargo → Hydrolysis inside vesicle → Color change	[12, 55]
$PMOXA_{20}$-$PDMS_{75}$-$PMOXA_{20}$	9.0	1.46	0.34	AqpZ	H_2O	X	1:25, 1:50, 1:200	V	PFR and biobeads	Vesicle size change	[56]
$PMOXA_{11}$-$PDMS_{76}$-$PMOXA_{11}$	7.8	1.48	0.25	BR	H^+	X	NA	V/Mc	MOr and SI	pH change	[57, 58]

http://dx.doi.org/10.5772/intechopen.71723

Polymer	M_n	PDI	*f*	Membrane protein	Transport Cargo	FI	mPAR	S	Incorporation method	Main functional incorporation measurement	References
$PMOXA_{11}$-$PDMS_{76}$-$PMOXA_{11}$	7.8	1.48	0.25	BR and ATPase	H^+	X	1:180	V	MOr and dialysis	pH change and bioluminescence assay	[15]
$PMOXA_{11}$-$PDMS_{76}$-$PMOXA_{11}$	7.8	1.48	0.25	BR and ATPase	H^+	X	1:20	V	PBR and dialysis	pH change	[59–61]
$PMOXA_{6}$-$PDMS_{90}$-$PMOXA_{6}$	9.5	NA	0.12	OmpF	L-ascorbic acid, CO, $Na_2S_2O_4$, $ONOO^-$	X	1:1300	V	PFR, dialysis, and SEC	Cargo → Absorbance change of encapsulated protein	[62]
$PMOXA_{21}$-$PDMS_{97}$-$PMOXA_{23}$	9.0	1.70	0.30	Hemaglutinin		X	1:3800	V	MAq and biobeads	MP → Fusion with fluorescence-labeled liposomes	[53]
$PMOXA_{9}$-$PDMS_{106}$-$PMOXA_{9}$	9.4	1.38	0.14	NADH reductase	e^-	X	1:1400	V	MAq, biobeads, and SEC	Cargo → Reduction of MP → EPR signal	[38]
$PMOXA_{13}$-$PDMS_{110}$-$PMOXA_{13}$	10.4	1.44	0.19	NADH reductase	e^-	X	1:1200	V	MAq, biobeads, and SEC	Cargo → Reduction of MP → EPR signal	[38]
$PMOXA_{14}$-$PDMS_{110}$-$PMOXA_{14}$	10.4	1.36	0.20	NADH reductase	e^-	X	1:1200	V	MAq, biobeads, and SEC	Cargo → Reduction of MP → EPR signal	[38]
$PMOXA_{15}$-$PDMS_{110}$-$PMOXA_{15}$	10.7	1.62	0.21	AqpZ	H_2O	X	1:25–1:500	V	PFR and SEC	Vesicle size change	[7, 19]
$PMOXA_{15}$-$PDMS_{110}$-$PMOXA_{15}$	10.7	1.62	0.21	OmpF		ND	NA	P	MAq		[53]
PMOXA-PDMS-PMOXA	20.0	NA		FhuA	Calcein	X	1:2,700,000	V	MOr, SI, and SEC	Cargo release → Color change	[54]

Polymer	M_n	PDI	*f*	Membrane protein	Transport Cargo	FI	mPAR	S	Incorporation method	Main functional incorporation measurement	References
$PMOXA_{65}$-$PDMS_{165}$-$PMOXA_{65}$		23.3	1.63	NADH reductase	e^-	X	1:550	V	MAq, biobeads, and SEC	Cargo release → Reduction of MP → EPR signal	[28]
PMOXA-PDMS-PMOXA	NA	NA	NA	BR	H^+	X	NA	P	MAq	pH change	[55, 56]
PMOXA-PDMS-PMOXA	NA	NA	NA	BR and CcO	H^+ and e^-	X	NA	V	MOr, SI, and SEC	Current and pH change	[56, 57]
PMOXA-PDMS-PMOXA	NA	NA	NA	CcO	e^-	X	NA	P	MOr, SI, and SEC	Current change	[55, 56]
PMOXA-PDMS-PMOXA	NA	NA	NA	OmpF	H^+	X	NA	P	MAq	Current change	[58]
$PMOXA_{110}$-$PDMS_{40}$-PEO_{25}	13.4	NA	0.75	AQP0	H_2O	ND	1:200	V	MOr, SI, and SEC		[15]
$PMOXA_{45}$-$PDMS_{40}$-$PMOXA_{67}$	10.6	NA	0.68	AQP0	H_2O	ND	1:200	V	MOr, SI, and SEC		[15]
MPEG-PVL	6.5	<1.2	0.00	Polymyxin B	Calcein	X	1:2	V	MAq	Cargo release → Color change	[63]
P2VP-PEO	NA	NA	NA	FhuA	NAD	–	NA	V	MOr, SI, and SEC	Cargo → Enzyme reaction inside vesicle → Absorbance change of cargo	[63]
PB_{12}-PEO_{10}	1.1	1.09	0.32	AQP0	H_2O	X	1:5–1:250	V	MAq and dialysis	Vesicle size change	[8]
PB_{12}-PEO_{10}	1.1	1.09	0.32	AQP0	H_2O	ND	1:1.3	Cr	MAq and dialysis		[8]
PB_{12}-PEO_{10}	1.1	1.09	0.32	AQP0	H_2O	ND	1:1–1:10	P	MAq and dialysis		[8]
PB_{12}-PEO_{10}	1.1	1.09	0.3	AqpZ	H_2O	X	1:50–1:1000	V	MAq and dialysis	Vesicle size change	[59]

http://dx.doi.org/10.5772/intechopen.71723

Polymer	M_n	PDI	f	Membrane protein	Transport Cargo	FI	mPAR	S	Incorporation method	Main functional incorporation measurement	References
PB_{12}-PEO_{10}	1.1	1.09	0.32	BR	H^+	X	1:500	V	MAq and biobeads	pH change	[60]
PB_{12}-PEO_{10}	1.1	1.09	0.3	SoPIP2;1	H_2O	—	1:200	V	MAq and biobeads	Vesicle size change	[59]
PB_{12}-PEO_{10}	1.1	NA	0.34	Hemolysin	Calcein	X	1:33,000	V	MAq and dialysis	Cargo release → Color change	[61]
PB_{22}-PEO_{14}	1.8	1.17	0.28	AQP0	H_2O	ND	2:1–1:300	P	MAq and dialysis		[8]
PB_{22}-PEO_{23}	2.2	1.09	0.39	AqpZ	H_2O	X	1:15–1:200	V	MAq and dialysis	Vesicle size change	[59]
PB_{22}-PEO_{23}	2.2	1.09	0.39	SoPIP2;1	H_2O	—	1:15–1:200	V	MAq and dialysis	Vesicle size change	[59]
PB_{29}-PEO_{16}	2.3	1.00	0.25	AQP10	H_2O	—	1:990	V	PFR and SE	Vesicle size change	—
PB_{35}-PEO_{14}	2.5	1.09	0.19	AqpZ	H_2O	—	1:15	V	MAq and dialysis	Vesicle size change	[59]
PB_{35}-PEO_{14}	2.5	1.09	0.19	SoPIP2;1	H_2O	—	1:15	V	MAq and dialysis	Vesicle size change	[59]
PB_{43}-PEO_{32}	3.7	1.03	0.31	AQP0	H_2O	X	1:600	V	PFR and SE	Vesicle size change	[62]
PB_{46}-PEO_{30}	3.8	1.04	0.28	AqpZ	H_2O	—	1:50, 1:100, 1:200	V	MAq and dialysis	Vesicle size change	[59]
PB_{46}-PEO_{32}	3.9	1.00	0.30	AQP0	H_2O	—	1:580	V	PFR and SE	Vesicle size change	—
PB52-PEO29	4.1	<1.1	0.25	Hemolysin	e^-	X	NA	P	MAq	Current change	[64]
PB52-PEO29	4.1	<1.1	0.25	Polymyxin B		X	NA	P	MAq	Current change	[65]
PB52-PEO29-LA	4.1	<1.1	0.25	Hemolysin	e^-	X	NA	P	MAq	Current change	[64]
PB52-PEO29-LA	4.1	<1.1	0.25	Polymyxin B		X	NA	P	MAq	Current change	[65]
PB92-PEO78	8.4	1.08	0.34	AQP0	H_2O	—	1:270	V	PFR and SE	Vesicle size change	—
PB125-PEO80	8.9	<1.1	0.28	Alamethicin	Calcein	—	1:2–1:8	V	MAq	Cargo release → Color change	[66]

Polymer	M_n	PDI	*f*	Membrane protein	Transport Cargo	FI	mPAR	S	Incorporation method	Main functional incorporation measurement	References
PHEMA25-PBMA25-PHEMA25	14.3	1.30	0.83	AqpZ		—	NA	P	MAq and biobeads	Current change	[67]
PHEMA25-PBMA25-PHEMA25	14.3	1.30	0.83	Hemolysin		X	NA	P	MAq	Current change	[67]
PHEMA25-PBMA25-PHEMA25	14.3	1.30	0.83	OmpF		—	1:70	P	MAq and biobeads	Current change	[67]
PEE37-PEO40	3.9	<1.1	0.39	Alamethicin	Calcein	X	1:2–1:8	V	MAq	Cargo release → Color change	[66]
PPO34-PGM14	6.5	1.30	0.66	Strepatividin-BSA		ND	1:5, 1:15, 1:50	V	PPFR		[68]
PI93-PEO87	10.2	1.00	0.31	FhuA	TMB	X	1:6700, 1:5,300,000	V	MOr, SI, and SEC	Cargo → Encapsulated enzyme activity → Color change	[63]
PEO136-PIB18-PEO136	8.0	1.86	0.90	Cecropin A	Calcein	X	1:30	V	MAq and SEC	Cargo release → Color change	[69]
P4MVP21-PS26-P4MVP21	13.1	NA	0.80	PR		X	1:10	V	MAq and precipitation	Absorbance change in membrane protein	[70]
P4MVP21-PS38-P4MVP21	14.3	1.19	0.74	PR		X	1:10	V	MAq and precipitation	Absorbance change in membrane protein	[70]
P4MVP29-PS42-P4MVP29	18.7	NA	0.78	PR		X	1:10	V	MAq and precipitation	Absorbance change in membrane protein	[70]
P4MVP22-PB28-P4MVP22	15.0	NA	0.92	PR		ND	1:10	V	MAq and precipitation		[71]

http://dx.doi.org/10.5772/intechopen.71723

Polymer	M_n	PDI	*f*	Membrane protein	Transport Cargo	FI	mPAR	S	Incorporation method	Main functional incorporation measurement	References
P4MVP22-PB28-P4MVP22	15.0	NA	0.92	RC	e^-	X	1:25	V	MAq and precipitation	Cargo → Reduction of MP → EPR signal	[72]
P4VP22-PB28-P4VP22	7.1	NA	0.82	PR		ND	1:10	V	MAq and precipitation		[71]
P4MVP29-PB56-P4MVP29	17.4	1.08	0.81	RC	e^-	X	1:25	V	MAq and precipitation	Cargo → Reduction of MP → EPR signal	[72]
P4MVP18-PB93-P4MVP18	13.9	1.06	0.62	PR		ND	1:10	V	MAq and precipitation		[71]

Table 1. Overview of studies of membrane protein incorporation into amphiphilic block copolymers. Most studies are done with the porn OmpF, followed by AqpZ.

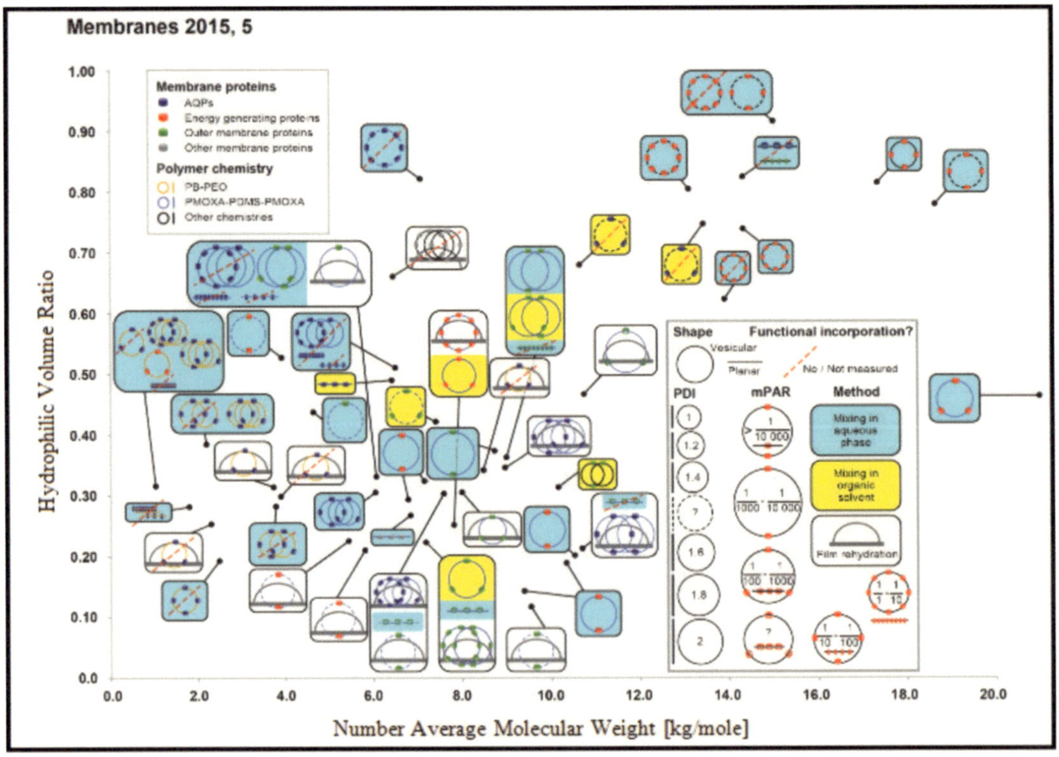

Figure 3. Summary of key parameters for membrane protein inclusion into amphiphilic block copolymers [81].

self-assembled morphology either planar or vesicular, *PDI* value of the polymer (rather than of the polymersomes), mPAR value, incorporation methodology, and if the incorporation process was functional, dysfunctional, or otherwise not measured. In instances where there are multiple sketches available in the box, then a variety of experimental runs have been conducted on the specific polymer example. Two crossing circles and two close lines, respectively, indicate that two different mPARs were researched under the same maintained circumstances. A dynamic with three crossing circles designates that three mPARs or more were examined. Whenever a parameter other than mPAR is investigated, such as incorporation method, incorporated membrane protein, or polymer chemistry, there is a new sketch available. In most cases, polymers that can engage in the process of functional incorporation require an f value somewhere between 0.2 and 0.35, and M_n value in the range of 2 and 12 kg/mol. In comparison to PB-PEO, PMOXA-PDMS-PMOXA has a significantly broader *PDI* [81] range, since its bilayer is water impermeable [7] and generally does not collapse in its dried form [82]. On the other hand, PB-PEO is noticeably more lipid-like because of its capacity to collapse easier and its greater water permeability potential [8]. Research suggests that these polymers that could not attain functional AQP incorporation are mostly PB-PEO polymers featuring small M_n and *PDI* values. The energy generating proteins (BR, CcO, NADH reductase, ATPase, RC, PR) and outer membrane proteins (OmpF, OmpG, FhuA, TsX) were integrated primarily into

http://dx.doi.org/10.5772/intechopen.71723

PMOXA-PDMS-PMOXA polymers. However, the outer membrane proteins have likewise been integrated into somewhat more exotic chemistries occurring in an *f* range where there is no expectation to locate vesicular structures. The vast majority of the experimental trials on functional incorporation were conducted with vesicular structures and with the mixing processes occurring in aqueous phase. In fact, most experimental cases at smaller *PDI* parameters showed that no functional membrane proteins could be integrated. This particular dynamic is in agreement with the research data released by Pata et al. [10]. Various types of mPARs have been actively implemented, only to arrive at the fact that no optimal ratios can be identified. Nevertheless, mPARs are grounded on the nominal or initial concentration values of polymers and membrane proteins, where the final mPAR data after the incorporation is completed may vary [83]. The section to follow will examine a number of approaches to quantifying membrane proteins, with particular focus on AQPs, and after the overview of the incorporation process.

2. Evaluation of AQP incorporation characterization methodologies

The process of identifying examples of functional AQPs incorporation may strike as potentially quite challenging, since the permeating solute is composed of neutral water molecules. The protein-mediated type of transport when it comes to neutral molecules, and specifically at the single protein levels, is consistently more difficult to assess than the transport parameters of charged molecules, such as ions or protons or in cases of specific chemical reactions, including the ATPase enzyme activity. While the deuterated water labeling was suggested as a measurement method using the Raman spectroscopy [84], researchers are concerned that these approaches to measurement can be further complicated because the water transport rate value in the AQP channel varies for deuterated water molecules when compared to the normalized water molecule rate [85]. SFLS is a common methodology used for calculating the functional integration. SFLS method relies on a dynamic where the proteopolymersomes are vigorously combined with an osmotically active agent (NaCl or sucrose) within a specifically defined amount of volume. If a hyperosmotic shock occurs, the proteopolymersomes will become smaller in size and this in turn will facilitate light scattering to increase. Once the content of incorporated AQPs is augmented, the overall shrinking rate will likewise begin to increase. Nonetheless, the SFLS approach is substantially influenced by the quality, or size distribution potential, of the polymersomes, the concentration of the osmolytes, and the AQP concentration within the polymersomes [35]. In theory, a visually based measurement can be accomplished using the freeze-fracture transmission electron microscopy (FF-TEM), however, the FF-TEM will not be able to provide sufficient data on the functional aspects. During the FF-TEM assessment, the proteopolymersomes are caught in their original shape with the aid of the quick-freezing process. After the proteopolymersomes are collected, the frozen sample is then fractured in such as manner that the fracture plane is located alongside the proteopolymersome bilayer, since this section is the most vulnerable point in the entire system. The experimental samples with integrated AQPs, or the cavities in which the AQPs were inserted within the bilayer, are subsequently exposed to the carbon/metal coating process. The replica formed during this procedure is then detached from the thawed out sample. As a result, the AQPs and cavities can be viewed and examined on the formed replica in the shape of separate spots on the proteopolymersomes content.

Alternative methodology available is the fluorescence correlation spectroscopy (FCS) process of the fluorescently labeled AQP. The FCS is based on the time-dependent fluctuations of fluorescence intensities within a defined microscopic space, otherwise known as the confocal volume, which are carefully observed and then exposed to autocorrelation function process. The specific number of particles within the confocal volume at given time interval can be calculated, however it depends on the diffusion times of particles spreading through the confocal volume. After the proper proteopolymersomes or proteoliposomes monitoring process, they are solubilized to micelles and monitored once more so as to calculate the proteins-per-vesicle-ratio, or the primary number of membrane proteins integrated into the bilayer of a single vesicle. Within this experimental scenario, it is presupposed that the micelles include only one AQP, and as a result the micelle-per-vesicle ratio is equivalent to the proteins-per-vesicle-ratio value. Additional information on the methodology is provided in Ref. [83]. On the other hand, it is possible to calculate the proteins-per-vesicle-ratio using a correlation between the proteopolymersome solution data and the AQP stock solution parameters.

In both of the outlined methodologies, the overall correlations of data have certain benefits as well as challenges, and these are outlined in greater detail in the FCS subsection. Small-angle X-ray scattering (SAXS) capacity to characterize the biological materials makes it an adaptive toolkit when it comes to particle structure. For instance, it can supply structural data about particles in a solution on a long-scale from 1 to 100 nm, and where the collected data is shown using scattering intensity values as a function of the magnitude parameter of the scattering vector q. The identified quantity is not dependent on experimental set-up's particular geometry and is linked directly to the scattering angle 2θ as $q = 4\pi \sin(\theta)/\lambda$, where λ is the wavelength value of the X-ray beam. The two scattering points that are separated by a distance d within a particle lead to an interference change that is signified by the scattering curve's increased intensity at $q = 2\pi/d$. The change in values implies that the larger sized features are probed at low q values and the smaller sized details are probed in the high-q region of the observed curves. Both the contrast and the strength with which a particle scatters is directly proportional to the particle's excess electron density, that is, the differentiation between the solvent used and the sample's electron density values. One of the SAXS issues is that this method demands access to extensive synchrotron radiation sources.

The upcoming section is an overview of SFLS, FF-TEM, FCS, and SAXS analyses featuring a variety of diblock copolymers containing optimal M_n and f range values for functional membrane protein integration processes, including PB29-PEO16, PB45-PEO14, PB33-PEO18, PB46-PEO32, PB92-PEO78, and PB43-PEO32. The PB-PEO was selected due to it functional AQP incorporation potential as reviewed earlier and the M_n and f range values that are simpler to manage in comparison to those of PMOXA-PDMS-PMOXA. For SFLS,

http://dx.doi.org/10.5772/intechopen.71723

SAXS, and FF-TEM, the AqpZ is applied as the integrated membrane protein, while the GFP-tagged human aquaglyceroporin AQP10 is used as part of the FCS experimental runs. The data relevant to these types of incorporation are listed in the supplemental material.

2.1. Stopped-flow light scattering

An illustration of the SFLS and its analysis, the information about PB45-PEO14 and PB33-PEO18 diblock copolymer proteo- and polymersomes, specifically with or without AqpZ, is outlined in **Figure 4**. In the case of PB33-PEO18, the rate constant value related to the augmentation of the light scattering intensity was somewhat greater when using AqpZ, while for PB45-PEO14, the rate it was even smaller. Such a dynamic demonstrates one of the major issues that exist with respect to the SFLS application. The distinct lack of substantial response to the alterations in extra vesicular osmolarity can be caused by the growth in the bilayer bending modulus caused by the existence of ApqZ, either blocked or non-functional. Analogous concerns have been noted in previous experimental runs using AqpZ as well as SoPIP2, and where only polymers of the smallest size (PB12-PEO10 and PB22-PEO23) showed a considerable change in SFLS between proteo- and polymersomes (the results are not shown). Additional explanation for the analogous SFLS signal could be found in the blockage of AqpZ channels with PEO chains. In this type of blockage, the AqpZ are situated directly in the bilayer and act as an impermeable hydrophobic blockage. Research conducted by Kumar et al. [8] indicates that this blockage dynamic is caused by the water permeation that is actively blocked by the sections corresponding to the integrated AqpZ, and where the proteopolymersomes' lower permeability values can be expected, if compared to the values

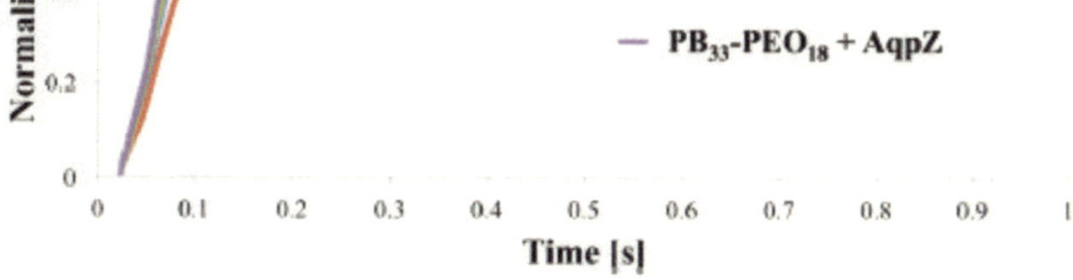

Figure 4. Normalized light scattering versus time for proteo- and polymersomes of PB45-PEO14 and PB33-PEO18, at an mPAR of 1:100. For PB45-PEO14 the apparent water permeability value is slightly lower for the proteopolymersomes *versus* polymersomes, while for PB33-PEO18 it is slightly increased [8].

occurring when using polymersomes. Alternatively, the incorporated AqpZ may continue being fully functional, while the polymer matrix remains resistant to various changes in volume parameters. These phenomena effectually undermine the idea that SFLS is not a stand-alone type of technique.

2.2. Freeze-fracture transmission electron microscopy

The summary of research data on FF-TEM for PB45-PEO14 proteopolymersomes is shown in **Figure 5**. Specifically, proteopolymersomes featuring an mPAR of 1:100 were created with the help of film rehydration (FR) approach, where they are frozen and then fractured in a Leica MED20 station. In the next step, the two planchets of the frozen sample are carefully separated, causing an intentional fracture that simulates something like a "crack" rather than a "cut" shape, which in turn lowers the smearing effects from the conventional FF procedures (for additional details refer to relevant supporting information). It is likely that because of the collapsed PB chains, all proteo- and polymersomes featured a distinguishable surface similar to a raspberry. In fact, the "typically" present spots that studies on proteoliposomes are usually associating with AQP [86] were not found. **Figure 5** outlines the bubble-like spots and their equal distribution among the polymersomes (**Figure 5a–d**) and proteopolymersomes (b, c, e, f). These bubble resembling spots may be PB chain accumulations (**Figure 5a–c**), or alternatively, phenomena produced by inferior fracturing (d–f) quality. Similar behavior was seen in proteo- and polymersomes in other PB-PEO polymers at different M_n and f values. The results collected during these experiments seem to indicate that FF-TEM sample preparation

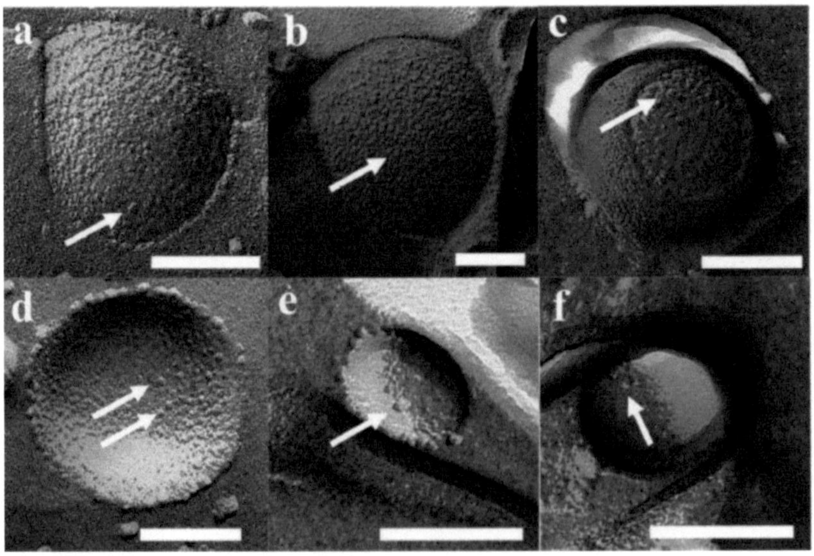

Figure 5. FF-TEM images of PB45-PEO14 proteosomes (b,c,e,f) and polymersomes (a,d). All vesicles showed spots that are potentially not from AqpZ, but instead from the collapsed PB chains (a–c) or failed fracturing phenomena (d–f). Scale bar is 100 nm [86].

process plays a significant role when it comes to false positive results. Dots were sporadically observed all over the samples, and since these spots were not AqpZ they could be polymer micelles. The observed spots could be removed whenever an augmented concentration step was omitted and the temperature was carefully controlled, after which the sample and cutting handling or metal coating parameters where managed (optimization protocol is provided in the supplementary information section). The existence of AqpZ could not be confirmed, even with polymers that have the shortest PB chains, such as PB45-PEO14 and PB32-PEO30. On the other hand, these experimental runs in themselves cannot reject the potential possibility that AqpZ tetramers could be present, since the hydrophilic PEO chains are comparatively large with respect to lipid head groups. As a consequence, some of the AqpZ may be hidden within the PB core structure.

2.3. Fluorescence correlation spectroscopy

FF-TEM and SFLS can result in several issues when applied as tools for potential evaluation of protein incorporation into polymersomes. As an alternative, FCS was examined as a possible method for collecting relevant quantitative data about the AQP incorporation process. The turn to FCS as a method was motivated by a recent Erbakan *et al.* research study published on the subject. In this paper, Erbakan *et al.* examine a range of AqpZ isoforms that are tagged with fluorophore in proteoliposomes, and where the protein-per-vesicle ratio was calculated and then validated with the help of SFLS [83]. One of the first steps to making this methodology more applicable is attempting to replicate Erbakan et al. and their proteoliposome experiments outlined in Ref. [83]. During an experimental run at an mPAR of 1:200, the measurements showed a proteins-per-vesicle-ratio of 5.35, a value comparable to the data found by Erbakan *et al.* at around 7.5. The variance in values could be created by the different AQP and the type of tagged fluorophore applied. Once the FCS instrument values for proteopolymersomes were optimized (further details available in the supporting information), FCS was implemented on the proteopolymersomes of PB45-PEO14 (mPAR 1:100), with AQP10- green fluorescent protein (GFP), and featuring OG-solubilized protein micelles. The data results from this application are outlined in **Figure 6**. In this case, a greater species number was achieved in the proteopolymersomes test sample than in the sample of protein micelle. This discrepancy in species number may be caused by to the same type of OG-induced accumulation. Greater autocorrelation signal value suggests a smaller number of particles in the confocal volume because of the slower diffusion time during the process. In this research attempt to simulate results obtained by Erbakan *et al.*, proteopolymersomes were also compared to the AQP10-GFP stock available. Researches Erbakan *et al.* were not able to do such a run, since the fluorophore applied (specifically, mBanana fluorescent protein) displayed a reduced fluorescence lifetime within pure OG environment (stock solution) when matched to the lipid/OG environment of solubilized protein micelles. On the other hand, GFP did not appear to change fluorescence lifetime values substantially even though AQP10-GFP was in OG (1.8 ns) and polymer/OG environmental parameters (1.97 ns, **Figure 6b**). These are similar to fluorophore implemented by Erbakan *et al.* (4 ns [83]) as part of the experiment and to the standardized GFP fluorescence lifetime values (3 ns). The variance in terms of the research project's GFP

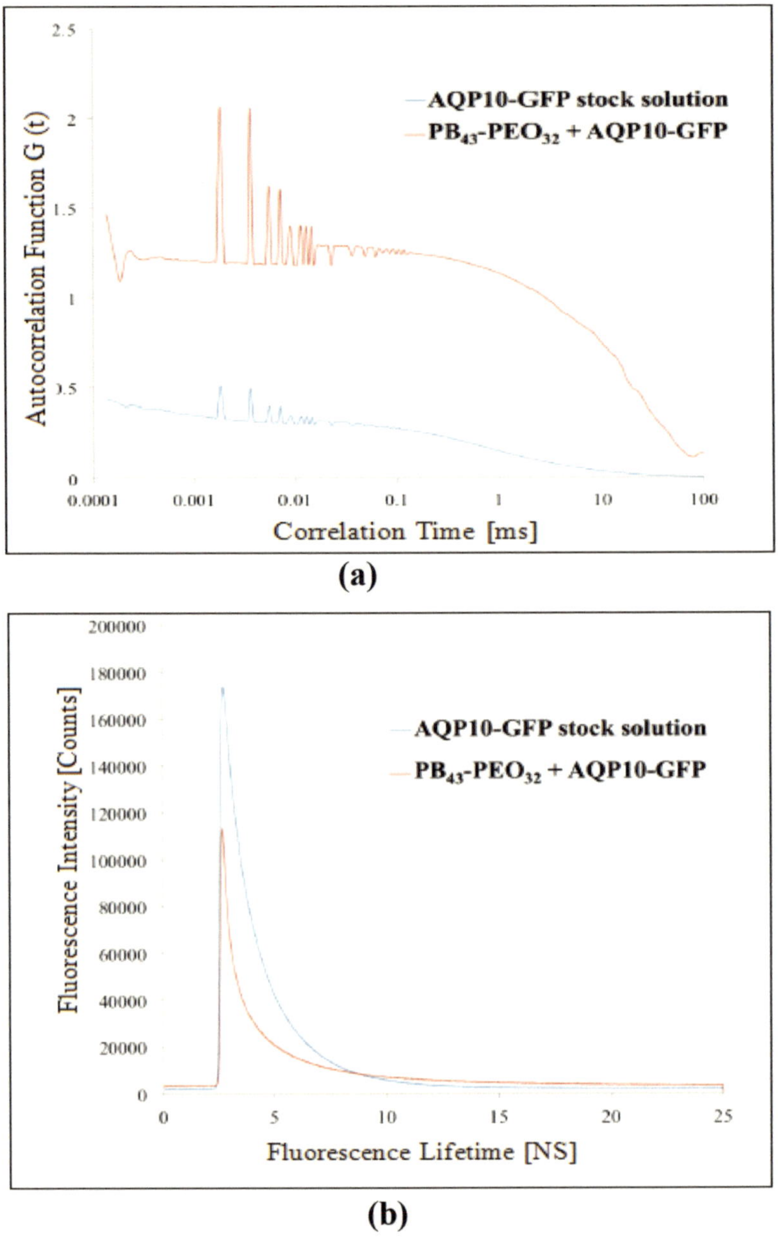

Figure 6. (a) Correlation diagram of proteopolymersomes and AQP10-GFP stock solution as a function of correlation time value τ against autocorrelation function $G(\tau)$. (b) Fluorescence lifetimes of the same batch of samples as a function of lifetime against intensity signal. In the instances where the intensities varied, the fluorescence lifetime was in a comparable range [83].

fluorescence lifetime values and the standardized data may be due to the shielding of the attached AQP10 and the OG environment, in addition to the possible influence of the instrument's fitting algorithm.

Research conducted suggests that the example correlation relies on the sole constituent of the system. For instance, when it comes to sensitive fluorophores, it is more important to comparatively analyze the AQP vesicles with the AQP micelles so as limit the potential impact on the fluorophore environment. With polymers, including the protein matrix, it is more advantageous to relate the AQP-fluorophore stock solution since the polymeric AQP micelles are capable of aggregating more easily. The difficulty caused by the correlation of AQP vesicles with AQP-fluorophore stock is that the resulting concentration value of AQP remains undetermined, and this seriously obscures the potential correlation with analogous AQP concentrations. Based on the species amount of pure AQP10-GFP in the confocal volume stock and the quantity from the proteopolymersome solution (**Figure 6a**), the proteins-per-vesicle-ratio was calculated as 2.87. The data obtained during these experimental runs show that FCS can in fact be applied as a dependable method for calculating AQPs in proteopolymersomes. This in turn invites a new opportunity for conducting a methodical research study where f and M_n are differentiated so as to locate relevant quantitative data about the wide range of polymers that can help obtain the highest proteins-per-vesicle-ratio values.

2.4. Small-angle X-ray scattering

Figure 7 showcases the scattering curves for FR prepared proteo- and polymersomes for PB33-PEO18 and PB45-PEO14. These examples went through a process of extruding and centrifuging before the actual measurements were taken. The typical linear slope was detected at low q values in the log-log plot, and featuring intensity that reflects the $q - 2$ power law. This type of response is characteristic in flat laminar structures. The extension of the slope far below the smallest detectable q-region indicates that there is a low curvature, or flat arrangement, even

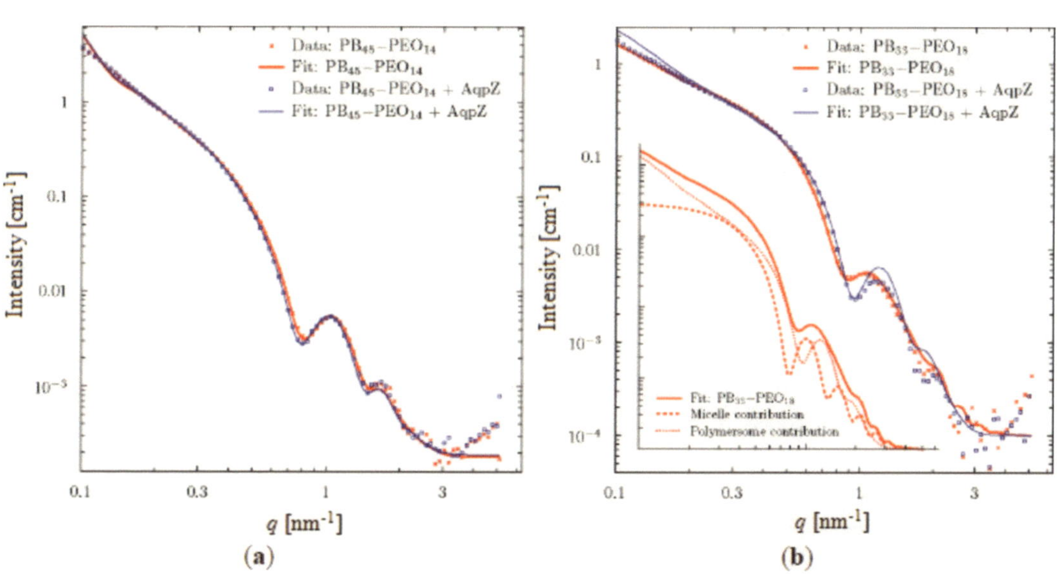

Figure 7. SAXS information for proteo- and polymersomes of (a) PB45-PEO14, and (b) PB33-PEO18 [83].

when it comes to the highest measurable length scale value of $q = 2\pi/0.1\ \text{nm} \approx 60\ \text{nm}^{-1}$. A typical oscillatory behavior occurs at higher q values, and could be caused by the problematic interference happening between positive contrast of PEO and negative contrast of PB. **Figure 7** shows that the fits were acquired with the help of a vesicle model involving three concentric spherical shells. In order to properly fit the PB33-PEO18 polymersomes, an additional contribution from block-copolymer micelles was necessary.

In cases focusing on the theoretical scattering in assorted straightforward geometrical objects, including spheres, ellipsoids, and cylinders of different contrast, the scattering values can be carefully calculated with relative ease. In fact, this data can be collected so as to form simplified models of the particles being studied. In this research model, the data were examined with the aid on a vesicle model based on three concentric spherical shells featuring interchanging contrast values and matched to shells of PEO, PB, and PEO. The thickness values varied in each individual shell so as to ensure the data fit and accuracy through the application of the least squares fitting routine. Exceptionally good fit correlations were found for the PB45-PEO14-system, further suggesting that the research data were in excellent agreement with the theoretical presupposition that diblock copolymers could form spherical vesicle structures. The correlated fits were shown to be especially sensitive to fluctuations in the factors that affect the central hydrophobic bilayer thickness founded using the PB-groups. These values were found to be in the ranges of 9.10 ± 0.1 and 8.94 ± 0.07 nm in the cases where AqpZ was either present or absent, respectively. With regards to the overall vesicle diameter value, the model suggests that it is greater than 60 nm, a result that aligns well with the initial analysis of the system. The collected data reflects that the fit parameters defining bilayer vesicles are created and that the inclusion of AqpZ incites very slight changes in the vesicles' structure.

In the case of the PB33-PEO18 proteopolymersomes, sufficiently suitable fits were obtained with the vesicle model based on a hydrophobic bilayer with the thickness values of 7.66 ± 0.05 nm. On the other hand, for experimental runs with polymersomes, there were no adequate fit options for the data that would ensure realistic physical parameters. In fact, for the data to fit the experimental approach needed to assume that the vesicles could coexist with a population of block copolymer micelles. The combined model fit parameters suggested that 76 wt% of the population was comprised of proteopolymersomes, and that 24 wt% were micelles with hydrophobic cores of diameter 11.7 ± 0.3 nm, thus showing a relatively good fit with the data overall. The **Figure 7** insert outlines the separate micelle and vesicle contributions.

To sum up, the SAXS inquiry exposes that in the case of for PB45-PEO14 the vesicles are created both with, as well as without, the AQP, where the AQP incorporation leads to a small variance in the average hydrophobic vesicle wall thickness value, and can imply a polymer puckering and dimpling near the integrated AQPs. For the PB33-PEO18 experimental runs, some micelle creation was seen, but this formation becomes lower once the AQP is successfully integrated. To sum up, this chapter examines the research that explored the characterization methods used for the functional integration of AQPs in PB-PEO diblock copolymers. The research results obtained suggest that both FF-TEM and

SFLS are in theory effective methods, however, when it comes to polymer systems the critical analysis of findings can provide ambiguous data that makes it problematic for applications. Alternatively, SAXS and FCS have been evaluated as capable of producing relevant information, with SAXS relying on access to large-scale facilities that can sustain synchrotron radiation resources.

Author details

Amira Abdelrasoul[1]*, Huu Doan[2] and Ali Lohi[2]

*Address all correspondence to: amira.abdelrasoul@usask.ca

1 Department of Chemical and Biological Engineering, University of Saskatchewan, Saskatoon, Saskatchewan, Canada

2 Department of Chemical Engineering, Ryerson University, Toronto, Ontario, Canada

References

[1] Kagawa Y, Racker E. Partial resolution of the enzymes catalyzing oxidative phosphorylation. The Journal of Biological Chemistry. 1971;**246**:5477-5487

[2] Nardin C, Thoeni S, Widmer J, Winterhalter M, Meier W. Nanoreactors based on (polymerized) ABA-triblock copolymer vesicles. Chemical Communications. 2000:1433-1434

[3] Zhang X, Tanner P, Graff A, Palivan CG, Meier W. Mimicking the cell membrane with block copolymer membranes. Journal of Polymer Science, Part A: Polymer Chemistry. 2012;**50**:2293-2318

[4] Tanner P, Baumann P, Enea R, Onaca O, Palivan C, Meier W. Polymeric vesicles: From drug carriers to nanoreactors and artificial organelles. Accounts of Chemical Research. 2011;**44**:1039-1049

[5] Choi HJ, Montemagno C. Artificial organelle: ATP synthesis from cellular mimetic polymersomes. Nano Letters. 2005;**5**:2538-2542

[6] Vriezema DM, Garcia PML, Sancho Oltra N, Hatzakis NS, Kuiper SM, Nolte RJM, Rowan AE, van Hest JCM. Positional assembly of enzymes in polymersome nanoreactors for cascade reactions. Angewandte Chemie, International Edition. 2007;**46**:7378-7382

[7] Kumar M, Grzelakowski M, Zilles J, Meier WP. Highly permeable polymeric membranes based on the incorporation of the functional water channel protein aquaporin Z. Proceedings of the National Academy of Sciences of the United States of America. 2007;**104**:20719-20724

[8] Kumar M, Habel J, Shen YX, Meier W, Walz T. High-density reconstitution of functional water channels into vesicular and planar block copolymer membranes. Journal of the American Chemical Society. 2012;**134**:18631-18637

[9] Andersen O. Bilayer thickness and membrane protein function: An energetic perspective. Annual Review of Biophysics and Biomolecular Structure. 2007;**36**:107-130

[10] Pata V, Dan N. The effect of chain length on protein solubilization in polymer-based vesicles (polymersomes). Biophysical Journal. 2003;**85**:2111-2118

[11] Srinivas G, Discher D, Klein M. Key roles for chain flexibility in block copolymer membranes that contain pores or make tubes. Nano Letters. 2005;**5**:2343-2349

[12] Discher DE, Ortiz V, Srinivas G, Klein ML, Kim Y, Christian D, Cai S, Photos P, Ahmed F. Emerging applications of polymersomes in delivery: From molecular dynamics to shrinkage of tumors. Progress in Polymer Science. 2007;**32**:838-857

[13] Aponte-Santamaría C, Briones R, Schenk A, Walz T, de Groot B. Molecular driving forces defining lipid positions around aquaporin-0. Proceedings of the National Academy of Sciences of the United States of America. 2012;**109**:9887-9892

[14] Hansen JS, Vararattanavech A, Plasencia I, Greisen PJ, Bomholt J, Torres J, Emnéus J, Hélix-Nielsen C. Interaction between sodium dodecyl sulfate and membrane reconstituted aquaporins: A comparative study of spinach SoPIP2;1 and *E. coli* AqpZ. Biochimica et Biophysica Acta (BBA) - Biomembranes. 2011;**1808**:2600-2607

[15] Stoenescu R, Graff A, Meier W. Asymmetric ABC-triblock copolymer membranes induce a directed insertion of membrane proteins. Macromolecular Bioscience. 2004;**4**:930-935

[16] Hite RK, Li Z, Walz T. Principles of membrane protein interactions with annular lipids deduced from aquaporin-0 2D crystals. The EMBO Journal. 2010;**29**:1652-1658

[17] Nehring R, Palivan CG, Casse O, Tanner P, Tüxen J, Meier W. Amphiphilic diblock copolymers for molecular recognition: Metal-nitrilotriacetic acid functionalized vesicles. Langmuir. 2009;**25**:1122-1130

[18] Kelly D, Abeyrathne P, Dukovski D. The affinity grid: A pre-fabricated EM grid for monolayer purification. Journal of Molecular Biology. 2008;**382**:423-433

[19] Kumar M. Biomimetic membranes as new materials for applications in environmental engineering and biology. Ph.D. Thesis. Champaign, IL, USA: University Illinois; 2010

[20] Kumar M, Meier W. Highly Permeable Polymeric Membranes. Patent WO 2009/078174; 18-06-2009

[21] Gonen T, Sliz P, Kistler J, Cheng Y, Walz T. Aquaporin-0 membrane junctions reveal the structure of a closed water pore. Nature. 2004;**429**:193-197

[22] Chandy G, Zampighi G, Kreman M, Hall J. Comparison of the water transporting properties of MIP and AQP1. The Journal of Membrane Biology. 1997;**159**:29-39

[23] Kumar M, Walz T. High Density Membrane Protein Membranes. Patent WO 2014/028923; 20-02-2014

[24] Habel J. Structural and functional characterization of aquaporin 0 incorporated in block copolymers and their resulting aggregate morphologies. Master Thesis. Basel, Switzerland: Universität Basel; 2011

[25] Hélix-Nielsen C. Biomimetic membranes for sensor and separation applications. Analytical and Bioanalytical Chemistry. 2009;**395**:697-718

[26] Pszon-Bartosz K, Hansen JS, Stibius KB, Groth JS, Emnéus J, Geschke O, Hélix-Nielsen C. Assessing the efficacy of vesicle fusion with planar membrane arrays using a mitochondrial porin as reporter. Biochemical and Biophysical Research Communications. 2011;**406**:96-100

[27] González-Pérez A, Stibius K, Vissing T. Biomimetic triblock copolymer membrane arrays: A stable template for functional membrane proteins. Langmuir. 2009;**25**:10447-10450

[28] Graff A. Amphiphilic copolymer membranes promote NADH: Ubiquinone oxidoreductase activity: Towards an electron-transfer nanodevice. Macromolecular Chemistry and Physics. 2010;**211**:229-238

[29] Wong D, Jeon TJ, Schmidt J. Single molecule measurements of channel proteins incorporated into biomimetic polymer membranes. Nanotechnology. 2006;**17**:3710-3717

[30] Uehlein N, Otto B, Eilingsfeld A, Itel F, Meier W, Kaldenhoff R. Gas-tight triblock-copolymer membranes are converted to CO2 permeable by insertion of plant aquaporins. Scientific Reports. 2012;**2**:1-4

[31] Wang H, Chung TS, Tong YW. Study on water transport through a mechanically robust aquaporin Z biomimetic membrane. Journal of Membrane Science. 2013;**445**:47-52

[32] Wang H, Chung TS, Tong YW, Jeyaseelan K, Armugam A, Chen Z, Hong M, Meier W. Highly permeable and selective pore-spanning biomimetic membrane embedded with aquaporin Z. Small. 2012;**8**:1185-1190

[33] Zhong PS, Chung TS, Jeyaseelan K, Armugam A. Aquaporin-embedded biomimetic membranes for nanofiltration. Journal of Membrane Science. 2012;**407-408**:27-33

[34] De Vocht C, Ranquin A, Willaert R, Van Ginderachter JA, Vanhaecke T, Rogiers V, Versées W, Van Gelder P, Steyaert J. Assessment of stability, toxicity and immunogenicity of new polymeric nanoreactors for use in enzyme replacement therapy of MNGIE. Journal of Controlled Release. 2009;**137**:246-254

[35] Grzelakowski M, Cherenet MF, Shen YX, Kumar M. A framework for accurate evaluation of the promise of aquaporin based biomimetic membranes. Journal of Membrane Science. 2015:1-32

[36] Grzelakowski M, Onaca O, Rigler P, Kumar M, Meier W. Immobilized protein-polymer nanoreactors. Small. 2009;**5**:2545-2548

[37] Ihle S, Onaca O, Rigler P, Hauer B, Rodríguez-Ropero F, Fioroni M, Schwaneberg U. Nanocompartments with a pH release system based on an engineered OmpF channel protein. Soft Matter. 2011;**7**:532-539

[38] Tanner P, Balasubramanian V, Palivan CG. Aiding nature's organelles: Artificial peroxisomes play their role. Nano Letters. 2013;**13**:2875-2883

[39] Tanner P, Onaca O, Balasubramanian V, Meier W, Palivan CG. Enzymatic cascade reactions inside polymeric nanocontainers: A means to combat oxidative stress. Chemistry - A European Journal. 2011;**17**:4552-4560

[40] Kowal JŁ, Kowal JK, Wu D, Stahlberg H, Palivan CG, Meier WP. Functional surface engineering by nucleotide-modulated potassium channel insertion into polymer membranes attached to solid supports. Biomaterials. 2014;**35**:7286-7294

[41] Winterhalter M, Hilty C, Bezrukov SM, Nardin C, Meier W, Fournier D. Controlling membrane permeability with bacterial porins: Application to encapsulated enzymes. Talanta. 2001;**55**:965-971

[42] Xie W, He F, Wang B, Chung TS, Jeyaseelan K, Armugam A, Tong YW. An aquaporin-based vesicle-embedded polymeric membrane for low energy water filtration. Journal of Materials Chemistry A. 2013;**1**:7592-7600

[43] Heinisch T, Langowska K, Tanner P, Reymond JL, Meier W, Palivan C, Ward TR. Fluorescence-based assay for the optimization of the activity of artificial transfer hydrogenase within a biocompatible compartment. ChemCatChem. 2013;**5**:720-723

[44] Broz P, Driamov S, Ziegler J, Ben-Haim N, Marsch S, Meier W, Hunziker P. Toward intelligent nanosize bioreactors: A pH-switchable, channel-equipped, functional polymer nanocontainer. Nano Letters. 2006;**6**:2349-2353

[45] Langowska K, Palivan CG, Meier W. Polymer nanoreactors shown to produce and release antibiotics locally. Chemical Communications. 2013;**49**:128-130

[46] Graff A, Sauer M, Meier W. Virus-assisted loading of polymer nanocontainer. Proceedings of the National Academy of Sciences of the United States of America. 2002;**99**:5064-5068

[47] Ranquin A, Versées W, Meier W, Steyaert J, Van Gelder P. Therapeutic nanoreactors: Combining chemistry and biology in a novel triblock copolymer drug delivery system. Nano Letters. 2005;**5**:2220-2224

[48] Lee H, Ho D, Kuo K, Montemagno CD. Vectorial insertion of bacteriorhodopsin for directed orientation assays in various polymeric biomembranes. Polymer. 2006;**47**: 2935-2941

[49] Choi HJ, Germain J. Effects of different reconstitution procedures on membrane protein activities in proteopolymersomes. Nanotechnology. 2006;**17**:1825-1830

[50] Choi HJ, Montemagno CD. Biosynthesis within a bubble architecture. Nanotechnology. 2006;**17**:2198-2202

[51] Choi HJ, Montemagno C. Light-driven hybrid bioreactor based on protein-incorporated polymer vesicles. IEEE Transactions on Nanotechnology. 2007;**6**:171-176

[52] Dobrunz D, Toma AC, Tanner P, Pfohl T, Palivan CG. Polymer Nanoreactors with dual functionality: Simultaneous detoxification of peroxynitrite and oxygen transport. Langmuir. 2012;**28**:15889-15899

[53] Thoma J, Belegrinou S, Rossbach P, Grzelakowski M, Kita-Tokarczyk K, Meier W. Membrane protein distribution in composite polymer-lipid thin films. Chemical Communications. 2012;**48**:8811-8813

[54] Onaca O, Sarkar P, Roccatano D, Friedrich T, Hauer B, Grzelakowski M, Güven A, Fioroni M, Schwaneberg U. Functionalized nanocompartments (synthosomes) with a reduction-triggered release system. Angewandte Chemie, International Edition. 2008;**47**:7029-7031

[55] Ho D, Chu B, Lee H, Montemagno C. Protein-driven energy transduction across polymeric biomembranes. Nanotechnology. 2004. DOI: 10.1088/0957-4484/15/8/038

[56] Ho D, Chu B, Lee H, Brooks EK, Kuo K, Montemagno CD. Fabrication of biomolecule–copolymer hybrid nanovesicles as energy conversion systems. Nanotechnology. 2005;**16**:3120-3132

[57] Xi JZ, Ho D, Chu B, Montemagno CD. Lessons learned from engineering biologically active hybrid nano/micro devices. Advanced Functional Materials. 2005;**15**:1233-1240

[58] Ho D, Chu B, Schmidt JJ, Brooks EK, Montemagno CD. Hybrid protein-polymer biomimetic membranes. IEEE Transactions on Nanotechnology. 2004;**3**:256-263

[59] Habel J. Functional and Chemical Characterization of Vesicular Diblock Copolymer Bilayers with Aquaporin Included. Technical Report, Aquaporin A/S, Copenhagen, Denmark; 2011

[60] Espina M. Barrier properties of biomimetic membranes. Master Thesis. Lyngby, Denmark: Danish Technical University (DTU) Kgs; 2012

[61] Nallani M, Andreasson-Ochsner M, Tan CWD, Sinner EK, Wisantoso Y, Geifman-Shochat S, Hunziker W. Proteopolymersomes: *In vitro* production of a membrane protein in polymersome membranes. Biointerfaces. 2011;**6**:153-157

[62] Bomholt J. Human Aquaporins—From *in vivo* detection to industrial scale production. Ph.D. Thesis. Copenhagen, Denmark: University Copenhagen; 2014

[63] Onaca O. Functionalized polymer vesicles and interactions with Polymyxin B and derivates. Ph.D. Thesis. Bremen, Germany: Universität Bremen; 2007

[64] Zhang X, Fu W, Palivan CG, Meier W. Natural channel protein inserts and functions in a completely artificial, solid-supported bilayer membrane. Scientific Reports. 2013;**3**. DOI: 10.1038/srep02196

[65] Dorn J, Belegrinou S, Kreiter M, Sinner EK, Meier W. Planar block copolymer membranes by vesicle spreading. Macromolecular Bioscience. 2011;**11**:514-525

[66] Vijayan K, Discher DE, Lal J, Janmey P, Goulian M. Interactions of membrane-active peptides with thick, neutral, nonzwitterionic bilayers. The Journal of Physical Chemistry. B. 2005;**109**:14356-14364

[67] Toughrai S. Functional surfaces through biomimetic block copolymer membranes. Ph.D. Thesis. Basel, Switzerland: Universität Basel; 2014

[68] Amado E, Schöps R, Brandt W, Kressler J. Spontaneous formation of giant bioactive protein-block copolymer vesicles in water. ACS Macro Letters. 2012;**1**:1016-1019

[69] Noor M, Dworeck T, Schenk A, Shinde P, Fioroni M, Schwaneberg U. Polymersome surface decoration by an EGFP fusion protein employing Cecropin a as peptide "anchor". Journal of Biotechnology. 2012;**157**:31-37

[70] Kuang L, Fernandes DA, O'Halloran M, Zheng W, Jiang Y, Ladizhansky V, Brown LS, Liang H. "Frozen" block copolymer nanomembranes with light-driven proton pumping performance. ACS Nano. 2014;**8**:537-545

[71] Hua D, Kuang L, Liang H. Self-directed reconstitution of proteorhodopsin with amphiphilic block copolymers induces the formation of hierarchically ordered proteopolymer membrane arrays. Journal of the American Chemical Society. 2011;**133**:2354-2357

[72] Kuang L, Olson TL, Lin S, Flores M, Jiang Y, Zheng W, Williams JC, Allen JP, Liang H. Interface for light-driven electron transfer by photosynthetic complexes across block copolymer membranes. Journal of Physical Chemistry Letters. 2014;**5**:787-791

[73] Andreasson-Ochsner M, Fu Z, May S, Ying Xiu L, Nallani M, Sinner EK. Selective deposition and self-assembly of triblock copolymers into matrix arrays for membrane protein production. Langmuir. 2012;**28**:2044-2048

[74] Gulati S, Jamshad M, Knowles TJ, Morrison KA, Downing R, Cant N, Collins R, Koenderink JB, Ford RC, Overduin M, et al. Detergent-free purification of ABC (ATPbinding-cassette) transporters. The Biochemical Journal. 2014;**461**:269-278

[75] Knowles TJ, Finka R, Smith C, Lin YP, Dafforn T, Overduin M. Membrane proteins solubilized intact in lipid containing nanoparticles bounded by styrene maleic acid copolymer. Journal of the American Chemical Society. 2009;**131**:7484-7485

[76] Hall AR, Scott A, Rotem D, Mehta KK, Bayley H, Dekker C. Hybrid pore formation by directed insertion of α-haemolysin into solid-state nanopores. Nature Nanotechnology. 2010;**5**:874-877

[77] Balme S, Janot JM, Berardo L, Henn F, Bonhenry D, Kraszewski S, Picaud F, Ramseyer C. New bioinspired membrane made of a biological ion channel confined into the cylindrical nanopore of a solid-state polymer. Nano Letters. 2011;**11**:712-716

[78] Ibragimova S. Supporting and stabilizing biomimetic membranes. Ph.D. Thesis. Lyngby, Denmark: Danish Technical University, Kgs; 2011

[79] Mech-Dorosz A, Heiskanen A, Bäckström S, Perry M, Muhammad HB, Hélix-Nielsen C, Emnéus J. A reusable device for electrochemical applications of hydrogel supported black lipid membranes. Biomedical Microdevices. 2015;**17**:21

[80] Bodor S, Zook JM, Lindner E, Tóth K, Gyurcsányi RE. Electrochemical methods for the determination of the diffusion coefficient of ionophores and ionophore-ion complexes in plasticized PVC membranes. Analyst. 2008;**133**:635-642

[81] Discher DE. Polymer vesicles. Science. 2002;**297**:967-973

[82] Nardin C, Hirt T, Leukel J, Meier W. Polymerized ABA triblock copolymer vesicles. Langmuir. 2000;**16**:1035-1041

[83] Erbakan M, Shen YX, Grzelakowski M, Butler PJ, Kumar M, Curtis WR. Molecular cloning, overexpression and characterization of a novel water channel protein from rhodobacter sphaeroides. PLoS One. 2014;**9**:e86830

[84] Ibata K, Takimoto S, Morisaku T, Miyawaki A, Yasui M. Analysis of aquaporin-mediated diffusional water permeability by coherent anti-stokes raman scattering microscopy. Biophysical Journal. 2011;**101**:2277-2283

[85] Mamonov AB, Coalson RD, Zeidel ML, Mathai JC. Water and deuterium oxide permeability through aquaporin 1: MD predictions and experimental verification. The Journal of General Physiology. 2007;**130**:111-116

[86] Itel F, Al-Samir S, Oberg F, Chami M, Kumar M, Supuran CT, Deen PMT, Meier W, Hedfalk K, Gros G, Endeward V. CO2 permeability of cell membranes is regulated by membrane cholesterol and protein gas channels. The FASEB Journal. 2012;**26**:5182-5191

Chapter 7

Recent Development in Aquaporin (AQP) Membrane Design

Amira Abdelrasoul, Huu Doan and Ali Lohi

Additional information is available at the end of the chapter

http://dx.doi.org/10.5772/intechopen.71724

Abstract

Development of Aquaporin Z (AqpZ) proteopolymersome has been substantial enough that it can obtain water separation membranes that feature fluxes of 11,000 L m^{-2} h^{-1}, a parameter value that is multiple orders of magnitude greater than the conventional industrial membranes available and possible only if the performance of AqpZ proteopolymersome can be properly scaled up. In fact, densely packed 2D AqpZ crystal arrays can in theory reach flux capacity of up to 16,000 L m^{-2} h^{-1}. On the other hand, these flux values may likely not be reached in practice, since various upscaling issues would prevent them from occurring. Nonetheless, AqpZ offers immense potentials benefits when it comes to biomimetic membranes. The research in membrane development is continuously ongoing, for example, only a few years ago in 2011, aquaporin-based biomimetic polymeric membranes (ABPMs) were viewed as the radically advanced membrane solution and, at the same time, removed from practical applications and commercial production. After 4 years of innovative thinking, ABPM membranes are produced for commercial consumption and with area dimensions of tens of m^2. Although it will take some time before this membrane technology becomes universally accessible, it has already gone beyond the confines of research theory and into practical application. The following chapter will explicitly outline the development of AQP biomimetic membrane technology.

Keywords: Aquaporin Z, membrane design, biomimetic structures, interfacial polymerization

1. Introduction

Development of Aquaporin Z (AqpZ) proteopolymersome has been substantial enough that it can obtain water separation membranes that feature fluxes of 11,000 L m^{-2} h^{-1}, a parameter value that is multiple orders of magnitude greater than the conventional industrial membranes available and possible only if the performance of AqpZ proteopolymersome can be properly scaled up. In fact, densely packed 2D AqpZ crystal arrays can in theory reach flux capacity of up to 16,000 L m^{-2} h^{-1} [1]. On the other hand, these flux values may likely not be reached in practice, since various upscaling issues would prevent them from occurring. Nonetheless, AqpZ offers immense potentials benefits when it comes to biomimetic membranes. The research in membrane development is continuously ongoing, for example, only a few years ago in 2011, aquaporin-based biomimetic polymeric membranes (ABPMs) were viewed as the radically advanced membrane solution and at the same time far removed from practical applications and commercial production [2]. After 4 years of innovative thinking, ABPM membranes are produced for commercial consumption and with area dimensions of tens of m^2 [3]. Although it will take some time before this membrane technology becomes universally accessible, it has already gone beyond the confines of research theory and into practical application. The following sections will explicitly outline the development of AQP biomimetic membrane technology and the research that helped to produce.

2. Planar biomimetic structures and membrane designs

The early industrial application approaches to membrane design were released by two Danish companies. They are the AquaZ (now Applied Biomimetic) and Aquaporin A/S. With the collaboration of the Danish Technical University (DTU), the University of Southern Denmark (SDU), DHI, Lund University in Sweden, Malaga University in Spain, Vilnius University in Lithuania, Ben-Gurion University of the Negev in Israel, and Veolia Water in France, the Aquaporin A/S joint research group became part of the EU-funded MEMBAQ project from 2006 to 2010, where they explored ways of using AQPs in industrial applications [4]. During the same period, AquaZ began to research on membrane development framed by the patent from Carlo Montemagno. In this patent, Montemagno conceptually outlined how AQPs that are embedded in lipid or polymeric bilayers can in theory function as a type of biomimetic membrane, despite offering a distinct design for the membrane [5].

The original Aquaporin A/S membrane design was constructed around the idea of an ethylene tetrafluoroethylene (ETFE) scaffold with 300 μm holes, created using laser-ablation, and directly inspired by the practice of painting or folding lipid chambers begun in the 1970s [6, 7]. In its earlier form, a freestanding lipid-bilayer film was created through the process of "painting" a two-phase solution over the area of the hole, specifically in places where the lipids transform from the organic solvent phase to the aqueous phase and then collect around the openings to form the bridging layers. A number of membrane peptides and proteins were integrated into the freestanding layers, such as porins [8]. Furthermore, freestanding polymethyloxazoline-polydimethylsiloxane-polymethyloxazoline (PMOXA-PDMS-PMOXA) polymer membranes with integrated gramicidin A channels were created [9] and then comprehensively categorized [10]. In the designs that followed, the membrane is maintained by PEO-dimethacrylate (PEO-DMA) type of hydrogels [11] or stabilized with the aid of surface plasma polymerization process [12]. In addition, an approach that was formulated during the continuous oil phase would create an interface lipid bilayer in between the lipid-coated water drops [13]. A later iteration of membrane design examined liquid membrane method with relatively small water flux for Spinach plasma membrane protein 2;1 (SoPIP2;1) proteoliposomes, which was placed between the NF membranes that were capable of providing an AQP fingerprint [14]. Designs such as these [14–18] helped to encourage innovative approaches to membranes and develop membrane-based biosensor projects [19]. In 2010, the hydrogel method created at Aquaporin A/S was introduced. This occurred once AquaZ and Montemagno came out with the ABLM design that featured internally cross-linked UV and PA-interconnected proteoliposomes that are positioned as motionless on the lipid-coated PA layer and maintained with a PEO hydrogel [20].

DHI Singapore and Aquaporin A/S began research collaboration with the SMTC on biomimetic membranes and their applications in 2009. Simultaneously, the Chung lab at NUS initiated biomimetic research project in conjunction with Wolfgang Meier research group. The collaboration at NUS continued to examine the Aquaporin's hydrogel method and attempted to create a planar proteobilayer, beginning with AqpZ proteoliposome fusion process conducted with pure and PEO-coated porous alumina. During the experimental runs, the research team recognized an increase in stability as the mPAR values became greater [21]. This method was outlined in 2012 as relying on a Langmuir-Blodgett film with nickel-chelated lipids that effectually binds to the histagged AqpZ. This method is similar in its framework to the approach developed by Kumar [22], however, based on lipids with consequent Langmuir-Schäffer deposition-mediated transfer onto the mica surface [23]. Kaufman et al. attempted to further the research by integrating spinach AQP (SoPIP2;1) into the positively charged bolalipid micelles that were fused on the negatively charged silica surface [24]. Research conducted by Tang et al. examined proteoliposomes fusion performance on polymer-coated and pure silica with the help of quartz crystal microbalance with dissipation (QCM-D) [25]. Tang et al. identified higher robustness as well as fusion resistance capabilities as the mPAR increased and additional proteoliposome stabilization potential with polyelectrolyte layers functioning at the maximum mPAR (1:25) in 1,2-diphytanoyl-sn-glycero-3-phosphocholine (DPhPC) liposomes [25]. SMTC research group likewise examined ABLMs in an attempt to further Kaufman's method for liposome fusion process on nanofiltration (NF) membranes, [26, 27] and fused AqpZ proteoliposomes on NF PA-polysulfone (PSf) membranes intentionally precoated with positively charged lipids using the spin-coating process [28]. In this case, the proteoliposomes were positioned onto the NF membrane and then slightly pressurized with 0.5 bar. The data obtained indicate that a linear correlation between the ABLM surface's roughness quality and mPAR shows AqpZ integration; however, an observable absence of influence from AqpZ onto the water flux Jv or the reverse salt flux J_s [28] was found.

3. Vesicular biomimetic structures in membrane design

Aquaporin A/S and SMTC launched a joint collaborative approach where AqpZ proteoliposomes were effectively inserted in the standard PA layer created using interfacial polymerization of trimesoyl chloride (TMC) and *m*-phenyl diamine (MPD) on a PSf support structure [29–31]. In this case, ABLMs were verified through testing with functional AqpZ proteoliposomes, PA-PSf membranes without any proteoliposomes, and proteoliposomes that include an inactive AqpZ mutation. The ABLMs and their qualifications were rigorously compared to the commercially available membranes with the aid of cross-flow RO tests on 42 cm^2 effective area. Although the *Js* values were similar in all the cases, the ABLMs including AqpZ proteoliposomes showed substantially higher *Jv* values than those measured for the ABLM with inactive AqpZ and PA-PSf membranes. Moreover, the ABLMs were capable of enduring 10 bar pressure, thus identifying them as suitable for applications with low-pressure RO. The ABLM *Jv* values were ~40% greater when compared to the commercially available brackish water RO membranes (BW30), as well as an order of magnitude higher if compared to the seawater RO membranes (SW30HR). The results of this analysis were further advanced through a systematic study showing that 1,2-dioleoyl-sn-glycero-3-phosphocholine (DOPC)–based proteoliposomes and proteoliposomes with mPAR of 1:200 offered the ideal water flux values as suggested by SFLS. The study also indicated that the addition of cholesterol might be able to seal the defects occurring on the proteoliposomes [32].

In order to obtain better sealing parameters and higher loading capacity, the SMTC research group tried to coat proteoliposomes with polydopamine (PDA) and then immobilized them on a 28-cm^2 NF polyamide imide (PAI) membrane by inserting them into the branched polyethyleneimine (PEI) that was crosslinked per PA bond at higher temperature values [33]. The data collected by SFLS indicated that the increased temperature values had greater adverse effects on the proteoliposomes' permeability potential than the PDA coating. Despite the potential issues, AqpZ functionality was shown and validated with optimized performance mPAR values of 1:200 once included and reconstituted into the PAI-PEI layer. Alternatively, the maximum SFLS results were reached at an mPAR value of 1:100 [32]. The reasons for this difference could be connected to the fact that AqpZ is influenced by the PEI branches or the PDA coating. Nevertheless, the *Jv* value was calculated as 36 $L\ m^{-2}\ h^{-1}\ bar^{-1}$, a result that qualifies it as the most advanced in all currently available biomimetic membranes [33].

Proteopolymersomes can also be further functionalized during the process of getting chemically bounded to a functional membrane counterpart. The process of liposomes and polymersomes functionalization has been widely researched for a number of decades [34–36]. The ABPMs featuring functionalized proteopolymersomes were originally introduced in a 2011 Montemagno patent. This patent established the conceptual framework for proteopolymersomes created out of polyethyloxazoline-polydimethylsiloxane-polyethyloxazoline (PEOXA-PDMS-PEOXA) triblock copolymers. In this type of proteopolymersomes, the methacrylate-functionalized PEOXA block is restraining the proteopolymersomes on a methacrylate functionalized cellulosic membrane [37]. After the patent was finalized, the initial experimental results collected during a practical application of this method were given by the NUS group [38]. This research group manufactured proteopolymersomes including AqpZ in methacrylate-functionalized PMOXA-PDMS-PMOXA and then tested them using SFLS. Unlike the data collected by Kumar and researchers [39], there were no substantial changes in SFLS signals when alternating mPAR values, a dynamic that reflects the potential concerns that exist with SFLS when it comes to rigid structures. Montemagno et al. reported that the proteopolymersomes were placed onto the acrylate-functionalized polycarbonate track-etched (PCTE) membranes and then immobilized using UV-crosslinking of the acrylate groups and implementing methacrylate of the PMOXA [37]. After this procedure, the proteopolymersomes were restrained even more through the process of pressure-assisted adsorption and potentially shattered using the "smooth extrusion". The AQP caused an augmentation in the *Jv* values as the mPAR values became higher, while there was an absence of *Jv*, when the polymersomes were applied by themselves. On the other hand, field emission-scanning electron microscopy (FE-SEM) and AFM exposed that the layer had a number of defects or imperfections [38]. In a research study conducted soon after, Montemagno et al. emulated a similar method that relied on acrylate-functionalized cellulose acetate membranes [40]. In this experimental dynamic, the researchers saw a surge in *Jv* values and a reduction in NaCl rejection potential with proteopolymersomes that have larger mPAR values. The marked growth in *Jv* values may be a sign of AQP activity; however, the NaCl rejection remained relatively low (33%) and the determined membrane area was quite small at only 7 mm^2 [40].

A different method suggested gold-disulfide binding for immobilization of disulfide-functionalized PMOXA-PDMS-PMOXA AqpZ proteopolymersomes on silicon and gold-coated porous alumina surfaces [41]. This FE-SEM approach indicated that complete pore coverage was attained at the pore diameter of 55 nm, and where greater size pores of 100-nm diameter continued to be open. As a result, despite observing the integration of AQP, the NaCl rejection potential was small [41]. In order to achieve an improved sealing capacity, the cysteamine was introduced with PDA as well as histidine coatings, once the proteopolymersome immobilization process on gold-coated PCTE was done [42]. The experimental data suggested that the *Jv* values increased, while the *Js* values decreased, as number of PDA-His layers increased. On the other hand, the ideal sealing capacity was achieved without the use of the proteopolymersomes. The pressure-reversed osmosis (PRO) mode testing (AL toward the water receiving draw side) caused a substantially higher *Js* values than the forward osmosis (FO) method of testing (AL toward the feed side) [42]. The mathematical simulations used for the ABPM suggest that when using the PRO mode, *Jv* value is dictated by permeability potential and the size of the vesicle. Alternatively, when applying the FO mode, the hydrostatic pressure value is shaped by the vesicle interior's solute concentration [43].

Innovation in membrane research helped initiate another variation of the method design that has been developed experimentally using AqpZ and methacrylate-functionalized and carboxyl-functionalized PMOXA-PDMS-PMOXA with amine-functionalized CA [44]. In this method, the proteopolymersomes are initially covalently connected to the CA, where the amine groups on CA and the carboxyl groups of PMOXA develop a PA bond. The approach then relies on the methacrylic cross-linking polymerization that is achieved by dipping the membrane into a mixture of ethylene glycol dimethacrylate, methyl methacrylate, and an initiator. As the polymerization time increases, the *Jv* value is linearly growing, while the NaCl rejection potential is reducing. A growth in *Jv* values and a drop in NaCl rejection of ABPMs were comparable to only polymersome-coated CA and methacrylated CA in both NF and FO modes and showed existence of AQP. Alternatively, the NaCl rejection rate of 61% suggested major defects and imperfections [44]. Additional illustration of methacrylate cross-linking is based on amine-functionalized AqpZ proteoliposomes with a PDA-precoated ultrafiltration (UF) polyacrylonitrile (PAN) membrane [45]. In this instance, the proteoliposomes are cross-linked internally using methacrylate and then slowly pressurized onto the PDA-PAN support, thus permitting the response of functionalized lipids and PDA amines. Additional stabilization is obtained with the aid of glutaraldehyde. The process of internal proteoliposomes cross-linking had a helpful influence on the overall stability. *Jv* and NaCl rejection potential between ABLMs and liposome-coated membranes indicated some effects that signal the presence of AqpZ. The low NaCl rejections and FE-SEM images reflected that the imperfections in the ABLM had a serious role in the membrane's performance [45]. Rather than rely on the process of chemical bonding, proteoliposomes or polymersomes have the capacity to be bound together by electrostatic forces. Kaufman et al. tried to use this approach so as to fuse positively charged bolamphiphilic proteoliposomes onto negatively charged NF PA and with sulfonate PSf (PSS) membranes [46]. In this experimental setup, the proteoliposome loading was improved due to the more negatively charged PSS membrane. The proteoliposome loading likewise contributed to a decrease in *Jv* values in addition to a raise in NaCl rejection potential, primarily because of the produced defects within the bolamphiphilic bilayer by SoPIP2;1 [46].

As researchers continued to expand alternative approaches, an electrostatic-binding-based approach was created and it relied on the process of embedding positively charged poly-L-lysine covered AqpZ proteoliposomes into the anionic section of a layer-by-layer (LbL) sandwich on a UF PAN membrane [47]. In this method, the anionic section consists of polyacrylic acid (PAA) and PSS and where the cationic counterpart is created out of the polyallylamine hydrochloride (PAH). When this method was applied experimentally, a distinct AqpZ effect was seen, *Jv* values were raised by 30–50% after the proteoliposomes addition, and the overall effect was greater once there were a larger number of negatively charged lipids available. In this experimental setup, the $MgCl_2$ rejection rate was comparable to the results obtained in the research data collected by Zhao et al. [29], but notably, there was no NaCl rejection available [47]. This research approach was further elaborated by encapsulating magnetic nanoparticles to incite more proteoliposomes to magnetically adsorb the polyanionic film. In this type of FO mode, the measured data showed a raise in *Jv* as well as *Js* values as the mPAR increase, and this accounts for the residual flaws, despite the attempts to introduce more vesicles onto the supporting substrate [48].

Together with other researchers from Ocean University in China, Wang continued to develop the earlier methods. In particular, they were able to immobilize AqpZ proteoliposomes using positively charged lipids on top of a negatively charged PSS layer, which was then followed by PEI on an UF PAN membrane [49]. Once the experimental data were collected, it became clear that the modest NaCl rejection capacity and *Jv* decrease suggest a highly imperfect membrane. A growth in *Jv* occurring between proteoliposomes and liposomes, in addition to an increase in *Jv* with greater mPAR values, may be indicative of the AQP presence. In this case, NaCl rejection continued to be unchanged for all membranes. The membrane performance was shown to be compromised, when the detergent treatment was used [49].

The method designs are comprehensively summarized [49], and the research data collected help to conclude that the embedment of proteopolymersomes or liposomes in a layer facilitates significantly more efficient membranes, if compared to layer-based immobilization approaches. One of the key benefits of the PA-embedment technique is that there is no need to precoat or functionalize processes that can severely limit any potential for upscaling and large applications [29]. The described performance data are relatively inefficient when compared to the theoretical predictions, since much more innovation is necessary. A critical setback is the fact that with an increasing mPAR values, *Jv* is likewise augmented; however, at the same time, the matrix layer becomes weaker in its structure and thus more susceptible to salt leakage. The inclusion of stabilizing and sealing polymer networks may help the rejection potential, but it might likewise compromise J_v [1]. In addition to ABPMs and ABLMs, two primary research directions are aimed at obtaining functioning biomimetic membranes with the aid of AQP-mimicking artificial channels, specifically organic building block nanochannels and carbon nanotubes (CNTs) [50]. The CNT method is used more frequently due to the fact that its faster water permeation has been proven in theoretical terms [51] as well as experimentally [52]. When it comes to the organic nanochannels and their applicability, five structures have been found promising as they can effectively compete with ABPMs, CNTs, and ABLMs. These five structures include zwitterionic coordination polymers based on zinc and N,N-diacetic acid imidazolium bromide [53], imidazole compounds with urea ribbons [54], helical pores of dendritic dipeptides [55], macrocycles of *m*-phenylene ethynylene [56], and hydrazine-appended pillar arenes [57]. When it comes to these structures, their primary benefit, if compared to other designs, is a smaller size featuring channel diameter values (3–10 Å) [50].

4. POSS: a new element in interfacial polymerization

Most of the FO and RO membranes are polyamide (PA-based) and are frequently referenced as the thin film composite (TFC) membranes because of their comparatively improved membrane design performance. The PA layer for these types of membranes primarily created using a reaction between an acyl chloride and an amine [58]. For this reaction to occur, there needs to be a process that effectively dissolves the amine group during the aqueous phase, as well as a process that dissolves the acyl chloride group during the organic phase [59]. In most cases, the membrane is fully wetted in the aqueous phase, which includes the amine group. After this initial process, the membrane becomes somewhat dry once the excessive visible liquid is removed, and the surface is kept as moist. In the next step, the organic phase featuring the acyl chloride group is introduced on the surface top area. It can be argued that the reaction's growth is directed into the organic phase [60] because of the preferential solubility that exists in amine group during the organic phase, if compared to the acyl chloride solubility values during the aqueous phase. In this experimental setup, the reaction growth causes the well-recognized valley and ridge formation in the PA layer. The standardized amine-acyl chloride grouping is Trimesoyl chloride (TMC) and *m*-phenyl diamine (MPD), and usually, they are complemented with molecules or additives with analogous chemistries and parameters in lower concentrations so as to improve chlorine resistance, rejection, or flux [58].

For the PA membrane, the ideal active layer (AL) suitable for water separation processes must show high levels of water permeability, while at the same time being able to reject a wide range of solutes and remain resilient during cleaning process. From a theoretical perspective, a model AL of an Aquaporin-based lipidic biomimetic membrane (ABPM) may even remain water impermeable, but only if they allow for adequate incorporation of proteopolymersomes where the water permits solely incorporated proteins to move through. As a result, the new AL components must be comprehensively researched so as to accommodate the existing requirements and parameters. For instance, an AL featuring homogenous thickness potential may be able to help proteopolymersome integration succeed. POSS (amine linker), TMC (acyl chloride linker), and polyhedral oligomeric silsesquioxane have been surveyed for possible applications when it comes the inclusion of proteopolymersomes in ABPMs. A research study recently outlined how POSS can be integrated as an AL layer component, with the data suggesting that POSS-TMC-layer distinctly showed a layer without valleys and ridges, but with elevated mechanical stability capacity in Polyacrylonitrile (PAN) membranes [61]. This experimental method can offer a potentially improved platform for the incorporation of proteopolymersomes, if compared to the valley and ridge prone *m*-phenyl diamine-Trimesoyl chloride MPD-TMC networks. A schematic overview of this experimental reaction is provided in **Figure 1**.

For the experimental trials, PA layers of POSS + TMC including polymersomes of PB29-PEO16 in the aqueous phase were created. During the experiment, the effects of vesicles on the AL capabilities were carefully assessed so as to serve as the basis for the integration of AQPs. In this instance, polybutadiene-polyethylene oxide PB29-PEO16 was chosen because of its capacity to create significant quantities of stable polymersomes during aqueous phase, if compared to other PB-PEO polymersomes [62]. When it comes to the microfluidic method, proteopolymersomes (AqpZ, PB33-PEO18, with molecular amphiphile-to-protein-ratio (mPAR) 1:100) were implemented. Specifically, PB33-PEO18 created significant quantities of stable polymersomes in aqueous phase and ensured effective AqpZ inclusion as shown by the small-angle X-ray

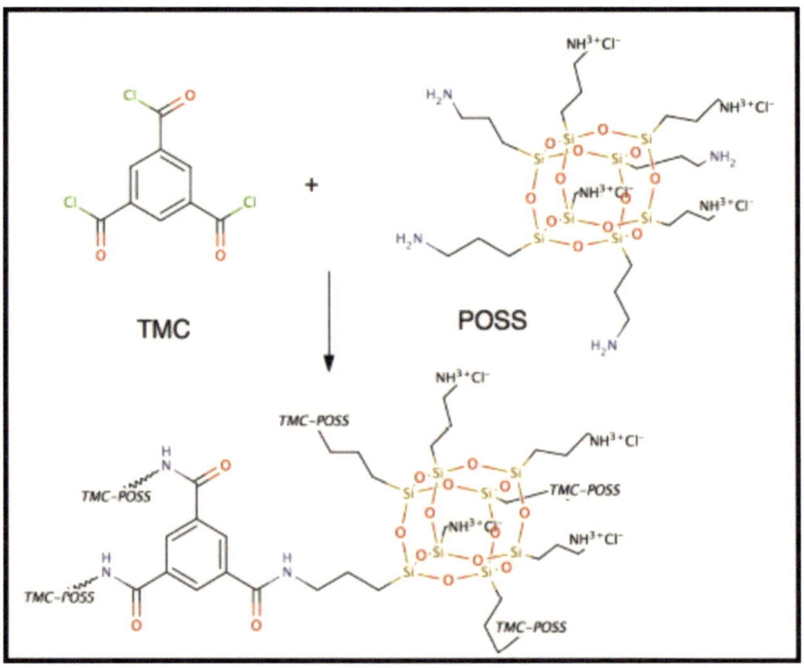

Figure 1. Chemical structure of POSS, TMC, and the resulting AL. POSS acting as the amine linker generates a highly stable and well-defined AL with TMC.

scattering (SAXS). Hexane was introduced as the organic phase, while milli-Q water was used as the aqueous phase. Furthermore, to obtain the smallest possible polydispersity, the polymersomes were sonicated. The process of sonications caused 95% of the polymersomes to show a diameter of 196 ± 83 nm, as the dynamic light scattering (DLS) was able to confirm.

To produce a nonsupported AL, the approach had to add both phases, one after another in a beaker, where an AL supported by microfiltration (MF) polyethersulfone (PES) layers was manufactured with the aid of various coating procedures. Next, the classification of the nonsupported AL was done with SEM, Fourier-transformed infrared spectroscopy (FTIR), and a new microfluidic method that permitted direct observation of the polymerization procedure [63]. For this approach, the characterization of supported AL was accomplished using SEM and FTIR. The supported AL was likewise verified in terms of functionality by implementing rejection and standard flux tests in FO mode and featuring methyl violet staining.

Once both phases are added, the sections of the formed nonsupported AL were first air dried and then effectively vacuum dried. During the process, these sections turned into structures resembling flakes. The FTIR examination of POSS + TMC, with inclusion of polymersomes, showcased the appearance of block copolymers in the AL, and this is visible in **Figure 2**. Specifically, the AL with polymersomes featured an absorption capacity peak at about 3000 cm^{-1} (C—H stretch), and this peak was likewise identified in spectra of PEO and PB [64, 65]. The data reflect that the polymersome-free AL displayed a broader peak at the similar wavelength value range; however, it did not allow for a specific maximum value as was seen in the polymersome-containing AL. These research data may be suggestive of an effective polymersome incorporation occurring in the AL. Moreover, polymersomes did not appear to prevent the PA creation, primarily due to the peaks of a PA bond, the C=O stretch

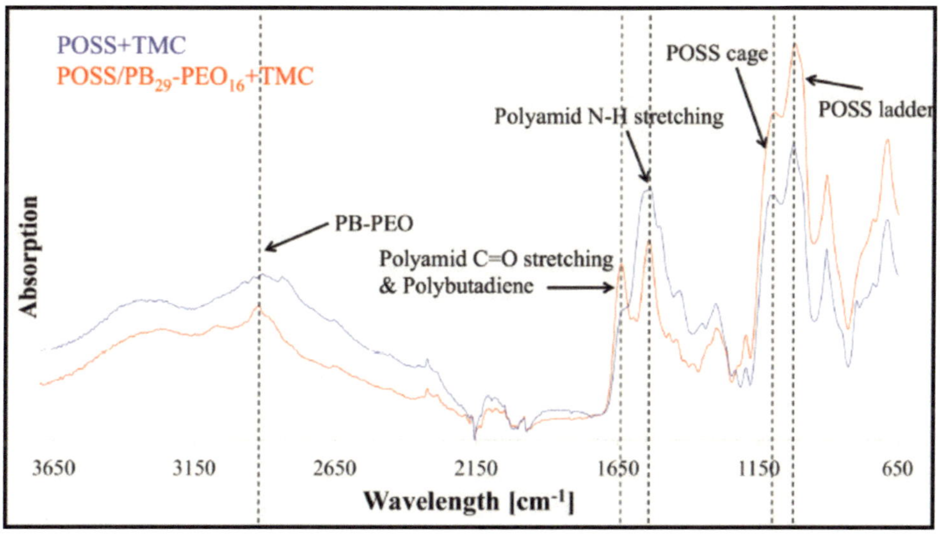

Figure 2. FTIR diagram of POSS/polymersomes + TMC AL (labeled red) and POSS + TMC control AL (labeled blue), as a function of wavelength mapped against absorption values. The AL with polymersomes had an absorption peak around 3000 cm^{-1}. This corresponds to PB and PEO and indicates their presence in the AL, where the characteristic absorption peaks for PA bonds and POSS were likewise present.

at 1636 cm^{-1} and the N—H stretch at 1545 cm^{-1} [61], being openly observable in the AL with polymersomes. Finally, partial hydrolysis of the POSS leading to the AL creation is not seriously disturbed by the occurrence of the polymersomes, as the usual peaks for the POSS-cage and ladder (1125 and 1040 cm^{-1} [61]) were observable in both of the AL. However, there was a noticeable influence of polymersomes on the TMC and its potential reactivity. Initially, Dalwani et al. research collaborative employed 2-g/L TMC for the supported as well as nonsupported AL [61]. During the testing, an alternative dynamic was observed where there was a nonsupported AL with 2 g/L, but with 0.5 g/L, TMC could not be incited to form. In theory, the TMC-POSS-stoichiometry was increased artificially due to the existence of other species in the aqueous phase. A surplus amount of TMC can delay the process of network structure construction, as TMC may not be able to connect the POSS cages and developing only in low molecular weight networks. In this case, 0.5-g/L TMC for the POSS/polymersomes + TMC AL and the nonsupported POSS + TMC AL was implemented.

Research data outlined in **Figure 3** show the FTIR outcomes supplemented by an SEM analysis of those samples. The fact that the POSS + TMC AL was seen as well-defined and smooth suggests that the data are in agreement with previously conducted research [61], as can be seen in **Figure 3a** and **b**. Once the polymersomes were introduced (**Figure 3c–e**) to the process, a visible discrepancy could be seen between the sides toward the organic phase. In particular, one of the sides does not show polymersomes presence (**Figure 3c**), while another side facing the aqueous layer is sufficiently coated with polymersomes (**Figure 3e**). The majority of polymersomes present seemed to be loosely stationed on the top AL. On the other hand, some of the polymersomes appeared to be enclosed by the AL at least to a certain, and their overall shape was much less sharp than the other types of shapes, as is shown by the dotted circles within the image in **Figure 3d**. A number of polymersomes were also implanted directly inside the AL, which can be observed when looking at the AL's cracked profile outline (shown by arrows in **Figure 3d**). This dynamic may indicate that the POSS method could be used for embedding the polymersomes in a manner that would make them helpful during the membrane fabrication process. A method based on the microfluidic field was recently published in a research study [63]. This particular approach suggests performing a visual type study of the exact evolution of interfacial polymerization reactions locations. During this assessment, the chip containing a hydrophobized microchamber becomes separated into two compartments with the aid of micropillars, each with a height of 50 μm and a diameter of 30 μm. In this case, the aqueous phase featuring amine linker was initiated using the microcapillary connections into one compartment and then created a water-air interface between the pillars. For the following step, the organic phase including the acyl chloride linker was inserted into the other compartment. The process showed that the AL formation occurring at the point of interface between the present solutions was seen with the aid of an optical microscope. The AL created as a result had a different formation time and morphology, since the parameters depend on the specific linkers. A practical application of this method indicates that the POSS + TMC creates a well-defined AL and has a formation time of within 4 seconds. Alternatively, the visible development of film in Jeffamine + TMC does not finalize even after 15 minutes, and the film itself exhibits the valley and ridge structural dynamic characteristic of AL created using interfacial polymerization [63].

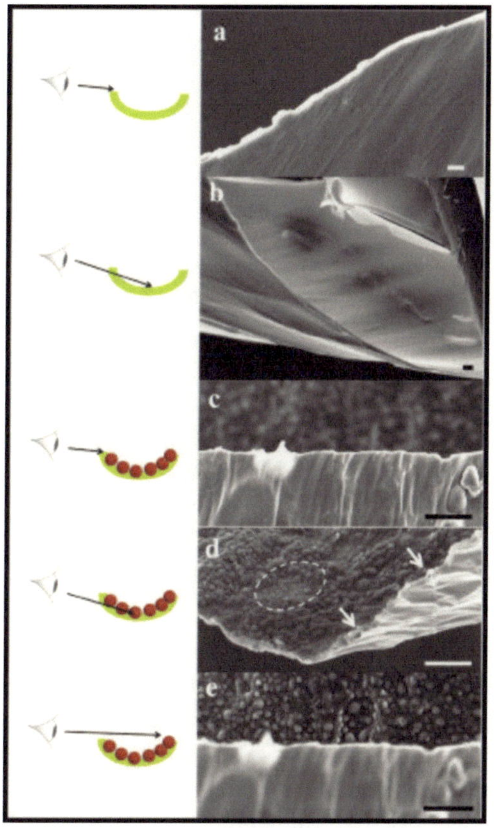

Figure 3. SEM images of POSS + TMC AL (a, b) and POSS/polymersomes + TMC AL (c–e) with schematic sketches, including the part of the layer that is being captured. Images of different flake parts were taken (labeled green in the sketch) for the AL test cases generated during the SEM preparation. The AL without polymersomes was smooth and well-defined and remained on the organic side once polymersomes were added. The aqueous side was covered with loosely attached and half-covered polymersomes (dotted circle in (d)). A few polymersomes could be observed inside the AL (arrows in (d)). Scale bar is 3 μm.

The following method was used to supervise POSS/proteopolymersomes + TMC AL (AqpZ & PB33-PEO18, mPAR 1:100) as shown in **Figure 4**. During the application of this approach, the chip used was not hydrophobized in an optimal manner and as a consequence forced a partial infusion of the aqueous phase directly into the channel space where the organic phase was occurring. However, the hydrophobization process remained sufficiently effective and was able to prevent the aqueous phase from transitioning into the other section in its entirety. Another possible explanation for the interface shift occurring from the pillar structures into the organic phase may be connected to the overpressure dynamic in the aqueous phase, since it is very difficult to regulate when the pressure range is around 104 Pa. The characteristic sharp AL was created on the aqueous-organic interface, once the organic phase containing TMC was introduced (shown in **Figure 4b** using dotted line 1). After this, the reaction proceeded as the diffused amine was added into the organic phase, while the created AL connecting the initial interfaces began to show the appearance of a new aqueous-organic interface

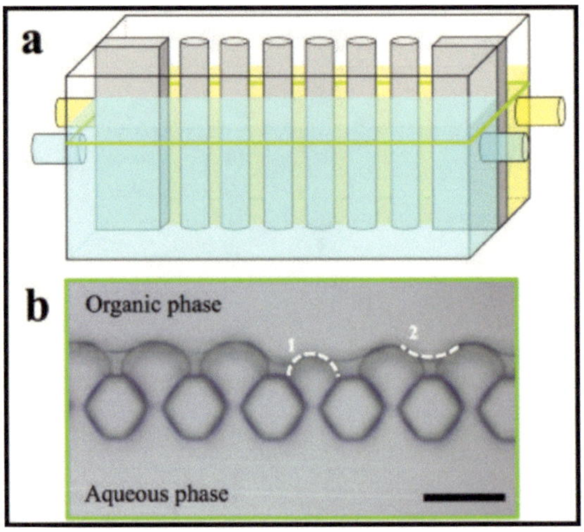

Figure 4. (a) Schematic sketch of the microfluidic chamber and micrographs of POSS/proteopolymersomes + TMC AL and (b) micrograph of the compartment. The aqueous phase entered into the other compartment. After introducing the organic phase, a well-defined AL was formed. The scale bar is 50 μm.

(indicated in **Figure 4b** using dotted line 2). The project's research data and subsequent observations suggest that a less dense AL was created by POSS/proteopolymersomes-TMC than the one created during the POSS-TMC reaction process. In this experimental setup, the formation time value was within the matter of seconds. Evidence likewise showed that the film quality continued to be acceptable and maintained the same shape, without any additional growth noted during the 12 hours that followed.

The POSS + TMC on MF PES support material coating was researched in response to the data produced by Dalwani et al. [61]. In this instance, the MF PES was supported by a nonwoven type of coating. The FTIR spectroscopy testing indicated that the polymersomes were present in the supported AL; on the other hand, the PA formation was substantially lower if compared to the nonsupported POSS/polymersomes + TMC AL, as seen in **Figure 5**. The primary issue that can hinder the analysis of this type of supported AL with FTIR is the possible absorption of the PES support, since it has a significant absorption possibility, specifically, in the sections located between 700 and 2000 cm^{-1}. In fact, the POSS absorption peaks can greatly interfere with the appearing PES peaks. For the supported POSS/polymersomes + TMC AL, a PB-PEO signal was apparent at 3000 cm^{-1}, and another minor peak around 1700 cm^{-1} also appeared in the FTIR spectra of PB [64]. In the case of nonsupported AL, however, peaks could not be identified. From a theoretical standpoint, the reason for this discrepancy is that it was overlayed because of the background signal within the 1600–3650 cm^{-1} region and that was more overpowering during the FTIR analysis spectra for the nonsupported AL. Furthermore, both the PA bonds were noted in the supported POSS + TMC AL, even though they were significantly smaller in the one including polymersomes. Specifically, the large peak (at 1580 cm^{-1}) near the N—H stretching peak is related to PES. The N—H stretching peak (1545 cm^{-1}) occurred solely in the supported POSS + TMC AL. Finally, the broad AL peak located in the range of 3150–3650 cm^{-1} is potentially connected to water and the unreacted amine groups.

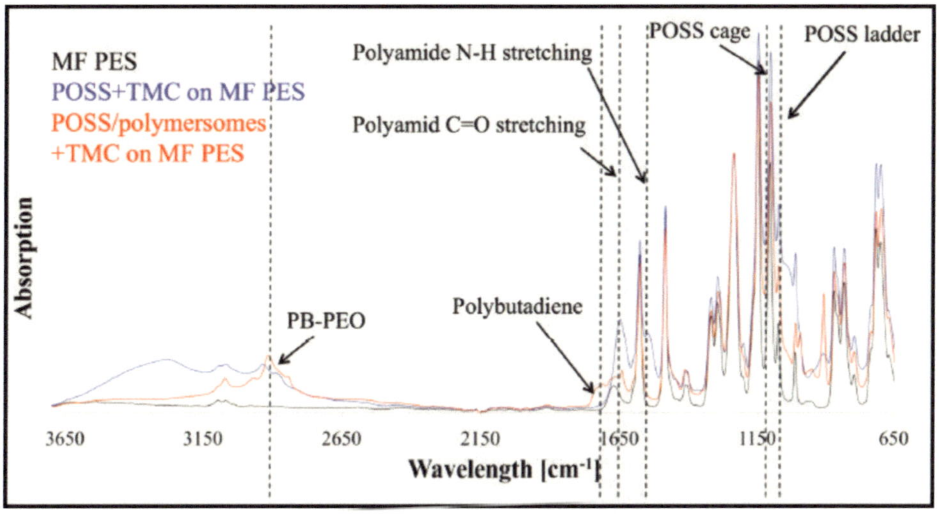

Figure 5. FTIR analysis of supported POSS/polymersomes + TMC AL (**red**) and POSS + TMC control AL (**blue**) on MF PES and pure MF PES (**black**). The PES-supporting material had high absorption values and interfered with numerous absorption peaks. A subtraction from the absorption spectra of pure PES resulted in negative peaks. As a result, only the spectra were normalized. PB-PEO was present in the AL with polymersomes; however, the PA formation was strongly suppressed.

Research data cannot with certainty provide the explanation for the subdual of the PA-signal in the supported POSS/polymersomes + TMC AL. There is a possibility that it might be connected to the TMC and its reactivity potential, a dynamic that has been overviewed earlier. In this experimental approach, 2 g/L TMC for the supported POSS + TMC AL as well as POSS/polymersomes + TMC AL were implemented, since there was no AL creation at 0.5 g/L value. Admittedly, an alternative TMC concentration value might prove to be more ideal for the supported POSS/polymersomes + TMC AL during the applications. Although the blockage of the PA formation incited by polymersomes was supposed to suppress the PA formation in the nonsupported AL, it was not able to do so. With the supported POSS + TMC AL, the AL formation was substantially lowered when altering from 2 to 0.5 g/L. A different hypothesis suggests that POSS + TMC do not in fact form effortlessly on MF PES. Recent research available on this has reported no former POSS + TMC AL formation on MF PES. Since the MF PES has considerably larger pore sizes than PAN, this may effectually impede the creation of a smooth type of layer. In comparison with the FTIR investigation, the SEM analysis revealed a totally covered POSS/polymersomes + TMC AL on the MF PES (shown in **Figure 6**). Moreover, TMC and POSS were able to cover the microporous PES structure in its entirety with a smooth surface layer, even though less distinct than found in PAN substrates [61]. Most likely, this can be attributed to the varying range of pore sizes as reported earlier. The addition of polymersomes facilitated a change and the AL began to exhibit submicron-sized bumps, 0.5–1 μm in height and 1.5–2 μm in length. In this case, with the 100 nm thickness covering AL [61] (**Figure 6** sketch in bottom left corner), there would appear to be groups of 6–9 polymersomes in a row with 1–3 layers. Unlike the nonsupported ALs arrangement, in this instance, only the side facing the organic phase can be effectually observed. When it

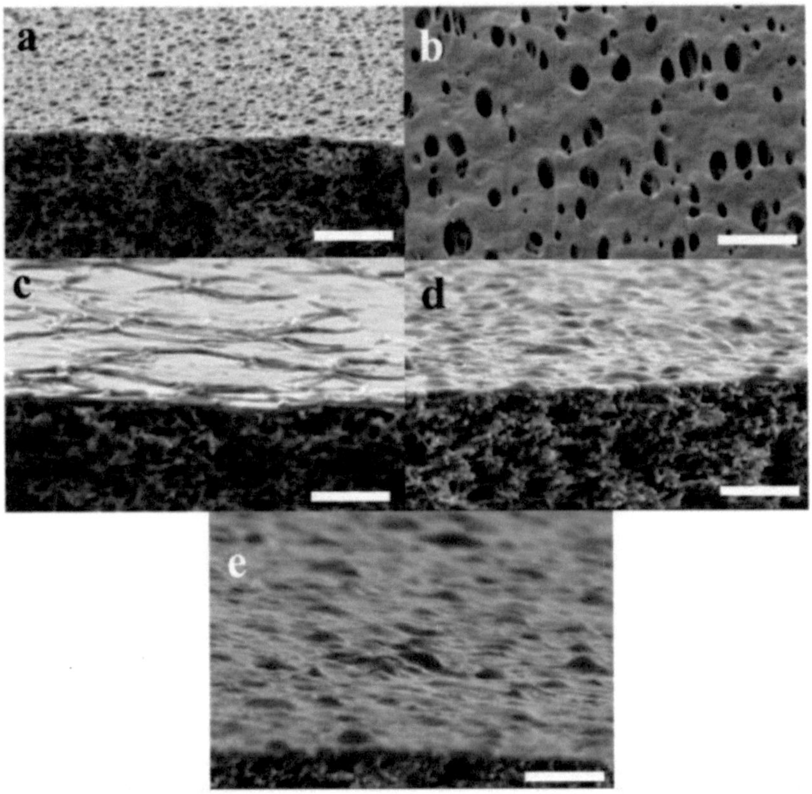

Figure 6. SEM images of MF PES (a, b) supported POSS + TMC AL on MF PES (c) and supported POSS/polymersomes+TMC on MF PES (d, e). Schematic sketch of polymersome coverage on the left (e). Micropores of the MF PES were covered completely by the POSS + TMC AL. After the addition of polymersomes, small bumps with dimensions similar to the polymersomes were observed on the side facing the organic in the AL. Greater bumps may be attributed to the accumulation of covered polymersomes. Scale bar is 3 μm.

comes to the supported POSS/polymersomes + TMC AL, polymersomes have a significantly greater effect on the shape of the AL side facing the organic phase than in examples with the nonsupported forms. Mostly this is caused by the variance in preparation and specifically with respect to POSS/polymersomes in solution at the nonsupported AL formation, occurring at the water-air interface or even being dried up on the MF PES at the supported AL formation. As a consequence, the overall probability of polymersomes being incorporated in the AL is greater for the supported AL rather than the nonsupported AL.

To conclude, the SEM analysis helped to successfully show the incorporation of polymersomes into a supported AL, even though the FTIR research information was potentially less reflexive of its internal dynamics and processes. The issue that limits the SEM and FTIR interrogation of the supported AL is that only a minor portion of the whole membrane is actually observed. Other concerns that can negatively affect a more in-depth study is that the AL samples can become brittle during the process of drying and then delaminate or possibly break off when exposed to liquid nitrogen currently being used for SEM sample preparation.

The new experimental directions included testing of POSS/polymersomes + TMC AL on MF PES with respect to flux potential and rejection values with FO mode. Notably, the collected data showed a lack of FO performance. An estimated one third of the membranes experimentally tested were determined to be impermeable to salt, as indicated by the small conductivity change occurring in the feed solution after 2 h. The remaining membranes were found to be leaky, which was reflected by an instant conductivity growth. Furthermore, only a minor fraction of the POSS/polymersomes + TMC and POSS + TMC leaky and sealed membranes could be assessed as comparable. Within the sealed type membranes, the pores were most likely obstructed by multiple collected layers of POSS + TMC AL. Moreover, the process of methyl blue staining showed an absence of scratches and pinholes on the surface, further indicating that the supporting PES was entirely covered with the AL. As research testing has already illustrated, MF PES is not capable of providing adequate general support for POSS + TMC ALs. Due to the smaller pore size between 5 and 30 nm, PAN support did not exhibit any flux without the added hydraulic pressure [66]. Such a dynamic may be suitable for POSS + TMC with a NF membrane; however, it is not sufficient for the FO. Research approach published by Lee et al. offers a potential compromise that would be based on employing UF PES membranes [67].

This subchapter focused on new insights with regard to the supported and nonsupported AL including POSS with polymersomes during the aqueous phase and TMC during the organic phase. The experimental runs showed that the nonsupported POSS/polymersomes + TMC AL could be successfully developed, with large number of polymersomes coated and a smaller number of them incorporated inside the AL. In particular, the supported POSS/polymersomes + TMC displayed a range of relevant properties. While the FTIR suggested a high suppression potential of the AL formations in cases where polymersomes were added, the SEM images indicated an entirely coated and substantially different AL after the addition of polymersomes. Experimental data showed that none of the created membranes that included TMC and POSS provided successful performances, most likely because of the incomplete coverage of the AL. Even though the experimental results were limited, the exploration of this approach was essential for future development of insight into how proteopolymersomes, TMC, and POSS are engaging with each other. There is a need to explore new avenues of research, and the next specific set of challenges will aim to develop a functional and effective water separation membrane based on these components and their potential strengths.

Author details

Amira Abdelrasoul[1]*, Huu Doan[2] and Ali Lohi[2]

*Address all correspondence to: amira.abdelrasoul@usask.ca

1 Department of Chemical and Biological Engineering, University of Saskatchewan, Saskatoon, Saskatchewan, Canada

2 Department of Chemical Engineering, Ryerson University, Toronto, Ontario, Canada

References

[1] Choi HJ, Montemagno C. Recent progress in advanced nanobiological materials for energy and environmental applications. Materials. 2013;**6**:5821-5856

[2] Pendergast MM, Hoek EMV. A review of water treatment membrane nanotechnologies. Energy & Environmental Science. 2011;**4**:1946-1971

[3] Ishibashi K. Aquaporin subfamily with unusual NPA boxes. Biochimica et Biophysica Acta. 2006;**1758**:989-993

[4] Verkman AS. More than just water channels: Unexpected cellular roles of aquaporins. Journal of Cell Science. 2005;**118**:3225-3232

[5] Montemagno C, Schmidt J, Tozzi S. Biomimetic Membranes. Patent WO 2004/011600; 5 February 2004

[6] Hansen JS, Perry M, Vogel J, Vissing T, Hansen CR, Geschke O, Emnéus J, Hélix-Nielsen C. Development of an automation technique for the establishment of functional lipid bilayer arrays. Journal of Micromechanics and Microengineering. 2009;**19**:2. DOI: 10.1088/0960-1317/19/2/025014

[7] Hansen JS. Development of supported biomimetic membranes for insertion of aquaporin protein water channels for novel water filtration applications. [Ph.D. thesis], Danish Technical University, Kgs. Lyngby: Denmark; 2010

[8] Pszon-Bartosz K, Hansen JS, Stibius KB, Groth JS, Emnéus J, Geschke O, Hélix-Nielsen C. Assessing the efficacy of vesicle fusion with planar membrane arrays using a mitochondrial porin as reporter. Biochemical and Biophysical Research Communications. 2011;**406**:96-100

[9] González-Pérez A, Stibius K, Vissing T. Biomimetic triblock copolymer membrane arrays: A stable template for functional membrane proteins. Langmuir. 2009;**25**:10447-10450

[10] Rein C, Pszon-Bartosz K, Stibius KB, Bjørnholm T, Hélix-Nielsen C. Free-standing biomimetic polymer membrane imaged with atomic force microscopy. Langmuir. 2011;**27**: 499-503

[11] Ibragimova S, Stibius K, Szewczykowski P, Perry M, Bohr H, Hélix-Nielsen C. Hydrogels for in situ encapsulation of biomimetic membrane arrays. Polymers for Advanced Technologies. 2010;**23**:182-189

[12] Perry M, Hansenz J, Stibius K. Surface modifications of support partitions for stabilizing biomimetic membrane arrays. Journal of Membrane Science and Technology. 2011. DOI: 10.4172/2155-9589.S1-001

[13] Rein C. Stabilization and characterization of 2D and 3D biomimetic membranes. [Ph.D. thesis], University of Copenhagen; Copenhagen, Denmark, 2011

[14] Aquaporin A/S. Biomimetic Membranes and Uses Thereof. Patent WO 2010/146365, 23 December 2010

[15] Aquaporin A/S. Membrane for Filtering of Water. Patent WO 2006/122566; 23 November 2006

[16] Aquaporin A/S. Biomimetic Water Membrane Comprising Aquaporins Used in the Production of Salinity Power. Patent WO 2007/033675; 29 March 2007

[17] Aquaporin A/S. Scaffold for Composite Biomimetic Membrane. Patent WO 2009/074155; 18 June 2009

[18] Aquaporin A/S. Assays Relating to Biomimetic Membranes and Their Uses. Patent WO 2010/146366; 23 December 2010

[19] Mech-Dorosz A, Heiskanen A, Bäckström S, Perry M, Muhammad HB, Hélix-Nielsen C, Emnéus J. A reusable device for electrochemical applications of hydrogel supported black lipid membranes. Biomedical Microdevices. 2015;**17**:21

[20] Montemagno C, Aquaz, Danfoss. Nanofabricated Membrane Using Polymerized Proteo-liposomes. Patent WO 2010/091078, 12 August 2010

[21] Wang H, Chung TS, Tong YW, Meier W, Chen Z, Hong M, Jeyaseelan K, Armugam A. Preparation and characterization of pore-suspending biomimetic membranes embedded with aquaporin Z on carboxylated polyethylene glycol polymer cushion. Soft Matter. 2011;**7**:7274

[22] Kumar M. Biomimetic membranes as new materials for applications in environmental engineering and biology. [Ph.D. thesis], University Illinois: Champaign, IL USA; 2010

[23] Sun G, Zhou H, Li Y, Jeyaseelan K, Armugam A, Chung TS. A novel method of AquaporinZ incorporation via binary-lipid Langmuir monolayers. Colloids and Surfaces B: Biointerfaces. 2012;**89**:283-288

[24] Kaufman Y, Grinberg S, Linder C, Heldman E, Gilron J, Freger V. Fusion of Bolaamphiphile micelles: A method to prepare stable supported biomimetic membranes. Langmuir. 2013;**29**:1152-1161

[25] Li X, Wang R, Wicaksana F, Zhao Y, Tang CY, Torres J, Fane AG. Fusion behaviour of aquaporin Z incorporated proteoliposomes investigated by quartz crystal microbalance with dissipation (QCM-D). Colloids and Surfaces B: Biointerfaces. 2013;**111**:446-452

[26] Kaufman Y, Berman A. Supported lipid bilayer membranes for water purification by reverse osmosis. Langmuir. 2010;**26**:7388-7395

[27] Freger V, Kaufman Y. Biomimetic Membranes, Their Production and Uses Thereof in Water Purification. Patent US 2011/0084026; 14 April 2011

[28] Li X, Wang R, Tang CY, Vararattanavech A, Zhao Y, Torres J, Fane AG. Preparation of supported lipid membranes for aquaporin Z incorporation. Colloids and Surfaces B: Biointerfaces. 2012;**94**:333-340

[29] Zhao Y, Qiu C, Li X, Vararattanavech A, Shen W, Torres J, Hélix-Nielsen C, Wang R, Hu X, Fane AG, et al. Synthesis of robust and high-performance aquaporin-based biomimetic membranes by interfacial polymerization-membrane preparation and RO performance characterization. Journal of Membrane Science. 2012;**423-424**:422-428

[30] Aquaporin AS. NTU. Aquaporin Based Thin Film Composite Membranes. Patent WO 2013/043118; 28 March 2013

[31] Aquaporin AS. Systems for Water Extraction. Patent WO 2014/128293; 28 August 2014

[32] Zhao Y, Vararattanavech A, Li X, Hélix-Nielsen C, Vissing T, Torres J, Wang R, Fane AG, Tang CY. Effects of Proteoliposome composition and draw solution types on separation performance of aquaporin-based Proteoliposomes: Implications for seawater desalination using aquaporin-based biomimetic membranes. Environmental Science & Technology. 2013;**5**:1496-1503

[33] Li X, Wang R, Wicaksana F, Tang CY, Torres J, Fane AG. Preparation of high performance nanofiltration (NF) membranes incorporated with aquaporin Z. Journal of Membrane Science. 2014;**450**:181-188

[34] Tanner P. Design and development of protein-polymer assemblies to engineer artificial organelles. [Ph.D. thesis], Universität Basel: Basel, Switzerland; 2013

[35] Pawar PV, Gohil SV, Jain JP, Kumar N. Functionalized polymersomes for biomedical applications. Polymer Chemistry. 2013;**4**:3160-3176

[36] Jesorka A, Orwar O. Liposomes: Technologies and analytical applications. Annual Review of Analytical Chemistry. 2008;**1**:801-832

[37] Montemagno C. Biomimetic Membrane Formed from a Vesicle-Thread Conjugate. Patent WO 2010/040353, 15 April 2010

[38] Wang H, Chung TS, Tong YW, Jeyaseelan K, Armugam A, Chen Z, Hong M, Meier W. Highly permeable and selective pore-spanning biomimetic membrane embedded with aquaporin Z. Small. 2012;**8**:1185-1190

[39] Kumar M, Grzelakowski M, Zilles J, Meier WP. Highly permeable polymeric membranes based on the incorporation of the functional water channel protein aquaporin Z. Proceedings of the National Academy of Sciences of the United States. 2007;**104**: 20719-20724

[40] Zhong PS, Chung TS, Jeyaseelan K, Armugam A. Aquaporin-embedded biomimetic membranes for nanofiltration. Journal of Membrane Science. 2012;**407-408**:27-33

[41] Duong PHH, Chung TS, Jeyaseelan K, Armugam A, Chen Z, Yang J, Hong M. Planar biomimetic aquaporin-incorporated triblock copolymer membranes on porous alumina supports for nanofiltration. Journal of Membrane Science. 2012;**409-410**:34-43

[42] Wang HL, Chung TS, Tong YW, Jeyaseelan K, Armugam A, Duong HHP, Fu F, Seah H, Yang J, Hong M. Mechanically robust and highly permeable AquaporinZ biomimetic membranes. Journal of Membrane Science. 2013;**434**:130-136

[43] Wang H, Chung TS, Tong YW. Study on water transport through a mechanically robust aquaporin Z biomimetic membrane. Journal of Membrane Science. 2013;**445**:47-52

[44] Xie W, He F, Wang B, Chung TS, Jeyaseelan K, Armugam A, Tong YW. An aquaporin-based vesicle-embedded polymeric membrane for low energy water filtration. Journal of Materials Chemistry A. 2013;**1**:7592-7600

[45] Sun G, Chung TS, Jeyaseelan K, Armugam A. Stabilization and immobilization of aquaporin reconstituted lipid vesicles for water purification. Colloids and Surfaces B: Biointerfaces. 2012;**102**:466-471

[46] Kaufman Y, Grinberg S, Linder C, Heldman E, Gilron J, Shen YX, Kumar M, Lammertink RGH, Freger V. Towards supported bolaamphiphile membranes for water filtration: Roles of lipid and substrate. Journal of Membrane Science. 2014;**457**:50-61

[47] Sun G, Chung TS, Jeyaseelan K, Armugam A. A layer-by-layer self-assembly approach to developing an aquaporin-embedded mixed matrix membrane. RSC Advances. 2013;**3**:473

[48] Sun G, Chung TS, Chen N, Lu X, Zhao Q. Highly permeable aquaporin-embedded biomimetic membranes featuring a magnetic-aided approach. RSC Advances. 2013;**3**:9178-9184

[49] Wang M, Wang Z, Wang X, Wang S, Ding W, Gao C. Layer-by-layer assembly of aquaporin Z-incorporated biomimetic membranes for water purification. Environmental Science & Technology. 2015;**49**:3761-3768

[50] Shen YX, Saboe PO, Sines IT, Erbakan M, Kumar M. Biomimetic membranes: A review. Journal of Membrane Science. 2014;**454**:359-381

[51] Thomas JA, McGaughey AJH. Reassessing fast water transport through carbon nanotubes. Nano Letters. 2008;**8**:2788-2793

[52] Holt J, Park H, Wang Y, Stadermann M, Artyukhin A, Grigoropoulos C, Noy A, Bakajin O. Fast mass transport through Sub-2-Nanometer carbon nanotubes. Science. 2006;**312**:1034-1037

[53] Fei Z, Zhao D, Geldbach TJ, Scopelliti R, Dyson PJ, Antonijevic S, Bodenhausen G. A synthetic zwitterionic water channel: Characterization in the solid state by X-ray crystallography and NMR spectroscopy. Angewandte Chemie, International Edition. 2005;**44**:5720-5725

[54] Barboiu M. Artificial water channels. Angewandte Chemie, International Edition. 2012;**51**:11674-11676

[55] Percec V, Dulcey AE, Balagurusamy VSK, Miura Y, Smidrkal J, Peterca M, Nummelin S, Edlund U, Hudson SD, Heiney PA, Duan H, Magonov SN, Vinogradov SA. Self-assembly of amphiphilic dendritic dipeptides into helical pores. Nature. 2004;**430**:764-768

[56] Zhou X, Liu G, Yamato K, Shen Y, Cheng R, Wei X, Bai W, Gao Y, Li H, Liu Y, et al. Self-assembling subnanometer pores with unusual mass-transport properties. Nature Communications. 2012;**3**. DOI: 10.1038/ncomms1949

[57] Tang CY, Zhao Y, Wang R, Hélix-Nielsen C, Fane AG. Desalination by biomimetic aquaporin membranes: Review of status and prospects. Desalination. 2013;**308**:34-40

[58] Alsvik I, Hägg MB. Pressure retarded osmosis and forward osmosis membranes: Materials and methods. Polymer. 2013;**5**:303-327

[59] Wittbecker EL, Morgan PW. Interfacial polycondensation. I. Journal of Polymer Science. 1959;**40**:289-297

[60] Ghosh AK, Jeong BH, Huang X, Hoek EMV. Impacts of reaction and curing conditions on polyamide composite reverse osmosis membrane properties. Journal of Membrane Science. 2008;**311**:34-45

[61] Dalwani M, Zheng J, Hempenius M, Raaijmakers MJT, Doherty CM, Hill AJ, Wessling M, Benes NE. Ultra-thin hybrid polyhedral silsesquioxane-polyamide films with potentially unlimited 2D dimensions. Journal of Materials Chemistry. 2012;**22**(30):14835-14838

[62] Zawada JF, Yin G, Steiner AR, Yang J, Naresh A, Roy SM, Gold DS, DS HHG, Murray CJ. Microscale to manufacturing scale-up of cell-free cytokine production—A new approach for shortening protein production development timelines. Biotechnology and Bioengineering. 2011;**108**:1570-1578

[63] Zhang Y, Benes NE, Lammertink RGH. Visualization and characterization of interfacial polymerization layer formation. Lab on a Chip. 2015;**15**:575-580

[64] Plasencia I, Survery S, Ibragimova S, Hansen JS, Kjellbom P, Helix-Nielsen C, Johanson U, Mouritsen OG. Structure and stability of the spinach aquaporin SoPIP2;1 in detergent micelles and lipid membranes. PLoS One. 2011;**6**:e14674

[65] Andersen OS, Nielsen C, Maer AM, Lundbæk JA, Goulian M, Koeppe RE II. Gramicidin channels as molecular force transducers in lipid bilayers. Biol. Skr. Dan. Vid. Selsk. 1998;**49**:75-82

[66] Germic L, Ebert K, Bouma RHB, Borneman Z, Mulder MHV, Strathmann H. Characterization of polyacrylonitrile ultrafiltration membranes. Journal of Membrane Science. 1997;**132**:131-145

[67] Lee KP, Zheng J, Bargeman G, Kemperman AJB, Benes NE. pH stable thin film composite polyamine nanofiltration membranes by interfacial polymerisation. Journal of Membrane Science. 2015;**478**:75-84

Chapter 8

Applications of Biomimetic and Bioinspired Membranes

Amira Abdelrasoul, Huu Doan and Ali Lohi

Additional information is available at the end of the chapter

http://dx.doi.org/10.5772/intechopen.71720

Abstract

Application fields of the biomimetic and bioinspired membranes are very similar to those of the existing synthetic membranes. Due to the hierarchical structures, as well as controlled selective transport and stability/resistance, the biomimetic and bioinspired membranes have extended applications in sustainable resources, environment, and energy aspects. The most important applications of biomimetic and bioinspired membranes are water treatment, clean energy, carbon capture, and health care.

Keywords: sustainability, water, clean energy, fuel cell, carbon capture

1. Introduction

The potential areas for the successful application of the bioinspired and biomimetic membranes are currently overlapping with the ones already taking advantage of and investing into existing synthetic membrane solutions. Because of their implicit hierarchical structures and the controlled selective transport and stability/resistance, bioinspired and biomimetic membranes attract new applications in areas including sustainable resources, energy management, and the environment. However, by far the most relevant areas in which biomimetic and bioinspired membranes can be used are water treatment, health care, carbon capture, and clean energy.

2. Water treatment

Global application of biometric membranes in water treatment processes is one of the primary current challenges. The difficulties associated with water treatment processes become a problem due to the growing amount of wastewater produced from both industrial and municipal sources, and the ways in which these sources have continuously incited concerns with various environmental issues on the global scale. The processes of water purification and wastewater treatment have been designed to allow for the reusing and recycling of discharged water. It is the ultimate aim of the future water reuse facilities and clean water access points to ensure that they can rely upon water drawn directly from the nontraditional sources, including industrial and municipal wastewaters [1]. In the last decade, the application of membrane technology production of high quality water from nontraditional water sources, including agricultural, industrial, municipal wastewater, brackish water, and seawater has been generally accepted. Successful use of filtration membranes has allowed creating the necessary alternative solutions for water scarcity, fresh water supplies, and beneficial reusing of discarded water. Most of the traditional pressure-driven membrane processes, including MF, UF, NF and RO are actively used in industrial and municipal wastewater treatment settings. Currently, there is an additional incentive to integrate next generation high-performance bioinspired and biomimetic membranes for water treatment for industrial and municipal dynamics. Both the diverse parameters and complexity of the water being filtered can incite additional concerns when it comes to membrane used for water and wastewater treatment. Membrane fouling is arguably the greatest issue for a widespread implementation of water treatment membranes, since it contributes to drastic flux decline, increases cleaning requirements, and dramatically raises energy consumption and operating costs. The issue of fouling frequently occurs in water filtration when the water going through the membrane contains foulants such as particles, colloids, macromolecules, hydrocarbon mixtures, microorganisms, and natural organic matter. The additional foulants can be deposited and then adsorbed onto the membrane's surface or its pore walls. This additional deposit can greatly reduce the water flux and detrimentally affect the separation performance of membranes. The bioinspired and biomimetic strategies introduced new ideas for better design and development of a range of antifouling membranes that have the potential to offer an improved separation capacity. Oil-contaminated wastewater produced as a result of hydrocarbon processing, oil-spill mixtures, and metallurgy creates extensive environmental pollution. As a result, it is essential to develop antifouling membranes that would be able to remove oil and other contaminants from water in an effective manner. Research based on the versatility of anchoring properties in mussel and their adhesive proteins has allowed to introduce hydrophilic Polydopamine (PDA) for an improved oil fouling resistance in MF, UF, NF, and RO membranes [2–7]. Freeman et al. [2, 7] placed PDA on PSf support polyester membranes in pressure-delayed osmosis and studied the antifouling ability of PDA-modified membranes used in oil/water filtration processes. The membranes modified with PDA for all dopamine concentrations, deposition times, and alkaline pH values were noticeably more resistant to oil fouling than the PDA-uncoated membranes during emulsified oil–water filtration. The advantages gained from the use of PDA also allowed for a range of reactions with functional organic molecules, such as Michael addition or Schiff base reactions between catechols and amines [3]. Amine-terminated poly(ethylene glycol) (PEG-NH_2) was anchored onto PDA-modified MF, UF, NF, and RO membranes and improved fouling resistance by taking advantage of the existing fouling resistance properties of PEG [3, 5, 6]. PDA and PDA-g-PEG-modified Polytetrafluoroethylene (PTFE) MF membranes had 20 and 56% higher flux, respectively, than the unmodified membranes 1 h after starting the process of filtering the emulsified oil/water. The PDA and PDA-g-PEG-modified PES UF membranes increased oil emulsion filtration flux by approximately 35%, if compared to the results obtained from their unmodified counterpart after 1 h of filtration time. NF and RO membranes featuring PDA-modified membranes showed roughly 30–50% greater flux parameters than the unmodified membranes after 1 day of oil emulsion filtration, while the PDA-g-PEG-modified membrane indicated no flux decline during the filtration process. The short-term BSA adhesion decrease was likewise observed in all instances of the PDA-g-PEG-modified membranes, and the overall tendency of the BSA adhesion was lowered with the raise of PEG graft molecular weight [3]. In a separate experimental trial, Freeman et al. [4] showed that PDA and PDA-g-PEG coatings might not be able to successfully control long-term membrane fouling. The traditional oil removing membranes are quickly fouled or often blocked by oil droplets due to their inherent oleophilic potential. The self-cleaning lotus leaves and their unique wettability properties can contribute to the wetting/antiwetting behavior of oil droplets, since they are essential to the design of membranes with low oil fouling potential. Research conducted by Tuteja et al. [8] effectively used oleophobic self-cleaning membranes with re-entrant texture and amphiphilic characteristics in oil–water separation procedures. The superoleophobic and superhydrophilic mesh membranes spin-coated with fluorodecyl POSS and x-PEGDA blend can selectively separate water from a range of oil–water mixtures using only the force of gravity. The membrane oleophobicity of water was the critical parameter for the separation of oil-in-water emulsions. Once the water phase of the emulsion reached the membrane, the PEGDA chains of amphiphilic surfaces would reconfigure and permit water to freely pass through the membrane, while the hexadecane droplets remained collected above the membrane (**Figure 1b**). Membrane oleophobicity occurring in both air and under water is essential for the separation process in water-in-oil emulsions. The PEGDA chains of amphiphilic surfaces begin the process of reconfiguring when the water droplets within the emulsion touch the membrane. In this case, the hexadecane phase remained above the membrane and the water droplets went through the membrane (**Figure 1d**). The test cases using these membranes suggest that they can separate different oil–water emulsions with great efficiency. In both instances, the permeate contained only ~0.1 wt% hexadecane, while the retentate contained only ~0.1 wt% water. Superoleophobic mesh membranes that were spin-coated with x-poly(dimethylsiloxane) (x-PDMS) blend and fluorodecyl POSS could likewise remove >99% of the emulsified oil droplets from a range of oil/water mixtures when triggered by the influence of electric field [9]. The low solid surface energy values and the re-entrant texture of the membrane permitted it to sustain water and hexadecane in the Cassie–Baxter state (superomniphobic) without an involvement of an electric field. Once the electric field was administered, the polar water in the Cassie–Baxter state under gravity transited into the Wenzel state. At the instance when the applied pressure became higher than the maximum pressure that the liquid–air interface could withstand, the water was able to permeate through the membrane whereas the oil became retained. The superoleophobic and superhydrophilic nanocomposite-coated membranes PDDA-PFO/SiO_2 prepared by Yang et al. showcased desirable water permeation and oil repellency behaviors. These membranes had the potential to selectively separate water from oil–water mixtures and at the same time offer good easy-recycling and antifouling parameter potential [10]. The advantages that can be derived from the bioinspired special wettability properties facilitated the successful application of various

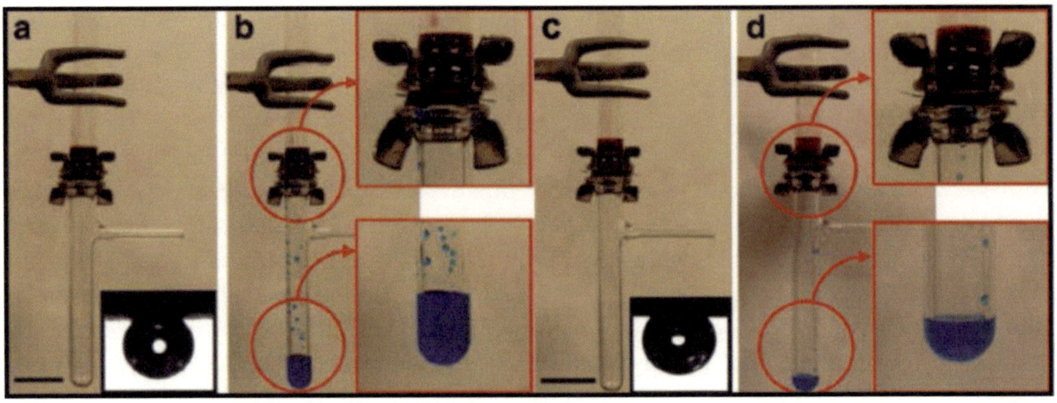

Figure 1. Separation of oil-in-water and water-in-oil emulsions. (a) Separation apparatus with a 50:50 (v:v) hexadecane-in-water emulsion ratio above the membrane. Inset, hexadecane droplet on a surface spin-coated with fluorodecyl POSS and x-PEGDA blend, submerged in water containing a dissolved nonionic surfactant. (b) Water-rich permeate passed through the membrane, while hexadecane-rich retentate was retained. (c) Separation apparatus with a 30:70 (v:v) water-in-hexadecane emulsion above the membrane. Inset, hexadecane droplet on a surface spin-coated with fluorodecyl POSS and x-PEGDA blend, submerged in water containing dissolved PS80. (d) Water-rich permeate passed through the membrane whereas hexadecane-rich retentate was retained. Water is dyed color blue and hexadecane is dyed color red. Ref. [8], Copyright 2012; reproduced with permission from the Nature Publishing Group.

superhydrophobic and superoleophilic membranes during the process of oil and water separation [11, 12]. Research group by Ding et al. [13–15] was able to achieve high-throughout separation of oil-water mixtures when they used fluorinated hybrid superhydrophobic and superoleophilic electrospun nanofibers. In this case, the oil–water emulsions or mixtures were poured onto the membranes and the oils quickly spread and permeated through the membranes with the water remaining on the membrane's surface. For instance, a promising flux of 3311 L m^{-2} h^{-1} and high separation efficiency potential was noted by researchers [14]. The hierarchical structures of fish scales that allow fish to maintain their bodies clean in oil-polluted water also provide potential ideas for solutions. To illustrate, underwater superoleophobic poly-acrylamide hydrogel-coated [16] and chitosan-coated [17] mesh membranes have been effectively applied in gravity-driven separation process of oil–water mixture where they showed separation efficiency potential higher than 99% for a range of oils. The hydrophilic coatings can absorb water as part of its balanced state in cases with water presence. Once the hydrogel coatings became in contact with the oil droplets, water can become trapped in the coarse nanostructures and the new oil–water–solid composite interface begins to showcase superoleophobic properties. This is caused by the trapped water molecules and the fact that they dramatically decrease the contact area between oil droplet and membrane's surface with discontinuous triple-phase contact line. Research group of Jin likewise proposed underwater superoleophobic polyelectrolyte-grafted Polyvinylidene difluoride (PVDF) membranes that can effectively separate surfactant-free oil-in-water emulsion with high separation success rate (>99.99%) and high flux potential (>1500 L m^{-2} h^{-1}, 0.01 MPs) [18, 19]. The amphiphilic self-cleaning membranes were created with compositional heterogeneity merging the fouling release property of low surface energy components with the fouling resistance of hydrophilic components on the surface. This self-cleaning membrane was then applied to oil/water emulsion separation [20–23].

The concerns that arise from membrane fouling can be effectively suppressed through dynamics that allow permeation flux to decline to an ultra low level with the smallest value of less than 3.4%, and ensure that after simple hydraulic washing process, the permeation flux recovery is retained at nearly 100%. As part of this vigorous filtration process, the fouling release properties of low surface energy micro domains stopped coalescence, spreading, and migration of the holistic hydrophobic oil droplets. This preventative measure remarkably reduced and even prevented the decline of the reversible flux. The hydrophilic domains improved the antifouling properties even more by creating compact hydration layer and opposing the nonspecific interaction between the membrane's surface and the incoming foulants. Medium levels of applied shear force effortlessly brushed the oil droplets away from the surface and then pushed them back into the rest of the feed solution (**Figure 2**). Alternatively, bioinspired superhydrophobic self-cleaning membranes featuring micro/nanoscale hierarchical structures and displaying very high levels of water repellence could also be used for membrane distillation. Researchers Wang et al. [24] introduced electrospun PS micro-/nanofibrous membranes displaying comparable micro/nanoscale hierarchical structures of silver ragwort leaf and lotus leaf during desalination with the aid of direct contact membrane distillation process. The properties of high superhydrophobicity (>150°) helped to prevent membrane wetting and guaranteed high liquid entry pressure for stable low permeate conductivity. The relatively high porosity values (~70%) improved the vapor permeation potential and allowed it to be about 4–5 times higher than the values found in commercial PVDF membranes. The application of the combined fabrication method ensured that the electrospun nanofibers were improved with silver nanoparticle and 1-dodecanethiol with the aid of poly-dopamine acting

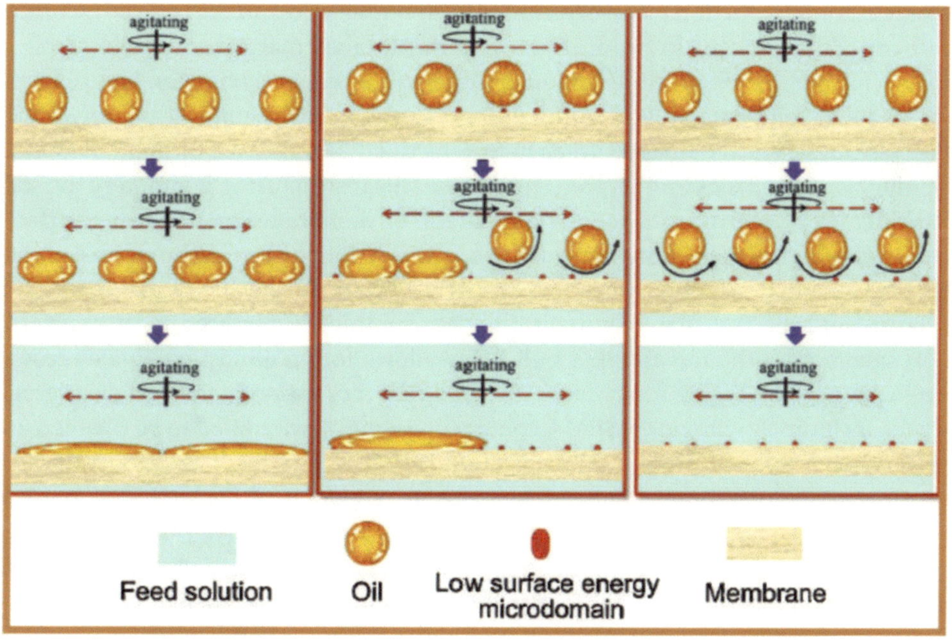

Figure 2. Illustration of the flux decline resistant mechanism in amphiphilic membranes with compositional heterogeneity [24].

as "bio-glue." This in turn gave the nanofiber membranes superhydrophobicity caused by the hierarchical structures, since they were advantageous for membrane distillation application process [25]. This substantial water flux enhancement of 2–3 times higher than a commercial PVDF membrane was credited to the open surface pore structure quality, while the stable and low conductivity (<5 μs cm^{-1}) was due to the lack of pathways for NaCl that would otherwise allow permeate to pass through the superhydrophobic membrane. Researchers Mansouri et al. [26] noted the heightened antifouling property of superhydrophobic fluorosilanized TiO_2 nanocomposite PVDF membranes when it came to various concentrations of both humic acid and NaCl solutions. This dynamic suggests a possibility of a long term antifouling performance potential in real seawater environment. A more successful treatment of industrial wastewater types can be achieved through the inclusion of nonfouling polymer brushes onto the membrane's surfaces. This approach offers a highly promising option for repelling various types of foulants, including hydrocarbons, microorganisms, colloids, and biomacromolecules. Membrane surfaces modified with phospholipid-like zwitterionic materials have been habitually implemented to repel proteins, cells, and other organic compounds from adhering to the membrane's surface. Research suggests that the zwitterionized PSf membrane based on the PSf/PDMAEMA-b-PSf-b-PDMAEMA blended membranes are almost without cell adhesion [27]. The PVDF-g-PCBMA and PVDF-g-PSBMA membranes based on PVDF/PVDF-g-PDMAEMA blend membranes showed an absence of protein deposition [28]. The PVDF membranes covered with amphiphilic PPO-b-PSBMA were able to successfully repel nonspecific protein surface adsorption throughout the dynamic filtration process and featuring minor irreversible flux decline ratio of 4.1% [29]. The zwitterionic colloid particles-coated PSf membranes have been reported as showing positive antifouling properties and facilitating a stable nanofiltration function when confronted with humic acid and BSA in a 30 h of filtration test scenario [30]. In order to produce durable membranes that can maintain stable effluent quality for long-duration water treatment operation, research has been focused towards membranes with self-healable antifouling properties. For instance, natural superhydrophobic plant leaves can self-heal damaged voids in the epicuticular wax layer through the process of rearranging wax molecules into layered structures. This restructuring guarantees the stability of wettability properties over the membrane's lifetime. Membranes created using surface segregation of amphiphilic copolymer may likewise include the rewarding properties of offering self-healing potential. These self-healing properties occur due to the damaged antifouling brush layers and the fact that eventually they are completely replaced by the surface segregation agents from the membrane's bulk. This allows for an almost complete recovery of antifouling properties in the membrane. The flux recovery ratio for the surface segregation membrane including zwitterionic SBMA content of 5.8 mol% was noted to be retained at 92%, even after three cycles of protein ultra-filtration [31]. The flux recovery ratio values of the surface segregation membrane featuring near-surface coverage of Pluronic F127 of 62 mol% showed flux recovery potential as high as 90% within three repetitive protein ultrafiltration operations [32]. The low surface free energy membrane surfaces produced using forced surface segregation of fluorine-containing copolymers could similarly maintain more successful antifouling performance capabilities with oily foulants after several cycles of ultrafiltration operation [20–22]. In a large number of water and wastewater treatment applications, contaminants such as metal ion, bacteria, proteins, and other contaminants polluting the water are removed primarily using the sieving mechanism that relies on the membrane's pore sizes and the size range of its contaminants. Research suggests that nanoporous membranes can be highly useful for the advantageous resolution but also for better capability throughout. Sophisticated selective transport properties of cell membranes helped develop a number of new strategies useful during fabrication of biomimetic and bioinspired membranes able to conduct effective water treatment.

The development of nanoporous biomimetic membranes containing Aquaporins (AQPs) holds a lot of possibilities for water purification applications. AQPs unique permeability and selectivity make these membranes a likely candidate for the design of high-performance membranes intended to be used for desalination. Research conducted by Kumar et al. [33] showed that AQP-incorporated triblock copolymer membranes can help develop significantly more controllable, sustainable, and productive water treatment membranes. These membranes would be able to provide a range of permeability levels, depending on the various concentrations of AQPs used. Research test cases showed that permeability peaked at a protein-to-polymer ratio of 1:50 and featuring permeability 3000 times greater than that of the pure polymer. Methodical studies have been conducted on active AQP-based composite membranes and have suggested relatively competitive water permeability potential and enhanced ion rejection values for existing RO, FO, and NF system [34–44]. An exceptional performance of AQP-based NF membrane can allow for the overall water flux of 36.6 L m^{-2} h^{-1} bar^{-1} and an $MgCl_2$ rejection of 95% (1 bar) [43]. In most cases, the AQP-based composite membranes had highly multivalent ion rejection capabilities, however, with a smaller NaCl rejection that restricted its applications for brackish and seawater desalination processes. Multiple innovative approaches for building AQP-based composite membranes with compatible NaCl rejection potential have been reported. Researches Chung et al. [35] cross-linked ruptured AQP-incorporated vesicles with acrylate-functionalized polycarbonate membrane support in order to facilitate a reduction in the number of uncovered pores to an insignificant level and to augment the NaCl rejection to a level above 98.5%. Chung et al. likewise implemented AQP-embedded vesicular membrane stabilized using an optimized layer-by-layer PDA-histidine coating during specific forward osmosis testing mode. This particular form osmosis testing relied on 6000 ppm NaCl as the feed, 0.8 M sucrose as the draw solute, and reporting high salt retention levels of 91.8% [40]. Alternatively, Tang et al. [38] determined that 1,2-dioleoyl-sn-glycero-3-phosphocholine-based proteoliposomes showed highly desirable osmotic water flux and NaCl reflection. They were also able to achieve adequate levels of NaCl rejection (~97%) and excellent water permeability parameters (4.0 L m^{-2} h^{-1} bar^{-1}) by using AQP-incorporated vesicles in well-established interfacial polymerization. Tang et al. offered a potential future of design criteria for AQP-based composite membranes for the application in desalination process [42]. Their research indicated that AQP-containing proteoliposomes were responsible for providing preferential water pathways in the ion rejection layer (**Figure 3a**), while fused AQP-containing lipid bilayers were in charge of NaCl rejection process (**Figure 3b**). The complex joining of lipid bilayer and proteoliposomes embedded matrix has been forecasted to ensure high water permeability levels and excellent salt rejection values (**Figure 3c**). Although there have been a number of successful accomplishments in the development of AQPs-based membranes, there are still key limitations to their proper scale up and larger scale applications because of the highly specialized requirements and excessively costly nanofabrication techniques [45]. Furthermore, the overall stability of the AQP-containing proteoliposomes during various extreme seawater conditions must be taken into account. On the other hand, carbon nanotubes (CNT) feature a much more rapid mass transport than the one calculated from continuum hydrodynamics models, such as the aquaporin water transport [46]. Prospective applications of CNT as a selective layer on the membrane's surface used for water treatment can be predicted through an array of high flux molecular channels. Membranes that have ultra high density of uniform molecular size pores and can be effortlessly produced on technical scale would constitute a considerable progress for membrane-based water treatment solutions [47]. The property of self-assembly in block polymers suggests new advantages in the production of nanoporous membranes with narrow pore size distributions, sharp molecular weight cut-off values, and high porosity. These features have influenced how nanoporous membranes have been fabricated using self-assembly of block copolymers to exhibit higher flux values if compared to commercial membranes, as reflected in **Table 1**. These fabricated membranes have superior abilities necessary for the competitive macromolecular separation platform due to the monodispersed pores' superior selectivity, smooth

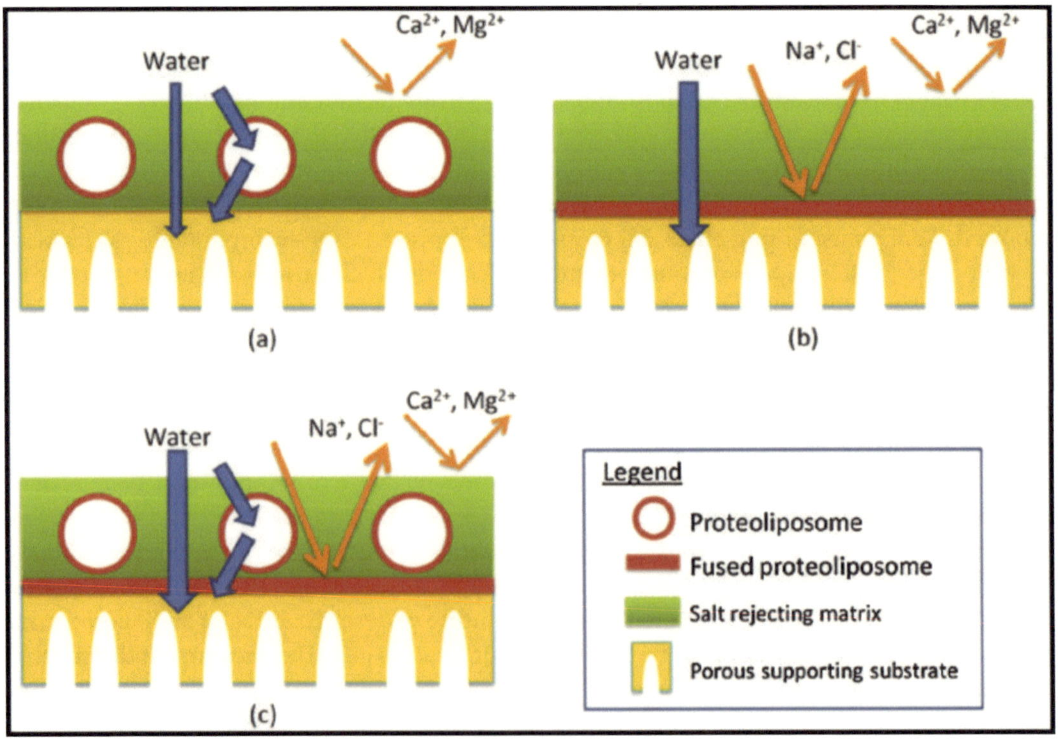

Figure 3. Conceptual designs of AQP-based composite biomimetic membranes. Ref. [42], Copyright 2013; reproduced with permission from the American Chemical Society.

http://dx.doi.org/10.5772/intechopen.71720

Assemblies	Effective pore diameter (nm)	Water permeability (L m^{-2} h^{-1} bar^{-1})
PS-b-PMMA	~15	~450
PS-b-PMMA	~17	~200
PS-b-PI-b-PLA	~22	~165
PS-b-P4VP	~8	~40
PS-b-P4VP	~19	~850
PS-b-P4VP	~50	~600
PS-b-P4VP	~25, 38	~450, 625
PS-b-PMMA	~1–2	~37
PI-b-PS-b-P4VP	~16–36	~150–850
PS-b-PEO	~20–30	~800
PS-b-P2VP	~8–25	~100–300
PS-b-P4VP	~100	~3200
PS-b-P4VP	~25	~600

Table 1. Summary of selective nanochannels in membranes based on block copolymer self-assembly.

surfaces that deter fouling, and high void fraction that permits higher fluxes [48]. Research conducted by Stamm et al. offer an illustration of this type of application. The nanoporous PS-b-P4VP membrane based on the self-assembly of PS-b-P4VP/HABA supramolecular complexes showed monodisperse pore radius of 12.3 nm and a high pore density of 2.43×10^{14} pores m^{-2}. These values are responsible for the high Congo red dye rejection (>98%) and rapid pure water flux (>600 L m^{-2} h^{-1} bar^{-1}). Additional zwitterionization and quaternization responses during aP4VP instances improve antibacterial and antifouling properties of membranes [49]. Despite the progress being made, there are still significant research strides when it comes to reducing the pore sizes down to the molecular dimensions for small molecule type of separations and to creating long range, dynamic, and highly selective nanochannels for large-scale production and industrial application.

3. Clean energy

The advancement of clean energy projects has invited a lot of interest because of the increasing demands for energy consumption and the growing environmental concerns that often accompany energy use and production. Membrane technology offers high efficiency, green conscious, and energy-saving qualities that can be actively used in clean energy production processes. In fact, membrane technologies show clear economic and technical advantages over many conventional technologies that are not environmentally conscious. Currently, biomimetic and bioinspired membranes are being incorporated into broader research and industrial applications and continue to promise further developments.

3.1. Fuel cells

Fuel cells are electrochemical devices that use the reaction between fuel, such as hydrogen, alcohol, or other hydrocarbon compound, and oxidant, or oxygen, to transform fuel's chemical energy into electric energy, but without being processed through the heat engine. For a wide range of fuel cells, the ion exchange membrane is the primary component that defines how well the fuel cell performs its job of conducting ion. Membranes' critical role in fuel cells is to prevent the diffusion of fuel from the anode to the cathode. If the membrane fails to effectively prevent the diffusion, it will dramatically reduce the performance of fuel cell because of the potential mixed effect and electrode poisoning [50, 51]. Different bioinspired and biomimetic tactics have been implemented in order to stimulate ion conduction process and inhibit fuel diffusion. Researchers Xu et al. [52] and Liu et al. [53] applied quaternized polymer to encourage mineralization of the silica precursor during the fabrication of hybrid anion exchange membranes. Once the process of hybridization was introduced, membrane's thermal stability and mechanical strength were enhanced by the stable structure of its inorganic component and the interactions between inorganic and organic phases. These improved parameters can offer a number of relevant benefits to the practical application of fuel cells in research and industrial settings. In this case, the methanol crossover was lowered from 4.1×10^{-10} to 8.45×10^{-11} $m^2\ s^{-1}$ because of the more tortuous methanol-transport pathway and the lowered free volume that heightened diffusion resistance properties of methanol [53]. Another promising variety of ion exchange membrane is the microphase-separated membrane that can form ion-cluster channels and improve the transport speed of ions. Despite its positive characteristics, this membrane has relatively high fuel permeability that restricts its application [50, 51]. There are a number of methods for reducing fuel permeability and the process of coating a fuel barrier layer has offered one such methodology due to its high efficiency and easy manipulation [54]. The barrier layer has the capacity to block the ion-rich hydrophilic domains in microphase-separated membranes, and as a result prevent the crossover of fuel [51]. However, the ion conductivity can decline if the coating layer is thick and nonconductive [50]. As a result, it is necessary to develop a fuel barrier layer with lower thickness and higher ion conduction capability. Wang et al. [54] relied on the bio adhesion phenomenon when modifying the Nafion membrane surface with PDA, one of the most often used cation exchange membranes with microphase-separated structure. The cross-linking structure of the PDA layer can block the ion-cluster channels and subdue the Nafion membrane swelling. Moreover, PDA's low hydrophilicity had negative effects on the solution of methanol at the membrane's surface. This caused a decrease of the membrane methanol crossover from 3.14×10^{-10} $m^2\ s^{-1}$ to 0.65×10^{-10} $m^2\ s^{-1}$. The ultra-thin quality of the PDA layer and the various proton conducting groups, such as amino, catechol, and imino, within the PDA layer forced Nafion membrane to sacrifice some of its conductivity. Recent research has focused on the block copolymers containing hydrophilic and hydrophobic segments as potentially useful for ion exchange membranes. These types of materials can be highly promising since they easily form ordered ion-cluster channels that resemble biological nanopores/nanochannels and can achieve high ion conductivity potential. A range of block copolymers with sulfonic acid and quaternary ammonium groups has been assembled for proton exchange membranes (PEMs) [55–57] and anion exchange membranes [58], respectively. During the process of creation, the hydrophilic segments form ordered channels with connected sulfonic acid or quaternary ammonium groups, and the hydrophobic segments form high-stable phases as a part of the supporting substrate. The membrane's morphology and the size of hydrophilic channel may be manipulated through a controlled management of the molecular weight, rigidity of each segment, and hydrophilicity/hydrophobicity [55, 57]. Balsara et al. [57] case studies designed PEMs employing poly (styrene sulfonate)-b-poly (methyl butylene) (PSS-b-PMB) with the width of dry hydrophilic phases ranging from 2.5 to 39 nm. This experimental study showed that the PEMs, with diameters of hydrophilic phases less than 5 nm, had higher water uptake and proton conductivity at greater temperatures if compared to Nafion membranes and the membranes with larger hydrophilic phases. For example, in an instance where the temperature was increased from 298 to 363 K, the water uptake was raised from 72.5 to 74.9 wt.%, while the proton conductivity increased from 11 to 19 S m^{-1} for membranes with hydrophilic phase of about 5 nm. Alternatively, membranes with 7 nm hydrophilic phase showed a lowered water uptake from 52.5 to 30.5 wt.%, and a decreased proton conductivity from 8 to 6 S m^{-1} using the same experimental parameters. This experimental phenomenon was caused by the capillary condensation that can occur in confined spaces where a suppression of evaporation water is happening. The obtained data offers a promising possibility for the future uses of PEMs at high temperature values. In addition to creating ordered channels through self-assembly of block copolymers, a number of top-down approaches have likewise been investigated for the construction of ordered channels for ion conduction. During these investigations, porous substrates were grafted [59, 60] or infiltrated [61] with poly electrolytes for ion conduction. Researchers Moghaddam et al. [60] constructed a silica membrane with pores sized 5–7 nm and grafted sulfonic acid groups onto the inner surface for better proton conduction. In this case, the maximum silica membrane conductivity is 11 S m^{-1}. In order to maintain high conductivity values at low humidity, an ultra-thin silica layer with the thickness and pore size of ~2 nm was placed at the mouth of the nanopores. The membrane's inner surface was likewise customized with sulfonic acid groups for improved proton conductivity. The pore's small size prevented water release process when used in fuel cells. In fact, the proton conductivity values of the two-layer silica membrane could remain constant while the humidity was below 20%; however, the proton conductivity of mono-layer silica membrane started to decrease once the humidity became 50–60%. The proton conductivity values in the two-layer silica membrane were two to three orders of magnitude greater than those of Nafion membranes at low humidity parameters. The rigidity and solvent resistance of porous substrates helped to effectively suppress the swelling of the poly electrolytes in pores, which if left untreated could lead to greater fuel crossover. As a result, the membrane was assessed as capable of advantageous durability in practical research and industrial applications.

3.2. Alcohol fuel

Over the last decade, alcohol fuels including methanol, ethanol, propanol, and butanol, produced from biomass have become a subject of research and investment interest because of their potential to offer an environmentally friendly and renewable alternative to fossil energy. When making alcohol-based fuels and specifically during the key process of fermentation, it is essential that the produced alcohol is timely removed because of its inhibitory effects on the yeast activity [62]. The process of pervaporation is a fitting method for highly efficient removal of alcohol during continuous fermentation because it is simple in operation, poison-free for microorganisms, energy saving, and easy to pair with the main reaction [62, 63]. During the fermentation process, water is likewise produced and is then removed as a part of the crude alcohol product that must be taken out to form high-purified alcohol. The process of pervaporation is a promising method for the separation of azeotropic mixtures, such as water–alcohol mixture, since it is not limited by the vapor–liquid equilibrium. Both water-permselective and alcohol pervaporation membranes can be implemented during the production of alcohol fuel. At this point in research, various biomimetic and bioinspired strategies have been actively used to generate membranes with greater separation performance that would reinforce the competitiveness of pervaporation method. When it comes to membrane separation processes, the permeation flux embodies the membrane's treatment capacity and effectually implies that high permeation flux values can lower the required membrane area and decrease investment membrane costs. Research has distinctly shown that the membrane's thickness can have a direct influence on the permeation flux capacity of pervaporation membranes. Specifically, thin membranes can reduce the diffusion pathway of the permeate molecules and augment membrane's permeation flux [64]. As a result, composite membranes involving a thin dense separation layer and a thick porous support layer formed using different materials are usually implemented in industrial-scale applications. To further reduce the separation layer's thickness so to enhance the permeate flux at a given applied pressure, the separation layer needs to be more durable; and a better bonding between the separation layer and the support layer is also desirable. For instance, bioadhesives and biomimetic adhesives, such as CP [65], polycarbophil calcium (PCP) [66], hyaluronic acid [67], dopamine [68], and gelatin [69] have been used for the creation of an intermediate layer or separation layer in composite membranes implemented for the process of ethanol (aqueous) solution dehydration. In research studies using CP and PCP as intermediate layers [65, 66], there was a notable change so far as the interfacial strength and interfacial compatibility were enhanced. Based on this research, composite membranes with thin and intact separation layers were successfully generated. Once an intermediate layer was added, the separation factor of composite membranes was raised by more than one order of magnitude and indicated the presence of advantageous long-term operational stability. Specifically, the membranes employing PCP as the intermediate layer exhibited an exceptional separation performance with the permeation flux of 1.39×10^3 g m^{-2} h^{-1} and the separation factor of 1279. Membranes generated using self-assembly of diblock or triblock copolymers has incited considerable interest in pervaporation because of their capability to provide continuous phase for the permeate transport process. PDMS [62] and polybutadiene (PB) [62, 70, 71] are commonly used for transporting blocks for ethanol-selective membranes. Researches Balsara et al. [62, 70] and Buonomenna et al. [71] examined the effects of molecular weight, as well as solvent and mass ratio of various blocks, on the ethanol/water separation performance and morphology of block copolymer membranes. A continuously growing molecular weight of block copolymers caused larger domain spacing and acted in a manner that improved the permselectivity of ethanol [62]. Both the solvent [71] and mass ratio of blocks [70] had a substantial effect on the self-assembled morphology, including spherical, lamellar, and cylindrical morphologies. Notably, the continuous cylindrical morphology showed the best separation performance. During a test featuring this type of membranes and fermentation broth as the feed solution, the results produce an enrichment of ethanol from 8 to 40 wt.% [70].

3.3. Clean gasoline

In the forthcoming years, fossil fuels will still continue to play a central role in the production of heat and power for daily life and industrial environments. As a result, the process of cleaning fossil fuels remains a critical strategy that can help curtail environmental pollution and its extensive implications. Sulfur compounds within the gasoline are the primary sources of atmospheric pollution and acid rain. These compounds are also poisonous catalysts of the vehicle exhaust gas convertor [72]. As a consequence, the sulfur compound content in gasoline must be strictly controlled and enforced. Over the last decade, the process of pervaporation desulfurization has received increased attention because of its various benefits over conventional hydrodesulfurization processes. Some of these advantages include higher selectivity, lower operating costs, decreased energy costs, facile scale up, easier maintenance of the octane number, as well as without hydrogen source and coproduct of H_2S gas [72–74]. Several bioinspired and biomimetic strategies have been implemented for the development of membrane materials with superior separation stability and performance potential. PDMS is used as the primary membrane material for pervaporation desulfurization of gasoline due to its superior processability, greater permeability, and high affinity for sulfur components. Despite these numerous advantages, pure PDMS membranes suffer from low strength and selectivity because of their highly flexible molecular chains [72, 73]. It has been noted that the appropriate incorporation of inorganic materials in polymeric matrix can in fact enhance the physicochemical stabilities and mechanical properties of the membrane. These hybrid membranes have the capacity to overcome the trade-off obstacle between selectivity and permeability effectively manipulating membrane's hydrophilicity/hydrophobicity and the arrangement of polymer chains that influence the entire chain spacing and chain rigidity values [75, 76]. PDMS–SiO_2 hybrid membranes were generated using the in situ biomimetic mineralization method [72, 73]. These membranes show that the utilization of silica precursor with greater reactivity values can contribute to the formation of smaller silica nanoparticles. As a result of this formation, a larger interfacial area caused more hydrogen bonds to occur between the silanol groups on the silica surface and the oxygen atoms on the polymer chains that in turn substantially improved the mechanical strength potential of the membranes. The infusion of silica into PDMS matrix is likewise helpful for increasing number and size of free volume cavities, a dynamic that actively supports lower diffusion resistance to the penetrant molecules. The prepared hybrid membrane displayed an exceptionally good desulfurization performance with a permeation flux of 10.8×10^3 g m^{-2} h^{-1} and a selectivity of 4.8 toward thiophene in gasoline model.

The high cohesive/adhesive energy properties of bioadhesives contributed to the preparation and incorporation of dopamine nanoaggregates into PDMS matrix [77]. Once the dopamine nanoaggregates were added, the chain rigidity and cohesive energy of PDMS were improved, as well as its thermal stability and swelling resistance. The free volume properties of PDMS membrane were similarly optimized though the intervening of dopamine nanoaggregates during the packing of PDMS polymer chains. The simultaneous improvement of permeation flux and enrichment factor was obtained when permeation flux increased from 2.78×10^3 to

6.90×10^3 g m^{-2} h^{-1} and enrichment factor grew from 4.3 to 4.5, thus surpassing the upper-bound curve of the PDMS control membrane. Researchers Jiang et al. generated an ultra-thin PDA coating on the PSf substrate. If compared to the PDMS membrane properties, the PDA membrane showcased higher hydrophilicity, higher cohesive energy, lower thickness, and higher adhesive strength with PSf substrate. Furthermore, this improved the PDA/PSf composite membrane and allowed it to have a more favorable pervaporation performance and long-term durability potential. Double amounts of PDA coating or even more than that were required for the pervaporation desulfurization process since the single-coated PDA showed relatively low selectivity toward thiophene. In addition to the PDMS, PEG can also act as an appropriate membrane material for pervaporation desulfurization of gasoline in accordance with the methodology outlined in the solubility parameter theory. Kong et al. [74, 78] created PEG-b-PAN membranes based on dispersed PEG micro domains and bulk PAN phases. In this case, an increase in the total flux and a decrease in sulfur enrichment factor were attained by either lowering PEG molecular weight or by increasing PEG mass content value that caused greater proportion of PEG micro domains. Additional research is necessary in order to comprehensively study pervaporation desulfurization permeability of PEG-based block copolymer membranes and their various applications in the industrial and environmental projects.

4. Carbon capture

Scientific research has confirmed that the global growth of greenhouse gas emissions in the atmosphere and particularly CO_2 emissions has acted as one of the key sources in the ongoing climate change over the past several decades. In 2014, CO_2 accounted for almost 80.9% of all U.S. greenhouse gas emissions caused by human activities. Carbon dioxide is naturally present in the Earth's atmosphere as a part of its carbon cycle, where the carbon naturally circulates between the atmosphere, oceans, soil, animals, and plants. Research prognosis suggests that the CO_2 emissions are going to severely increase in the foreseeable future [79, 80]. In fact, the International Panel on Climate Change predicts that ice-melt will offer drastically higher projections of between 2.4 and 6.2 feet sea level rise by 2100, because of the increasing of CO_2 emissions [80]. As part of the solution for resolving this global dilemma, technologies for obtaining CO_2 from gaseous mixtures can be divided into three specific categories, namely liquid absorption, solid adsorption, and membrane separation [80]. If compared with liquid- and solid-oriented types of technologies, the membrane separation technology offers a series of inherent advantages. Membrane separation is an environmentally friendly process without hazardous chemicals, simple in operation, requiring small equipment and low energy consumption. Substantial research projects have been dedicated to manufacturing high-performance robust membranes using biomimetic and bioinspired strategies for the development of membrane-based carbon capture technology. In fact, the adhesive capacity and iron-fortified property of marine adhesive proteins have been used for membrane design. For instance, Fe^{3+}–dopamine organometallic nanoaggregates (Fe^{3+}–DA) were integrated into polymeric matrix as a filler in order to construct defect-free and ultra-thin hybrid membranes [81]. In this instance, the interfacial interaction between filler and polymer matrix was optimized by varying the molar ratio of Fe^{3+} to DA. This gave the membrane favorable free volume that was beneficial to the selective diffusion of CO_2 molecules within the membrane. There was a substantial improvement in CO_2/CH_4 selectivity values from 21 to 72, while the comprehensive performance exceeded the most recent upper bound line. Moreover, the membrane attained a positive suppression of CO_2-induced plasticization at high operating pressures. The process of self-assembly of block copolymer into a membrane with highly ordered and continuous structure can be an advantage for mass transport through the membrane and has been researched for applications in CO_2 separation and capture. Researchers Cohen et al. [82] interrogated the permeation of gases through PS-b-PB block copolymer membranes featuring highly oriented lamellar microstructure. The results obtained by Cohen et al. suggest that the highest permeability was achieved when the lamellae were oriented in a position parallel to the permeation direction. An analogous dependence of permeability on the direction of lamellae orientation was noted by Kofinaset al. [83]. Recent research project by Gao et al. [80, 83] synthesized block copolymers with PS segments and linear PEO or brush-type PEO (poly [oligo (ethylene glycol) methyl ether methacrylate], POEGMA) segments. After the process of self-assembly, cylindrical structures were formed for both BCPs with PEO cylinders oriented perpendicularly to the surface. The perpendicular channel and the ether oxygen linkages in PEO segments were in support of CO_2 permeation. Specifically, the brush-type PEO segments featured lower crystallinity properties if compared to linear PEO segments. These properties afforded the cylindrical channels with high CO_2 affinity and free volume, thus aiding the solution and diffusion properties of the CO_2 molecules. As a consequence, the membrane showed an exceedingly high CO_2 permeance levels of 5.92×10^{-7} m^3 m^{-2} s^{-1} Pa^{-1} and an extensive separation performance that surpassed the Robeson's upper bound line. As an enzyme that exists in many organisms, carbonic anhydrase (CA) plays a critical role in the transport of CO_2 in vivo. It is by far the quickest catalyst for CO_2 hydration and dehydration processes as it features a turnover rate of 106 mol CO_2 mol^{-1} CA s^{-1} and can be used at low CO_2 concentration values [84]. With high levels of catalysis effectiveness, CA-immobilized membranes have been widely implemented for the CO_2 separation process [85, 86]. Researchers Zhang et al. [85] have created a hollow fiber membrane reactor by embedding CA in hydrogel that showed great performance with CO_2/N_2 selectivity of 820, CO_2/O_2 selectivity of 330, and CO_2 permeance values of 3.70×10^{-10} m^3 m^{-2} s^{-1} Pa^{-1}. To circumvent the CA inactivation at the industrial-scale production level, Wang et al. [87] generated a biomimetic poly(N-viny limidazole)–zinc (PVI–Zn(II)) complex so as to simulate the active site of CA (a Zn(II) tetrahedral center bound to three imidazole residues and a hydroxyl) and obtain a high-performance membrane for the separation of CO_2/N_2. The essential effect of PVI–Zn(II) complex on CO_2 hydration reversibly was carefully verified using a varying molar ratio of PVI/Zn(II), pH value of the PVI–Zn(II) solution, and by replacing PVI with polyvinyl pyrrolid (PVP), which significantly changed the structure and the amount of the complex. The CO_2 permeance and CO_2/N_2 selectivity of PVI–Zn(II) complex membrane were determined to be close to three and two times higher than those of pure PVI membranes, respectively.

5. Health care

Health care is becoming one of the most viable fields for the application of innovative membrane technology. New membrane technologies are being used in a wide range of health care areas, including in diagnosis, prevention, treatment of disease, injury, and other physical and mental impairments in humans. In particular, bioinspired and biomimetic membranes are being implemented in health care and research centers during the design and creation of effective artificial organs, such as artificial kidney, lung, and liver. In addition to excellent selective permeation properties and antifouling performance, the specific requirement for membranes utilized in artificial organs is the compatibility of the membrane's surfaces with its immediate surroundings, such as cytocompatibility and hemocompatibility [88, 89]. Currently, multiple biomimetic and bioinspired methods have been used to construct highly compatible surfaces by immobilizing natural or synthesized macromolecules on the membrane's surface. PVP is a hydrophilic polymer with a highly advantageous biocompatibility that exceeds all the ones available for use as blood plasma substitutes. Zhao et al. [88] altered the PES membrane surfaces with PVP by applying surface segregation of amphiphilic triblock copolymer PVP-b-PMMA-b-PVP. If compared to the PES membranes, the membranes modified by Zhao et al. offered an improved hemocompatibility (BSA adsorption decreased from 19 to 10 $\mu g/cm^2$, platelet adhesion decreased from 25×10^7 cells/cm^2 to nearly zero, blood coagulation time increased from about 55 to 90 s) and cytocompatibility, with more flat cell morphology, higher surface coverage, and favorable cell viability. These enhancements invested the membranes with a promising potential in the field of blood purification, including hemodialysis and artificial liver production and maintenance. Heparin is a widely used blood anticoagulant with remarkable cytocompatibility, hemocompatibility, and cell proliferation properties [90, 91]. For the grafting of heparin [90–92] and heparin analogs [91] onto the membrane's surfaces, biomimetic adhesion method was employed as a relatively quick, green conscious, and highly effective option relying on coating dopamine-anchored heparins or precoating PDA followed by grafting heparin onto the surfaces. As an alternative to the PVP-modified membranes, the dopamine-heparin-modified membranes showed a noticeably greater improvement in the anticoagulation performance, with clotting time amplified from 10 to more than 60 min for PE/dopamine-heparin membranes [90]. Furthermore, the improved water flux (from 371.4 to 644.9 L m^{-2} h^{-1}) [90], discernably suppressed adhesion, activation, and transmutation of platelets, and helped cell attachment and growth on membrane's surface [90, 91]. These results were the result of an increased surface hydrophilicity and biological activities of heparin molecules. In comparison to the highly complex processes and excessive costs of extracting heparin from the animal body, synthesized heparin analogs have comparable functions and structure of heparin, making it definitively more appealing for practical applications in health care. For example, phosphoryl choline is a biological zwitterion situated on the cell membrane's outside surface and that can offer critical antifouling properties to the cell membrane. It also features encouraging biocompatibility because of its zwitterionic nature and electrostatically induced hydration properties [89]. During the last decade, surface modifications using phosphorylcholine and other zwitterions have been widely examined in an attempt to enhance biocompatibility of the membrane's surfaces [27, 28, 93–99]. Chang et al. [93, 94, 96] examined the hemocompatibility of zwitterionic surfaces by grafting poly(sulfobetainemethacrylate) (PSBMA) onto the membranes. In this case, the optimum antifouling (low protein adsorption), anticoagulant (long plasma-clotting time), and antithrombogenic (low hemolysis of red blood cells solution) properties were obtained once the membrane's surface had the greatest hydration capacity and the lowest charge bias values. To sum up, the PSBMA-grafted membranes can provide highly desirable nonbioadhesive characteristics when in contact with tissue cells and bacterial medium. They also offered properties that make it a suitable micro environment for skin wound healing, and have been identified as having great potential applications in the rational design and quick preparation of advanced wound dressings [96]. Although more research needs to be conducted on the long term applicability of these membrane approaches, the surface zwitterionization appears as one of the most advantageous strategies for constructing biocompatible surfaces because of its consistently high-rate performance, great diversity, and easy processability properties.

Author details

Amira Abdelrasoul[1]*, Huu Doan[2] and Ali Lohi[2]

*Address all correspondence to: amira.abdelrasoul@usask.ca

1 Department of Chemical and Biological Engineering, University of Saskatchewan, Saskatoon, Saskatchewan, Canada

2 Department of Chemical Engineering, Ryerson University, Toronto, Ontario, Canada

References

[1] Shannon MA, Bohn PW, Elimelech M, Georgiadis JG, Marinas BJ, Mayes AM. Science and technology for water purification in the coming decades. Nature. 2008;**452**:301-310

[2] Kasemset S, Lee A, Miller DJ, Freeman BD, Sharma MM. Effect of polydopamine deposition conditions on fouling resistance, physical properties, and permeation properties of reverse osmosis mem-branes in oil/water separation. Journal of Membrane Science. 2013;**425-426**:208-216

[3] McCloskey BD, Park HB, Ju H, Rowe BW, Miller DJ, Chun BJ, Kin K, Freeman BD. Influence of polydopamine deposition conditions on pure water flux and foulant adhesion resistance of reverse osmosis, ultrafiltration, and microfiltration membranes. Polymer. 2010; **51**:3472-3485

[4] Miller DJ, Araújo PA, Correia PB, Ramsey MM, Kruithof JC, vanLoosdrecht MCM, Freeman BD, Paul DR, Whiteley M, Vrouwen-velder JS. Short-term adhesion and long-term biofouling testing of polydopamine and poly(ethylene glycol) surface modifications of membranes and feed spacers for biofouling control. Water Research. 2012;**46**:3737-3753

[5] McCloskey BD, Park HB, Ju H, Rowe BW, Miller DJ, Freeman BD. A bioinspired fouling-resistant surface modification for water purification membranes. Journal of Membrane Science. 2012;**413-414**:82-90

[6] Miller DJ, Kasemset S, Wang L, Paul DR, Freeman BD. Constan tflux crossflow filtration evaluation of surface-modified fouling-resistant membranes. Journal of Membrane Science. 2014;**452**:171-183

[7] Arena JT, McCloskey B, Freeman BD, McCutcheon JR. Surface modification of thin film composite membrane support layers with polydopamine: Enabling use of reverse osmosis membranes in pressure retarded osmosis. Journal of Membrane Science. 2011;**375**:55-62

[8] Kota AK, Kwon G, Choi W, Mabry JM, Tuteja A. Hygro-responsive membranes for effective oil–water separation. Nature Communications. 2012;**3**:1025

[9] Kwon G, Kota AK, Li Y, Sohani A, Mabry JM, Tuteja A. On-demand separation of oil–water mixtures. Advanced Materials. 2012;**24**:3666-3671

[10] Yang J, Zhang Z, Xu X, Zhu X, Men X, Zhou X. Supe rhydrophilic–super oleophobic coatings. Journal of Materials Chemistry. 2012;**22**(7):2834

[11] Li J, Shi L, Chen Y, Zhang Y, Guo Z, Su B, Liu W. Stable superhydropho-bic coatings from thiolligand nano crystals and their application in oil/water separation. Journal of Materials Chemistry. 2012;**22**:9774-9781

[12] Zhang W, Shi Z, Zhang F, Liu X, Jin J, Jiang L. Super hydrophobic and super oleophilic PVDF membranes for effective separation of water-in-oil emulsions with high flux. Advanced Materials. 2013;**25**:2071-2076

[13] Shang Y, Si Y, Raza A, Yang L, Mao X, Ding B, Yu J. An in situ polymerization approach for the synthesis of super hydrophobic and super oleophilic nanofibrous membranes for oil–water separation. Nanoscale. 2012;**4**:7847-7854

[14] Tang X, Si Y, Ge J, Ding B, Liu L, Zheng G, Luo W, Yu J. In situ polymerized super hydrophobic and supe roleophilic nanofibrous membranes for gravity driven oil–water separation. Nanoscale. 2013;**5**:11657-11664

[15] Huang M, Si Y, Tang X, Zhu Z, Ding B, Liu L, Zheng G, Luo W, Yu J. Gravity driven separation of emulsified oil–water mixtures utilizing in situ polymerized super hydrophobic and super oleophilic nanofibrous membranes. Journal of Materials Chemistry A. 2013;**1**:14071-14074

[16] Xue Z, Wang S, Lin L, Chen L, Liu M, Feng L, Jiang L. A novel super hydrophilic and underwater super oleophobic hydrogelcoated mesh for oil/water separation. Advanced Materials. 2011;**23**:4270-4273

[17] Zhang S, Lu F, Tao L, Liu N, Gao C, Feng L, Wei Y. Bio-inspired anti-oil-fouling chitosan-coated mesh for oil/water separation suitable for broad pH range and hyper-saline environments. ACS Appl MaterInterfaces. 2013;**5**:11971-11976

[18] Zhu Y, Zhang F, Wang D, Pei XF, Zhang W, Jin J. A novel zwitterionic polyelectrolyte grafted PVDF membrane for thoroughly separating oil from water with ultra high efficiency. Journal of Materials Chemistry A. 2013;**1**:5758-5765

[19] Zhang W, Zhu Y, Liu X, Wang D, Li J, Jiang L, Jin J. Salt-induced fabrication of super hydrophilic and underwater superoleophobic PAA-g-PVDF membranes for effective separation of oil-in-water emulsions. Angewandte Chemie (International Ed. in English). 2014;**53**: 856-860

[20] Zhao XT, Chen WJ, YL S, Zhu W, Peng JM, Jiang ZY, Kong L, Li YF, Liu JZ. Hierarchically engineered membrane surfaces with superior antifouling and self-cleaning properties. Journal of Membrane Science. 2013;**441**:93-101

[21] Chen WJ, Su YL, Peng JM, Zhao XT, Jiang ZY, Dong YN, Zhang Y, LiangYG LJZ. Efficient wastewater treatment by membranes through constructing tunable antifouling membrane surfaces. Environmental Science & Technology. 2011;**45**:6545-6552

[22] Chen WJ, YL S, Peng JM, Dong YN, Zhao XT, Jiang ZY. Engineering a robust, versatile amphiphilic membrane surface through forced surface segregation for ultra low flux-decline. Advanced Functional Materials. 2011;**21**:191-198

[23] Li YF, YL S, Zhao XT, He X, Zhang RN, Zhao JJ, Fan XC, Jiang ZY. Antifouling, high-flux nanofiltration membranes enabledby dual functional polydopamine. ACS Applied Materials & Interfaces. 2014;**6**:5548-5557

[24] Li X, Wang C, Yang Y, Wang X, Zhu M, Hsiao BS. Dual-biomimetic superhydrophobic electrospun polystyrene nanofibrous membranes for membrane distillation. ACS Applied Materials & Interfaces. 2014;**6**:2423-2430

[25] Liao Y, Wang R, Fane AG. Engineering super hydrophobic surface on poly (vinylidene fluoride) nanofiber membranes for direct contact membrane distillation. Journal of Membrane Science. 2013;**440**:77-87

[26] Razmjou A, Arifin E, Dong G, Mansouri J, Chen V. Super hydrophobic modification of TiO_2 nanocomposite PVDF membranes for applications in membrane distillation. Journal of Membrane Science. 2012;**415-416**:850-863

[27] Zhao YF, Zhu LP, Yi Z, Zhu BK, YY X. Improving the hydrophilicity and fouling-resistance of polysulfone ultrafiltration membranes via surface zwitterionic alization mediated by polysulfone-based triblock copolymer additive. Journal of Membrane Science. 2013;**440**:40-47

[28] Yi Z, Zhu LP, YY X, Gong XN, Zhu BK. Surface zwitter ionicalization of poly(vinylidene fluoride) porous membranes by post-reaction of the amphiphilic precursor. Journal of Membrane Science. 2011;**385-386**:57-66

[29] Venault A, Chang Y, Yang HS, Lin PY, Shih YJ, Higuchi A. Surface self-assembled zwitterionization of poly(vinylidene fluoride) microfiltration membranes via hydrophobic-driven coating for improved blood compatibility. Journal of Membrane Science. 2014;**454**:253-263

[30] Ji YL, Zhao Q, An QF, Shao LL, Lee KR, ZK X, Gao CJ. Novel separation membranes based on zwitterionic colloid particles: Tunable selectivity and enhanced antifouling property. Journal of Materials Chemistry A. 2013;**1**:12213-12220

[31] Sun Q, YL S, Ma XL, Wang YQ, Jiang ZY. Improved antifouling property of zwitterionic ultrafiltration membrane composed of acrylonitrile and sulfobetaine copolymer. Journal of Membrane Science. 2006;**285**:299-305

[32] Zhao W, YL S, Li C, Shi Q, Ning X, Jiang ZY. Fabrication of antifouling polyethersulfone ultrafiltration membranes using Pluronic F127 as both surface modifier and pore-forming agent. Journal of Membrane Science. 2008;**318**(12):405

[33] Kumar M, Grzelakowski M, Zilles J, Clark M, Meier W. Highly permeable polymeric membranes based on the incorporation of the functional water channel protein Aquaporin Z. Proceedings of the National Academy of Sciences of the United States of America. 2007;**104**:20719-20724

[34] Duong PHH, Chung TS, Jeyaseelan K, Armugam A, Chen Z, Yang J, Hong M. Planar biomimetic aquaporin-incorporated triblock copolymer membranes on porous alumina supports for nanofiltration. Journal of Membrane Science. 2012;**409-410**:34-43

[35] Wang H, Chung TS, Tong YW, Jeyaseelan K, Armugam A, Chen Z, Hong M, Meier W. Highly permeable and selective pore-spanning biomimetic membrane embedded with aquaporin Z. Small. 2012;**8**:1185-1190

[36] Zhong PS, Chung TS, Jeyaseelan K, Armugam A. Aquaporin-embedded biomimetic membranes for nanofiltration. Journal of Membrane Science. 2012;**407-408**:27-33

[37] Sun G, Chung TS, Jeyaseelan K, Armugam A. A layer-by-layer self-assembly approach to developing an aquaporin-embedded mixed matrix membrane. RSC Advances. 2013;**3**: 473-481

[38] Zhao Y, Qiu C, Li X, Vararattanavech A, Shen W, Torres J, Hélix-Nielsen C, Wang R, Hu X, Fane AG, Tang CY. Synthesis of robust and high-performance aquaporin-based biomimetic membranes by interfacial polymerization-membrane preparation and RO performance characterization. Journal of Membrane Science. 2012;**423-424**:422-428

[39] Sun G, Chung TS, Jeyaseelan K, Armugam A. Stabilization and immobilization of aquaporin reconstituted lipid vesicles for water purification. Colloids and Surfaces, B: Biointerfaces. 2013;**102**:466-471

[40] Wang H, Chung TS, Tong YW, Jeyaseelan K, Armugam A, DuongHHP FF, Seah H, Yang J, Hong M. Mechanically robust and highly permeable Aquaporin Z biomimetic membranes. Journal of Membrane Science. 2013;**434**:130-136

[41] Wang H, Chung TS, Tong YW. Study on water transport through a mechanically robust Aquaporin Z biomimetic membrane. Journal of Membrane Science. 2013;**445**:47-52

[42] Zhao Y, Vararattanavech A, Li X, Hélixnielsen C, Vissing T, Torres J, Wang R, Fane AG, Tang CY. Effects of proteoliposome composition and draw solution types on separation performance of aquaporin-based proteoliposomes: Implications for seawater desalination using aquaporin-based biomimetic membranes. Environmental Science & Technology. 2013;**47**:1496-1503

[43] Li X, Wang R, Wicaksana F, Tang C, Torres J, Fane AG. Preparation of high performance nanofiltration (NF) membranes incorporated with aquaporin Z. Journal of Membrane Science. 2014;**450**:181-188

[44] Xie W, He F, Wang B, Chung TS, Jeyaseelan K, Armugam A, Tong YW. An aquaporin-based vesicle-embedded polymeric membrane for low energy water filtration. Journal of Materials Chemistry A. 2013;**1**:7592-7600

[45] Tang CY, Zhao Y, Wang R, Hélix-Nielsen C, Fane AG. Desalination by biomimetic aquaporin membranes: Review of status and prospects. Desalination. 2013;**308**:34-40

[46] Holt JK, Park HG, Wang Y, Stadermann M, Artyukhin AB, Grigoropoulos CP, Noy A, Bakajin O. Fast mass transport through sub-2-nanometer carbon nanotubes. Science. 2006;**312**:1034-1037

[47] Nunes SP, Sougrat R, Hooghan B, Anjum DH, Behzad AR, Zhao L, Pradeep N, Pinnau I, Vainio U, Peinemann KV. Ultraporous films with uniform nanochannels by block copolymer micelles assembly. Macromolecules. 2010;**43**:8079-8085

[48] Phillip WA, O'Neill B, Rodwogin M, Hillmyer MA, Cussler EL. Self-assembled block copolymer thin films as water filtration membranes. ACS Applied Materials & Interfaces. 2010;**2**:847-853

[49] Tripathi BP, Dubey NC, Choudhury S, Simon F, Stamm M. Antifouling and anti biofouling pH responsive block copolymer based membranes by selective surface modification. Journal of Materials Chemistry B. 2013;**1**:3397-3409

[50] Lin H, Zhao C, Ma W, Li H, Na H. Low water swelling and high methanol resistant proton exchange membrane fabricated by cross-linking of multilayered polyelectrolyte complexes. Journal of Membrane Science. 2009;**345**:242-248

[51] Zhao C, Lin H, Zhang Q, Na H. Layer-by-layer self-assembly of polyaniline on sulfonated poly(arylene ether ketone) membrane with high proton conductivity and low methanol crossover. International Journal of Hydrogen Energy. 2010;**35**:10482-10488

[52] Lin X, Wu C, Wu Y, Xu T. Free-standing hybrid anion-exchange membranes for application in fuel cells. Journal of Applied Polymer Science. 2012;**123**:3644-3651

[53] Xiong Y, Liu QL, Zhu AM, Huang SM, Zeng QH. Performance of organic–inorganic hybrid anion-exchange membranes for alkaline direct methanol fuel cells. Journal of Power Sources. 2009;**186**:328-333

[54] Wang JT, Xiao LL, Zhao YN, Wu H, Jiang ZY, Hou WQ. A facile surface modification of Nafion membrane by the formation of self-polymerized dopamine nano-layer to enhance the methanol barrier property. Journal of Power Sources. 2009;**192**:336-343

[55] Uehara H, Kakiage M, Sekiya M, Yamagishi T, Yamanobe T, Nakajima K, Watanabe T, Nomura K, Hase K, Matsuda M. Novel design solving the conductivity vs. water-uptake trade-off for polymer electrolyte membrane by bicontinuous crystalline/amorphous morphology of block copolymer. Macromolecules. 2009;**42**:7627-7630

[56] Park MJ, Downing KH, Jackson A, Gomez ED, Minor AM, Cookson D, Weber AZ, Balsara NP. Increased water retention in polymer electrolyte membranes at elevated temperatures assisted by capillary condensation. Nano Letters. 2007;**7**:3547-3552

[57] Moore HD, Saito T, Hickner MA. Morphology and transport properties of mid block-sulfonated tri block copolymers. Journal of Materials Chemistry. 2010;**20**:6316-6321

[58] Sun L, Guo J, Zhou J, Xu Q, Chu D, Chen R. Novel nano structured high-performance anion exchange ionomers for anion exchange membrane fuel cells. Journal of Power Sources. 2012;**202**:70-77

[59] Yameen B, Kaltbeitzel A, Langner A, Duran H, Muller F, Gosele U, Azzaroni O, Knoll W. Facile large-scale fabrication of proton conducting channels. Journal of the American Chemical Society. 2008;**130**:13140-13144

[60] Moghaddam S, Pengwang E, Jiang YB, Garcia AR, Burnett DJ, Brinker CJ, Masel RL, Shanno MA. An inorganic–organic proton exchange membrane for fuel cells with a controlled nanoscale pore structure. Nature Nanotechnology. 2010;**5**:230-236

[61] Yamaguchi T, Zhou H, Nakazawa S, Hara N. An extremely low methanol crossover and highly durable aromatic pore-filling electrolyte membrane for direct methanol fuel cells. Advanced Materials. 2007;**19**:592-596

[62] Jha AK, Chen L, Offeman RD, Balsara NP. Effect of nanoscale morphology on selective ethanol transport through block copolymer membranes. Journal of Membrane Science. 2011;**373**:112-120

[63] Vane LMA. Review of pervaporation for product recovery from biomass fermentation processes. Journal of Chemical Technology and Biotechnology. 2005;**80**:603-629

[64] Raisi A, Aroujalian A. Aroma compound recovery by hydrophobic pervaporation: The effect of membrane thickness and coupling phenomena. Separation and Purification Technology. 2011;**82**:53-62

[65] Ma J, Zhang M, Wu H, Yin X, Chen J, Jiang ZY. Mussle-inspired fabrication of structurally stable chitosan/polyacrylonitrile composite membrane for pervaporation dehydration. Journal of Membrane Science. 2010;**348**:150-159

[66] Zhao CH, Wu H, Li XS, Pan FS, Li YF, Zhao J, Jiang ZY, Zhang P, Cao XZ, Wang BY. High performance composite membranes with a poly-carbophil calcium transition layer for pervaporation dehydrationof ethanol. Journal of Membrane Science. 2013;**429**:409-417

[67] Ma J, Zhang MH, Jiang ZY, Nie MC, Liu GX. Facile fabrication of structurally stable hyaluronic acid-based composite membranesinspired by bioadhesion. Journal of Membrane Science. 2010;**364**:290-297

[68] Chen J, Chen X, Yin X, Ma J, Jiang ZY. Bioinspired fabrication of composite pervaporation membranes with high permeation flux and structural stability. Journal of Membrane Science. 2009;**344**:136-143

[69] Zhao J, Ma J, Chen J, Pan FS, Jiang ZY. Experimental and molecular simulation investigations on interfacial characteristics of gelatin/polyacrylonitrile composite pervaporation membrane. Chemical Engineering Journal. 2011;**178**:1-7

[70] Jha AK, Tsang SL, Ozcam AE, Offeman RD, Balsara NP. Mastercurve captures the effect of domain morphology on ethanol pervaporation through block copolymer membranes. Journal of Membrane Science. 2012;**401-402**:125-131

[71] Buonomenna MG, Golemme G, Tone CM, De Santo MP, Ciuchi F, Perrotta E, Zappone B, Galiano F, Figoli A. Ordering phenomena in nanostructured poly(styrene-b-butadiene-b-styrene) (SBS) membranes for selective ethanol transport. Journal of Membrane Science. 2011;**385-386**:162-170

[72] Li B, Yu SN, Jiang ZY, Liu WP, Cao RJ, Wu H. Efficient desulfurization by polymer–inorganic nano composite membranes fabricated in reverse micro emulsion. Journal of Hazardous Materials. 2012;**211-212**:296-303

[73] Li B, Liu WP, Wu H, SN Y, Cao RJ, Jiang ZY. Desulfurization of model gasoline by bioinspired oleophilic nanocomposite membranes. Journal of Membrane Science. 2012;**415-416**:278-287

[74] Lu F, Kong Y, Lv H, Yang J. The removal of thiophene from n-heptane/thiophene mixtures by polyethylene glycol-block-poly acrylonitrile membranes. Petroleum Science and Technology. 2012;**30**:1232-1238

[75] Peng FB, LY L, Sun HL, Wang YQ, Liu JQ, Jiang ZY. Hybridorganic–inorganic membrane: Solving the tradeoff between permeability and selectivity. Chemistry of Materials. 2005;**17**:6790-6796

[76] Freeman BD. Basis of permeability/selectivity tradeoff relations in polymeric gas separation membranes. Macro molecules. 1999;**32**:375-380

[77] Liu WP, Li YF, Meng XX, Liu GH, Hu S, Pan FS, Wu H, Jiang ZY, Wang BY, Li ZX, Cao XZ. Embedding dopamine nano aggregates into a poly(dimethylsiloxane) membrane to confer controlled interactions and free volume for enhanced separation performance. Journal of Materials Chemistry A. 2013;**1**:3713-3723

[78] Lu F, Kong Y, Lv H, Yang J, Feng Z. The correlation between solvent treatment and the microstructure of PAN-b-PEG copolymer membranes. Polymer Journal. 2011;**43**:378-384

[79] MacDowell N, Florin N, Buchard A, Hallett J, Galindo A, Jackson G, Adjiman CS, Williams CK, Shah N, Fennell P. An overview of CO_2 capture technologies. Energy & Environmental Science. 2010;**3**:1645-1669

[80] Yang HQ, ZH X, Fan MH, Gupta R, Slimane RB, Bland AE, Wright I. Progress in carbon dioxide separation and capture: A review. Journal of Environmental Sciences. 2008;**20**:14-27

[81] Li YF, Wang SF, Wu H, Wang JT, Jiang ZY. Bio adhesion-inspiredpolymer–inorganic nanohybrid membranes with enhanced CO_2 capture properties. Journal of Materials Chemistry. 2012;**22**:19617-19620

[82] Csernica J, Baddour RF, Cohen RE. Gas permeability of a polystyrene/polybutadiene block copolymer possessing a mis oriented lamellar morphology. Macromolecules. 1989;**22**: 1493-1496

[83] Drzala PL, Halasab AF, Kofinasa P. Microstructure orientation and nanoporous gas transport in semi crystalline block copolymer membranes. Polymer. 2000;**41**:4671-4677

[84] Bao L, Trachtenberg MC. Facilitated transport of CO_2 across a liquid membrane: Comparing enzyme, amine, and alkaline. Journal of Membrane Science. 2006;**280**:330-334

[85] Favre N, Pierre AC. Synthesis and behaviour of hybrid polymer–silica membranes made by sol gel process with adsorbed carbonic anhydrase enzyme, in the capture of CO_2. Journal of Sol-Gel Science and Technology. 2011;**60**:177-188

[86] Zhang YT, Zhang L, Chen HL, Zhang HM. Selective separation of low concentration CO_2 using hydrogel immobilized CA enzyme based hollow fiber membrane reactors. Chemical Engineering Science. 2010;**65**:3199-3207

[87] Yao K, Wang Z, Wang J, Wang S. Biomimetic material-poly(N-vinylimidazole)-zinc complex for CO_2 separation. Chemical Communications. 2012;**48**:1766-1768

[88] Ran F, Nie S, Zhao W, Li J, Su B, Sun S, Zhao C. Biocompatibility of modified polyethersulfone membranes by blending anamphiphilic triblock co-polymer of poly(vinyl pyrrolidone)-b-poly(methyl methacrylate)-b-poly(vinyl pyrrolidone). Acta Biomaterialia. 2011;**7**:3370-3381

[89] Wang ZG, Wan LS, Xu ZK. Surface engineerings of polyacrylonitrile-based asymmetric membranes towards biomedical applications: An overview. Journal of Membrane Science. 2007;**304**:8-23

[90] Jiang JH, Zhu LP, Li XL, YY X, Zhu BK. Surface modification of PE porous membranes based on the strong adhesion of poly-dopamine and covalent immobilization of heparin. Journal of Membrane Science. 2010;**364**:194-202

[91] Ma L, Qin H, Cheng C, Xia Y, He C, Nie C, Wang L, Zhao C. Mussel-inspired self-coating at macro-interface with improved biocompatibility and bioactivity via dopamine grafted heparin-like polymers and heparin. Journal of Materials Chemistry B. 2014;**2**:363-375

[92] Gao A, Liu F, Xue L. Preparation and evaluation of heparin-immobilized poly(lactic acid) (PLA) membrane for hemodialysis. Journal of Membrane Science. 2014;**452**:390-399

[93] Chang Y, Chang WJ, Shih YJ, Wei TC, Hsiue GH. Zwitterionic sulfo betaine-grafted poly(vinylidene fluoride) membrane with highly effective blood compatibility via atmospheric plasma-induced surface copolymerization. ACS Applied Materials & Interfaces. 2011;**3**:1228-1237

[94] Chen SH, Chang Y, Lee KR, Wei TC, Higuchi A, Ho FM, Tsou CC, Ho HT, Lai JY. Hemocompatible control of sulfobetaine-grafted polypropyl-ene fibrous membranes in human whole blood via plasma-induced surface zwitterionization. Langmuir. 2012;**28**: 17733-17742

[95] Zhao J, Song L, Shi Q, Luan S, Yin J. Antibacterial and hemocompatibility switchable polypropylene nonwoven fabric membrane surface. ACS Applied Materials & Interfaces. 2013;**5**:5260-5268

[96] Jhong JF, Venault A, Hou CC, Chen SH, Wei TC, Zheng J, Huang J, Chang Y. Surface zwitterionization of expandedpoly(tetrafluoroethylene) membranes via atmospheric plasma-induced polymerization for enhanced skin wound healing. ACS Applied Materials & Interfaces. 2013;**5**:6732-6742

[97] Yuan J, Huang X, Li P, Li L, Shen J. Surface-initiated RAFT polymerization of sulfobetaine from cellulose membranes to improve hemocompatibility and antibiofouling property. Polymer Chemistry. 2013;**4**:5074-5085

[98] Yue WW, Li HJ, Xiang T, Qin H, Sun SD, Zhao CS. Grafting of zwitter ion from polysulfone membrane via surface-initiated ATRP with enhanced antifouling property and biocompatibility. Journal of Membrane Science. 2013;**446**:79-91

[99] Liu PS, Chen Q, SS W, Shen J, Lin SC. Surface modification of cellulose membranes with zwitterionic polymers for resistance to protein adsorption and platelet adhesion. Journal of Membrane Science. 2010;**350**:387-394

Chapter 9

Challenges and Opportunities for Biomimetic Membranes

Amira Abdelrasoul, Huu Doan and Ali Lohi

Additional information is available at the end of the chapter

http://dx.doi.org/10.5772/intechopen.71721

Abstract

A brief overview of the fundamental and practical challenges as well as of the current status of biomimetic membrane technologies is presented.

Keywords: development, scale up, antifouling, nanopores, cost, challenge

1. Introduction

This chapter presents a brief overview of the fundamental and practical challenges as well as of the current status of biomimetic membrane technologies. An accompanying summary is also given in **Table 1**. The specific focus of this chapter is aquaporin-based membranes that have initiated a substantial increase in research activities and several recent and ongoing commercialization attempts [1, 2].

2. Biomimetic nanopores

The fabrication of biomimetic nanopores, including those created with solid-state materials such as silicon, is a relatively new research area. As a result, the scale-up of biomimetic nanopore to practical and implementable dimensions for separation applications could face several challenges. The making of nanoscale pores using current methods, such as i-beam and e-beam lithography, is currently a lab scale process which requires expensive infrastructure. Furthermore, one of the fundamental challenges still being explored is the functionalization of biomimetic nanopores using specialized biological molecules and chemistry, which

	Practical challenges	Fundamental challenges	Current status
Biomimetic nanopores (solid state)	Scale-up at reasonable cost	Selection of functionalization ligands to provide selectivity	Commercialization attempts for DNA sequencing; no separation applications yet
Carrier-mediated biomimetic membranes	Stability; process configuration; low transport rates would require large membrane areas	Overcome low transport rates for separation applications	Not commercialized Sensing electrode commercialization; no separation applications
Protein-mediated biomimetic membranes	Membrane protein production scale-up; large membrane scale-up; leakage prevention	Limited range of proteins and polymers	Commercialization attempts ongoing
Artificial channel membranes	Scale-up	Designing specificity into channels, packing channels in membranes, increasing permeability	A new research area for water channels, ion channels studied but not commercialized
Antifouling strategies	Cost and efficacy	Not substantial for most applications	Various stages of research and commercialization

Table 1. Challenges to development and scale-up of various biomimetic membranes.

may be difficult to implement on large scales. Questions regarding the ligands that can be used to functionalize pores are also difficult to address, in particular if discrimination such as that is seen in potassium channels, is desirable. On the other hand, their application to DNA sequencing has reached commercialization level with several technologies licensed to start-up organizations [3].

3. Carrier-mediated biomimetic membranes

Carrier-mediated biomimetic membranes include liquid membranes (LMs) and ionophore-based membranes. LMs have gained interest in researchers in the last several decades, and as a result, an excellent understanding of the transport process has been developed. However, commercialization efforts in separation processes have been hindered by the poor stability and other practical difficulties in the implementation of these membranes. The practical difficulties include unstable immobilization of LMs in supported liquid membranes (SLMs) and process inefficiencies in separation of the recovered materials from the emulsion phase in emulsion liquid membranes (ELMs). Nevertheless, some applications have been scaled up to the necessary pilot and larger scales in recent years. ELMs have been used for zinc, phenol, and cyanide removal from industrial waste streams [4]. Ionophore-based membranes are widely used in ion-selective electrodes. Ion-selective membranes are the gold standard for this type of application. However, their application in separation membranes has not yet progressed sufficiently due to the low transport rates of ions in polymeric matrices [5]. In order to provide fluidity to the polymer matrix, plasticizers can be used, but these still do not improve the transport to reasonable levels necessary for separation applications.

4. Membrane protein-mediated biomimetic membranes

Protein-based biomimetic membranes, and in particular aquaporin-based membranes, have gained significant interest in recent years leading to multiple attempts at their commercialization. However, there are several fundamental and practical challenges that need to be addressed before large-scale membranes suitable for industrial applications can be developed. Applications of aquaporin biomimetic membranes face many critical challenges, primarily because of the limited scope of research studies conducted in this area. In particular, block copolymers (BCPs) that have been used for inserting membrane proteins have been limited a single polymer type with polydimethyl siloxane (PDMS) hydrophobic block [6]. Recent reports have shown that the mammalian eye lens aquaporin (AQP0) was successfully incorporated into poly(butadiene)-block-poly(ethylene oxide) block copolymer (PB-PEOBCP) membranes [7]. While these polymers have been shown to insert membrane proteins, it is not well understood what dictates membrane protein polymer interactions and compatibilities. Perhaps other polymers with superior characteristics have not been explored because a rational basis for polymers election does not yet exist. More experimental and theoretical explorations are required to develop this rapidly growing field. A related question is how to quantify the insertion efficiency of membrane proteins in BCPs in order to determine the best and most compatible polymer for a particular membrane protein. No effective method currently exists for quantifying the amount of protein inserted per unit membrane area with sufficient accuracy. Biochemistry-based methods such as stern blotting [8, 9], antibody-gold labeling [10], and freeze fracture [11] are difficult to implement and do not provide relevant quantitative information. A new method is needed to accurately determine insertion efficiency and compatibility of membrane proteins in various polymers and to provide a rational basis for BCP selection. A successful biomimetic membrane would require a high level of protein packing in the membrane. In most studies, full function of aquaporins in BCP membranes has only been demonstrated at packing densities that are relatively low and when the concentration of membrane proteins in native membrane systems, such as eye lenses, retina, and bacterial photosynthetic membranes, is considered. A typical packing density showing the expected function was demonstrated for AqpZ reconstituted into poly-(2-methyloxazoline)-block-poly-(dimethyl siloxane)-block-poly-(2-methyloxazoline) (PMOXA-PDMS-PMOXA) triblock copolymer membranes at a molar polymer-to-protein ratio (PoPR) adjusted for triblock architecture of 50–100, beyond which permeability has been shown to decrease [12]. In a recent study, AQP0 function was shown to persist for PoPR of 15 in a PB-PEO polymer [7]. This study also indicated that, while the constitution methodology is critical, polymer block length and chemistry may also be the important factors that determine how much protein could be functionally reconstituted into BCP membranes. The possibility of obtaining a high density of functional membrane proteins in BCP membranes has significant implications for applications of such systems. High level of protein packing density has been shown in lipid bilayers of protein-based membrane protein, using 2D crystallization [13–15] and several native membranes described earlier [13, 16, 17]. A more comprehensive understanding and characterization of membrane protein-BCP compatibility will also assist in making highly packed aquaporin-based membranes, similar to lipid-based membrane protein 2D crystals. There is also a need to explore aquaporin membranes beyond the traditionally used *Escherichia coli*

AqpZ, which may feature higher permeability and better insertion efficiency in BCPs. Another key challenge is the use of systems for expressing large amounts of membrane proteins. As has been reported, AqpZ is well expressed in *E. coli* and yields up to 200 mg L^{-1} of culture in a fusion form [18]. Results like these can be promising for scale-up of this particular aquaporin. In general, yield values are much lower (typically 1 mg L^{-1} of culture) for membrane proteins. The major membrane protein expression systems that have been developed and used widely in laboratory research setting are *E. coli*, yeast, and mammalian cells (Chinese hamster ovary cells in particular). However, membrane protein production is limited by both the cells' ability to survive membrane protein overproduction and the lack of coordination between membrane protein production and cell membrane production to accommodate membrane proteins. Alternative approaches to produce membrane proteins in general and aquaporins in particular should be further explored. The main practical challenge is in the scale-up of defect free membranes. So far, the sizes that are being realized are in the scale of square millimeter, even though rapid strides are being made in this research direction [19–30]. The methods used for membrane fabrication-vesicle position, monolayer formation, and pore suspended bilayers are all challenging to replicate at larger scales. Furthermore, most substrates used to support or immobilize active AqpZ are highly specialized and range from gold-coated track-etched membranes [23, 29] to polymer-based membranes [22, 24–28, 30]. The more scalable approach may be achieved through the use of polymer membranes, if technical concerns with regards to sealing around deposited vesicles and bilayers are solved. The economic aspects of making such membranes can likewise be problematic. Membrane protein purification is expensive, primarily due to the necessary process of disrupting cell membranes, use of specialty nonionic detergents ultracentrifugation, and chromatography. A thorough analysis of membrane protein scale-up has not been conducted before and should be a critical step forward if this class of membrane progresses to larger production scales and commercialization. Another challenge may be the unknown landscape of regulations regarding the use of membrane proteins, particularly in water treatment and industrial applications. The possibility of release of these materials is real and must be regulated. This issue is similar to the one faced by membranes that incorporate nanomaterials and are under specific regulations.

5. Artificial channel-based membranes

The research of artificial channel-based biomimetic membranes is relatively new, and so far most of the work has focused on synthesis and characterization. Transport measurements are still rudimentary in this field [31], and more studies are necessary to be able to compare their efficiency to membrane protein channels. Artificial water channels attract significant interest since they might prove to be the key materials for water purification. The challenges of carbon nanotubes (CNTs) for desalination applications, where it could have the most impact, include insufficient salt rejection levels and the inability to be used in manufacturing large-scale aligned CNT membranes [32]. Organic nanochannels-based water channels, in particular, are just in the early stages of being explored [33–37]. The only semiempirical principle available is the mimicking of natural selective filters. However, the current structures are still far from the perfect design models. Current data indicate that they suffer from low permeability (43 orders of magnitude lower than the aquaporins) and possibly imperfect rejection of solutes in some cases where channel diameters are large [36]. As mentioned by LeDuc in 2011, extensive hydrogen bonding helps encapsulate water within the channel, but also reduces the mobility of water molecules. This is probably the reason why the channels showed very low water permeability values (44 orders of magnitude lower than the lipid background permeability). This also leads to another challenge of finding a way to measure the permeability of low-permeable channels. A systematic platform for water permeability measurement needs to be established [31]. The next generation of water channels is expected to improve the design of the pore structure in order to increase water permeability while maintaining or improving solute rejection values. The geometry of the channels is also one possible area of improvement, as this will assist in packing these channels with very high density in lipid or polymer matrix for membrane fabrication. None of these ion or water channels have been tested in a practical membrane form since they are currently being studied in lipid vesicles. However, they hold great promise for separation applications due to their higher stability, properties potentially matching natural channels, scalability of their production, and ability of immobilization in a membrane-like support in a scalable manner.

6. Biomimetic antifouling strategies

Bioinspired antifouling strategies proposed for existing membranes are also generating greater interest in this research field. Many of the approaches proposed, and specifically surface modification, have the potential of being technically feasible. A cost-benefit analysis and their practical implementation may be important to consider prior to advancing them to the application level, in particular, because some of these approaches may actually decrease the initial permeability of membranes.

Author details

Amira Abdelrasoul[1]*, Huu Doan[2] and Ali Lohi[2]

*Address all correspondence to: amira.abdelrasoul@usask.ca

1 Department of Chemical and Biological Engineering, University of Saskatchewan, Saskatoon, Saskatchewan, Canada

2 Department of Chemical Engineering, Ryerson University, Toronto, Ontario, Canada

References

[1] Shannon MA, Bohn PW, Elimelech M, Georgiadis JG, Marinas BJ, Mayes AM. Science and technology for water purification in the coming decades. Nature. 2008;**452**:301-310

[2] Kasemset S, Lee A, Miller DJ, Freeman BD, Sharma MM. Effect of polydopamine deposition conditions on fouling resistance, physical properties, and permeation properties of reverse osmosis membranes in oil/water separation. Journal of Membrane Science. 2013;**425-426**:208-216

[3] McCloskey BD, Park HB, Ju H, Rowe BW, Miller DJ, Chun BJ, Kin K, Freeman BD. Influence of polydopamine deposition conditions on pure water flux and foulant adhesion resistance of reverse osmosis, ultrafiltration, and microfiltration membranes. Polymer. 2010;**51**: 3472-3485

[4] Kislik VS. Liquid Membranes: Principles and Applications in Chemical Separations and Wastewater Treatment. UK: Elsevier Science; 2009

[5] Armstrong RD, Todd M. Ionic mobilities in PVC membranes. Electrochimica Acta. 1987;**32**:155-157

[6] Kumar M, Payne M, Poust S. Polymer-based biomimetic membranes for desalination. In: Hélix-Nielsen C, editor. Zilles J, Biomimetic Membranes for Sensor and Separation Applications. Netherlands: Springer; 2012. p. 43-62

[7] Kumar M, Habel JEO, Shen YX, Meier WP, Walz T. High-density reconstitution of functional water channels into vesicular and planar block copolymer membranes. Journal of the American Chemical Society. 2012;**134**:18631-18637

[8] Gershoni JM, Palade GE. Protein blotting: Principles and applications. Analytical Biochemistry. 1983;**131**:1-15

[9] Nakamura K, Tanaka T, Kuwahara A, Takeo K. Micro assay for proteins on nitrocellulose filter using protein dye-staining procedure. Analytical Biochemistry. 1985;**148**:311-319

[10] Stoenescu R, Graff A, Meier W. Asymmetric ABC-triblock copolymer membranes induce a directed insertion of membrane proteins. Macromolecular Bioscience. 2004;**4**:930-935

[11] Severs NJ. Freeze-fracture electron microscopy. Nature Protocols. 2007;**2**:547-576

[12] Kumar M, Grzelakowski M, Zilles J, Clark M, Meier W. Highly permeable polymeric membranes based on the incorporation of the functional water channel protein Aquaporin Z. Proceedings of the National Academy of Sciences of the United States of America. 2007;**104**:20719-20724

[13] Jap BK, Zulauf M, Scheybani T, Hefti A, Baumeister W, Aebi U, Engel A. 2D crystallization: From art to science. Ultramicroscopy. 1992;**46**:45-84

[14] Hasler L, Heymann JB, Engel A, Kistler J, Walz T. 2D crystallization of membrane proteins: Rationales and examples. Journal of Structural Biology. 1998;**121**:162-171

[15] Abeyrathne PD, Chami M, Pantelic RS, Goldie KN, Stahlberg H. Preparation of 2D crystals of membrane proteins for high-resolution electron crystallography data collection. In: Grant JJ, editor. Methods in Enzymology. Vol. 1. Academic Press; 2010. p. 25-43. DOI: 10.1016/S0076-6879(10)81001-8

[16] Raunser S, Walz T. Electron crystallography as a technique to study the structure on membrane proteins in a lipidic environment. Annual Review of Biophysics. 2009;**38**:89-105

[17] Renault L, Chou HT, Chiu PL, Hill R, Zeng X, Gipson B, Zhang Z, Cheng A, Unger V, Stahlberg H. Milestones in electron crystallography. Journal of Computer-Aided Molecular Design. 2006;**20**:519-527

[18] Lian J, Ding S, Cai J, Zhang D, Xu Z, Wang X. Improving aquaporin Z expression in *Escherichia coli* by fusion partners and subsequent condition optimization. Applied Microbiology and Biotechnology. 2009;**82**:463-470

[19] Kaufman Y, Berman A, Freger V. Supported lipid bilayer membranes for water purification by reverse osmosis. Langmuir. 2010;**26**:7388-7395

[20] Wang H, Chung TS, Tong YW, Meier W, Chen Z, Hong M, Jeyaseelan K, Armugam A. Preparation and characterization of pore-suspending biomimetic membranes embedded with Aquaporin Z on carboxylated polyethyleneglycol polymer cushion. Soft Matter. 2011;**7**:7274-7280

[21] Duong PHH, Chung TS, Jeyaseelan K, Armugam A, Chen Z, Yang J, Hong M. Planar biomimetic aquaporin-incorporated triblock copolymer membranes on porous alumina supports for nano filtration. Journal of Membrane Science. 2012;**409-410**:34-43

[22] Li XS, Wang R, Tang CYT, Vararattanavech A, Zhao Y, Torres J, Fane T. Preparation of supported lipid membranes for aquaporin Z incorporation. Colloids and Surfaces B: Biointerfaces. 2012;**94**:333-340

[23] Wang HL, Chung TS, Tong YW, Jeyaseelan K, Armugam A, Chen ZC, Hong MH, Meier W. Highly permeable and selective pore-spanning biomimetic membrane embedded with aquaporin Z. Small. 2012;**8**:1185-1190

[24] Zhao Y, Qiu C, Li X, Vararattanavech A, Shen W, Torres J, Hélix-Nielsen C, Wang R, Hu X, Fane AG, Tang CY. Synthesis of robust and high- performance aquaporin-based biomimetic membranes by interfacial polymerization-membrane preparation and RO performance characterization. Journal of Membrane Science. 2012;**423-424**:422-428

[25] Zhong PS, Chung TS, Jeyaseelan K, Armugam A. Aquaporin-embedded biomimetic membranes for nano filtration. Journal of Membrane Science. 2012;**407-408**:27-33

[26] Sun G, Chung TS, Jeyaseelan K, Armugam A. A layer-by-layer self-assembly approach to developing an aquaporin-embedded mixed matrix membrane. RSC Advances. 2013;**3**: 473-481

[27] Sun G, Chung TS, Chen N, Lu X, Zhao Q. Highly permeable aquaporin-embedded biomimetic membranes featuring a magnetic-aided approach. RSC Advances. 2013;**3**:9178-9184

[28] Sun G, Chung TS, Jeyaseelan K, Armugam A. Stabilization and immobilization of aquaporin reconstituted lipid vesicles for water purification. Colloids and Surfaces B: Biointerfaces. 2013;**102**:466-471

[29] Wang HL, Chung TS, Tong YW, Jeyaseelan K, Armugam A, Duong HHP, Fu F, Seah H, Yang J, Hong M. Mechanically robust and highly permeable Aquaporin Z biomimetic membranes. Journal of Membrane Science. 2013;**434**:130-136

[30] Xie W, He F, Wang B, Chung TS, Jeyaseelan K, Armugam A, Tong YW. An aquaporin-based vesicle-embedded polymeric membrane for low energy water filtration. Journal of Materials Chemistry A1. 2013;**26**:7592-7600

[31] Sakai N, Matile S. Synthetic ion channels. Langmuir. 2013;**29**:9031-9040

[32] Pendergast MM, Hoek EMV. A review of water treatment membrane nanotechnologies. Energy Environmental Sciences. 2011;**4**:1946-1971

[33] Fei Z, Zhao D, Geldbach TJ, Scopelliti R, Dyson PJ, Antonijevic S, Bodenhausen G. A synthetic zwitterionic water channel: characterization in the solid state by X-ray crystallography and NMR spectroscopy. Angewandte Chemie, International Edition. 2005;**44**:5720-5725

[34] Kaucher MS, Peterca M, Dulcey AE, Kim AJ, Vinogradov SA, Hammer DA, Heiney PA, Percec V. Selective transport of water mediated by porous dendritic dipeptides. Journal of the American Chemical Society. 2007;**129**:11698-11699

[35] LeDuc Y, Michau M, Gilles A, Gence V, Legrand YM, vanderLee A, Tingry S, Barboiu M. Imidazole-quartet water and proton dipolar channels. Angewandte Chemie, International Edition. 2011;**50**:11366-11372

[36] XB H, Chen Z, Tang G, Hou JL, Li ZT. Single-molecular artificial transmembrane water channels. Journal of the American Chemical Society. 2012;**134**:8384-8387

[37] Zhou X, Liu G, Yamato K, Shen Y, Cheng R, Wei X, Bai W, Gao Y, Li H, Liu Y, Liu F, Czajkowsky DM, Wang J, Dabney MJ, Cai Z, Hu J, Bright FV, He L, Zeng XC, Shao Z, Gong B. Self-assembling sub nanometer pores with unusual mass-transport properties. Nature Communications. 2012;**3**:949